Mobile NMR and MRI

Developments and Applications

New Developments in NMR

Titles in the Series:

1: Contemporary Computer-Assisted Approaches to Molecular Structure Elucidation
2: New Applications of NMR in Drug Discovery and Development
3: Advances in Biological Solid-State NMR
4: Hyperpolarized Xenon-129 Magnetic Resonance: Concepts, Production, Techniques and Applications
5: Mobile NMR and MRI: Developments and Applications

How to obtain future titles on publication:
A standing order plan is available for this series. A standing order will bring delivery of each new volume immediately on publication.

For further information please contact:
Book Sales Department, Royal Society of Chemistry, Thomas Graham House, Science Park, Milton Road, Cambridge, CB4 0WF, UK
Telephone: +44 (0)1223 420066, Fax: +44 (0)1223 420247
Email: booksales@rsc.org
Visit our website at www.rsc.org/books

Mobile NMR and MRI
Developments and Applications

Edited by

Michael L. Johns
University of Western Australia, Crawley, Australia
Email: michael.johns@uwa.edu.au

Einar O. Fridjonsson
University of Western Australia
Email: einar.fridjonsson@uwa.edu.au

Sarah J. Vogt
University of Western Australia
Email: sarah.vogt@uwa.edu.au

and

Agnes Haber
University of Western Australia
Email: agnes.haber@uwa.edu.au

ROYAL SOCIETY OF CHEMISTRY

THE QUEEN'S AWARDS FOR ENTERPRISE: INTERNATIONAL TRADE 2013

New Developments in NMR No. 5

Print ISBN: 978-1-84973-915-3
PDF eISBN: 978-1-78262-809-5
ISSN: 2044-253X

A catalogue record for this book is available from the British Library

Published by The Royal Society of Chemistry,
Thomas Graham House, Science Park, Milton Road,
Cambridge CB4 0WF, UK

Registered Charity Number 207890

For further information see our web site at www.rsc.org

Printed in the United Kingdom by CPI Group (UK) Ltd, Croydon, CR0 4YY, UK

Preface

When asked by Prof. Bill Price and Prof. Bruce Balcom to put this book together I was somewhat surprised given my research to that point at the University of Cambridge as part of Prof. Lynn Gladden's MRRC had largely been using superconducting magnets. However, following a relocation to the University of Western Australia, I had started to work on applications using low field, portable NMR/MRI instruments; initially out of financial necessity! However it soon became apparent that a lot of my fears, and indeed prejudices, regarding low field were (in context) unfounded and in fact there was a suite of potential applications for which such NMR instrumentation was eminently suitable, although not necessarily optimal in terms of robustness and cost. I also got to interact with a suite of low field NMR companies, and associated applied academics in this space, who were collectively hugely impressive in terms of both their commitment and the developments that they had achieved. Who would have thought we could routinely acquire chemically resolved spectra on small portable benchtop systems a decade ago!

NMR and MRI obviously are extremely useful in medicine and in chemical characterisation. They are also extremely powerful in terms of their use as a laboratory research tool in a very broad range of disciplines. Penetration into other commercial/industrial applications has however been, I think we need to be honest, rather modest. There are a whole suite of reasons for this, dost and sensitivity being arguably the two main ones. Sensitivity is an interesting issue. NMR by comparison with other spectroscopic techniques is inherently insensitive. It really is not ideal for measuring at ppm levels; however, that is where the industry demand often is. NMR/MRI is however incredibly sensitive to a range of sample physical/chemical characteristics, usually in terms of how these impact on relaxation and diffusion measurements. This opens up enormous opportunities. However, this sensitivity to multiple characteristics and parameters also makes interpreting acquired signals quantitatively

New Developments in NMR No. 5
Mobile NMR and MRI: Developments and Applications
Edited by Michael L. Johns, Einar O. Fridjonsson, Sarah J. Vogt, and Agnes Haber

Published by the Royal Society of Chemistry, www.rsc.org

often inherently difficult. Being quantitative is an absolute necessity—this was really hammered home for me in my formative years by Prof. Lynn Gladden and later by Prof. Bruce Balcom. Yes, 'pretty MRI pictures' are often good at convincing 'management' to fund further research work; however, to be really useful for industry at large, techniques really need to be quantitative. What quantitative actually means in this context, is a conversation best left for another forum!

With the above in mind, the field of low field mobile NMR and MRI is clearly currently offering some significant developments that show promise to enable its true potential to be fully exploited. The confluence of developments in hardware, signal enhancement, pulse sequence design and data processing are all encouraging, and collectively will be greater than the sum of their parts. We have done our best to capture some of these in the current textbook. We hope it will enable or even inspire some new practical applications of mobile NMR/MRI and open up new product opportunities. I suspect that a few new applications might 'break the dam wall'. Let's see.

Finally, I would like to thank the chapter authors for their efforts in helping put this book together. I am in awe of their intellect, dedication and achievements in their respective fields. I would also like to thank my fellow editors—Einar Fridjonsson, Sarah Vogt and Agnes Haber. I recruited all of them to Perth, the most isolated city in the world, credit to them that they were still willing to contribute. I would particularly like to single out Sarah Vogt who essentially performed all the tedious editing required and a lot of the required author correspondence. Her efforts were immense and are greatly appreciated.

Mike Johns
University of Western Australia

Contents

New Developments in NMR No. 5
Mobile NMR and MRI: Developments and Applications
Edited by Michael L. Johns, Einar O. Fridjonsson, Sarah J. Vogt, and Agnes Haber

Published by the Royal Society of Chemistry, www.rsc.org

CHAPTER 1

Introduction

EIICHI FUKUSHIMA*[a]

[a]ABQMR, 2301 Yale Blvd., SE (Suite C2), Albuquerque, NM 87106, USA
*E-mail: eiichi@abqmr.com

Until recently, the thought of an NMR or MRI instrument that is compact and mobile was not congruent with laboratory or clinical NMR and MRI. The reason for this discrepancy is easy to discern: laboratory/clinical NMR/MRI instruments use well-characterized and reasonably strong magnetic fields that are difficult to achieve with portable (mostly permanent) magnets. In addition, the complexity of typical NMR experiments in terms of electronic requirements meant that the electronics modules occupied significant real estate, albeit often hidden in electronic racks and enclosures, and required nontrivial power consumption. This monograph describes many new developments that overcome these traditional barriers to mobile NMR.

Because of the lack of small and lightweight magnets with field strengths and homogeneity comparable to those used in laboratories, most applications of compact and mobile NMR were for specific and limited parameters, using magnets with lesser performance than that of laboratory NMR magnets, or no magnets at all. Such applications include simply measuring the magnetic field strength and progressing to detecting the presence or mobility of certain materials in samples of interest to oil, agricultural, and other industries. Many of the advances were realized in situations where the limitations of the magnet are accepted and the measurements tailored to a limited number of parameters that are important to that application but otherwise unobtainable.

New Developments in NMR No. 5
Mobile NMR and MRI: Developments and Applications
Edited by Michael L. Johns, Einar O. Fridjonsson, Sarah J. Vogt, and Agnes Haber

Published by the Royal Society of Chemistry, www.rsc.org

However, there have been recent advances in magnet technology that have significantly helped push the field of mobile NMR forward. Chapters 4 and 5 cover single sided and Halbach magnet developments, respectively, while there is also a recent review on low field magnets for industrial use.[1] A key development is the successful adaptation of mechanical shimming to small single sided magnets, described in Section 4.2.2, opening the possibility of resolving moderate resolution chemical shifts. Another exciting recent advance, that of using bulk high T_c superconductors, is introduced near the end of this chapter and is more fully described in Chapter 11.

Another factor that is contributing to the advance of compact mobile NMR and MRI is the amazing evolution that is still taking place in electronics that was unimaginable a few decades ago. Not only can the electronics be made much smaller than they were in the past, a given size of circuitry can perform at a much higher level, enabling not only miniaturization but also enhanced performance. Consequently, both complex experimental protocols as well as rapid and high-capacity data acquisition contribute to the progress of this field. NMR-on-a-board devices have now been on the market for a number of years and have contributed to the development of many mobile NMR systems. Chapter 6 contains a detailed description of these developments.

Throughout the history of NMR, there have been some examples of mobile or compact NMR that were outside the mainstream of the discipline. Probably the earliest example of mobile NMR was proton magnetometry, which simply measured the Earth's magnetic field strength. A 1955 article on the subject[2] even uses the word portable in the quotation "The apparatus consists of a rugged and portable nuclear magnetometer head, requiring no careful orientation in setting up..." It appears that the author was far ahead of his time. It is known that such proton magnetometers were used to successfully survey archeological sites as early as 1962.[3,4]

Strong[5] described an unusual proton magnetometer developed by Wadsworth[6] in his Scientific American Amateur Scientist column in 1968. This hand-carried device found buried marker magnets that defined areas in a field to study the encroachment of certain grasses into those areas over a period of a few years. The ingenious battery-powered device was operated *via* a spring-loaded manual plunger that created the "pulse sequence" and the free induction decay (FID) was detected in a set of earphones worn by the operator as he/she walked the field. The presence of a marker magnet was detected by a beat signal between two identical water samples spaced 2.6 m apart on a pole carried by the operator. Wadsworth stated that the device cost approximately £10 in parts, although the recently inflated price of copper has increased the present day cost many-fold!

Unbeknownst to most of us NMR practitioners, mobile proton magnetometry has continued to evolve, especially in the miniaturization of electronics as well as the development and application of modern software. There is at least one book that describes the construction and use of proton magnetometers[7] and articles on the modern uses of such devices also exist online[8,9] with major emphasis on studying the temporal variation of Earth's magnetic field. These inexpensive devices enable extremely accurate measurements of the temporal variation of Earth's magnetic field intensity by practitioners all over the world, and their results can be correlated with sunspot activity and rapidly communicated *via* the internet.

Until the mid-1990s, all Earth's field NMR operated in a manner similar to the proton magnetometer, *i.e.*, the spins were initially polarized along the axis of a coil that was energized for that purpose and then allowed to precess around the Earth's magnetic field, generating a FID that was analyzed, mostly for its Larmor frequency. Then, Stepišnik's group succeeded in obtaining magnetic resonance images in Earth's magnetic field. These workers realized that when the pre-polarization field decreases adiabatically, the magnetization follows the effective applied oscillating magnetic field to align with Earth's magnetic field. At that point, the standard spin-warp pulse sequence could be performed using spin echoes to yield images. This was a significant insight that made Earth's field NMR useful for diffusion measurements with pulsed-gradient spin-echo (PGSE) sequences. In contrast, it is still difficult to make magnetic resonance images in the Earth's magnetic field. This field is reviewed in detail by Mohorič and Stepišnik.[10]

One of the best-known research projects carried out with this technique is an experiment to measure the characteristics of brine in the pore structure of Antarctic sea ice by Callaghan's group, as described in another review.[11] The ability of these instruments to carry out standard pulse NMR experiments allowed the measurement of diffusion in the pores of ice cores with PGSE sequences, resulting in information about pore geometries. The harsh experimental conditions were compensated for by the near absence of electromagnetic noise and the presence of feathered observers.

While we are on the subject of Earth's field NMR, it is worth noting that, quite unexpectedly to most of us, ultra high-resolution heteronuclear *J*-coupling data have been obtained at Earth's field.[12] In subsequent experiments, it has been shown that even chemical shift information can be obtained under some circumstances in Earth's magnetic field.[13] This method is not required to be mobile but the hardware used for such experiments is quite minimal and any NMR experiment that does not use an artificial magnet is likely to be mobile. Above all, such an apparatus that does not use high quality artificial magnets is likely to be inexpensive, opening up the possibility of sophisticated experiments that are also affordable. In this experiment, a permanent magnet is used to pre-polarize the spins before the sample is physically transported to the NMR detector for the experiment. However, because the only function of pre-polarization is to enhance the signal, the magnetic field for this purpose need not be homogeneous.

The aforementioned sophisticated data processing, of course, affects all areas of NMR. Chapter 8 discusses two particular aspects of data processing, specifically, compressed sensing and Bayesian techniques. Such advances in data processing are relevant to mobile and compact NMR and MRI, even if for no other reason than usual mobile and compact NMR and MRI can always use more S/N. Of course, the additional information content arising from these new techniques represents an extra benefit to the field.

Figure 1.1 (a) Photo of a mine detector that is mounted in front of a tractor. The trailer carried the batteries. (b) Photo of a mail bomb detector based on ^{14}N nuclear quadrupole resonance (NQR). NQR is well-suited for mobile instrumentation because it requires no magnetic fields. (c) Photo of a soil moisture detector and (d) photo of a cement core moisture detector probe, presaging the modern rock core analyzers, such as those described in Chapter 10.

Another isolated effort to take NMR technology out of the laboratory and into the industrial atmosphere was initiated by Bill Rollwitz at Southwest Research Institute (SwRI) in 1953(!) and flourished in the 1970s through to the 1990s. Their numerous projects included landmine detection, moisture measurement in concrete and soil, studies of asphalt aging, inspection of solid rocket motors, tissue measurements, and dynamite detection in checked airline baggage. Because the NMR community-at-large was unaware of SwRI's efforts in NMR, it might be of some interest here to view a few of their mobile devices (Figure 1.1).

A major enabling technology utilized by SwRI, for example, in its landmine detector, is unilateral NMR whereby samples are positioned outside the magnet rather than between poles of electric or permanent dipole magnets or in axial holes of superconducting solenoids. Blümich and his colleagues at RWTH-Aachen led the movement to develop portable and industrial NMR with an emphasis on unilateral NMR. Nowadays, unilateral NMR is available commercially and this technology has been used for such tasks as analyzing frescos, determining the condition of restored paintings, and even an examination of a mummy![14]

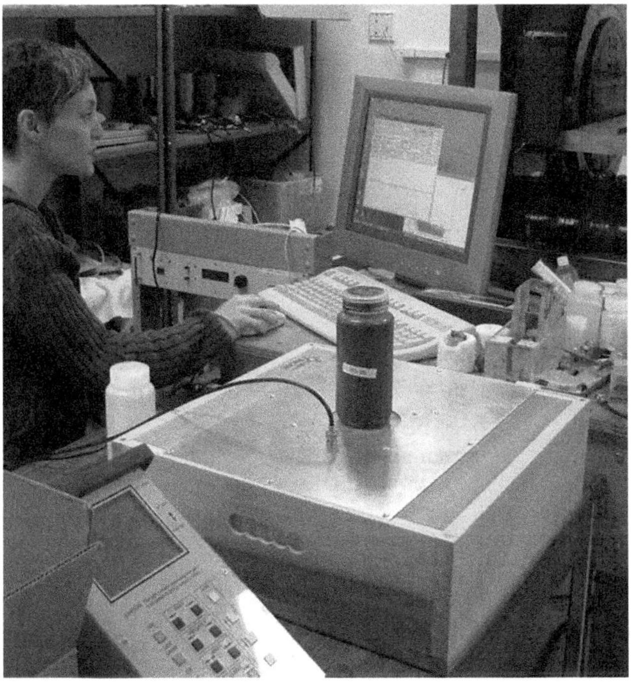

Figure 1.2 Unilateral NMR device with which a signal is obtained from a sample of crude oil in a 500 mL plastic bottle on top of the device.

A compact unilateral NMR device at ABQMR that looks at fairly large samples, for example, coconuts, watermelon, and large bottles, is shown in Figure 1.2. It has a favorable spot, *i.e.*, a sweet spot, at a distance of 12–15% of the magnet's diameter (10 cm), in contrast to the more common devices without a sweet spot that look at shallower depths, albeit with much better S/N. The magnet for this scheme is discussed more fully in Chapter 4.

These developments are expected to advance applications in many other areas, such as agricultural products,[15] trees and plants,[16–18] and foods (Chapter 3), not to mention medical applications with possibly large societal consequences (Chapter 11). The unusual magnet designs involved in this technology will be covered in Chapters 4 and 5.

There is also a push towards smaller permanent dipole magnets that are lighter, cheaper, and more portable, largely facilitated by discoveries of rare-earth magnetic materials, such as SmCo and NdFeB. Figure 1.3 shows three 1 T permanent NdFeB-based magnets with 40, 25, and 5 mm gaps, the first two made by Neomax and the smallest made by Aster. The 5 mm gap magnet, also shown in Figure 11.2d, is suitable for performing NMR in capillaries, utilizing microcoils, as discussed in Chapter 6.

Prior to the discovery of these modern magnetic materials, the most common permanent magnets were made of alnico, an alloy of iron, aluminum, nickel, and cobalt. Possibly the earliest commercial "desk-top" NMR

Figure 1.3 Three 1 T permanent NdFeB-based magnets with 40, 25, and 5 mm gaps. The first two were made by Neomax (Japan) and the third by Aster Enterprises in the US.

apparatus was the Spinloc pulse-NMR that used an alnico magnet covered by a stainless steel tea cozy. The operator would drop a seed into the sample chamber through a hole at the top of the tea cozy for the relaxation measurements to determine oil content and then blow it back out by squeezing a rubber bulb that injected air at the bottom of the sample tube.

Another technology that relies on seeing signals from outside the magnet is downhole well-logging, used by the oil industry to characterize the strata and determine the presence, distribution, and fluid permeability of oil. Initially, attempts were made to do these NMR experiments in Earth's magnetic field but the scheme became more practical with the idea of lowering magnets into the borehole in order to enhance the magnetic field strength outside the borehole.[19] It has quietly, at least to the laboratory NMR world, become a field unto itself over the past 25–30 years.[20] The presence of oil is distinguished from that of water by a distinct range of values of T_2 compared with water. In addition, the diffusion coefficient could be measured in order to distinguish liquid oil from gas, and a relatively new development of significance is to combine the diffusion information with T_2. This field will be described in Chapter 2.

An activity that started in the 1980s, termed NMR sounding, uses an ~100 m diameter loop coils for Earth's field NMR detection of water and other proton-containing fluids 50–100 m underground. Such technology is mobile but hardly qualifies as "compact" NMR because it requires a large truckload of wire to form the NMR coil. The large scale is needed because the technique depends on seeing an enormous volume of sample to overcome the very poor specific sensitivity of NMR in the very weak Earth's magnetic field. This is especially true here because there is no pre-polarization to enhance S/N as in the previously mentioned Earth's field NMR experiments. One of the major challenges in using such large coils is the sensitivity to long- and

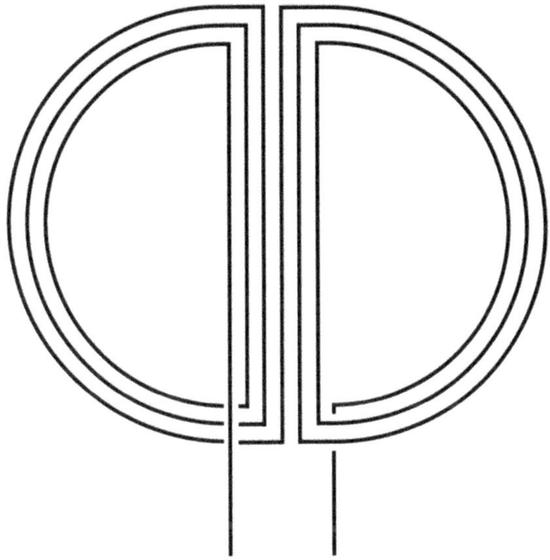

Figure 1.4 Schematic of a flat gradiometer coil used in Earth's field NMR that is sensitive to a flat region parallel to the plane of the coil. Coils with length scales from 61 cm to 6 m have been used to detect proton signals from water without pre-polarization.

short-range interference that exists near man-made objects. Thus, a major topic of research in this area concerns mitigation of such interference by analog and digital means, often needing to use an extra coil to obtain a reference signal. There are several commercially available systems and the discipline will be described in Chapter 9.

Recent experiments at ABQMR demonstrated that it is possible to obtain NMR signals in Earth's magnetic field from shallower depths over smaller areas without prepolarization. Prototype two-dimensional coils obtained excellent signals from a ~10 cm deep water over an area of ~400 cm^2 in 32 scans. The essential components of these coils are: (1) a gradiometer configuration to attenuate long-range interference, (2) use of a sufficient amount of wire to enhance the sensitivity, and (3) geometry that fits the shape of the sample.[21]

Figure 1.4 shows a sketch of a flat gradiometer coil showing, for simplicity, only six turns of parallel wires in the center that form an array that is sensitive to a flat region that is parallel to the plane of the coil. An actual coil wound in this way has 992 turns of #18 wire in the 61 cm long, 10 cm wide, center array. The signal-to-noise ratio (S/N) depends on the coil parameters as:

$$\mathrm{S/N} \propto \frac{B_1/i}{\sqrt{r}} \tag{1.1}$$

where B_1/i is the magnetic field strength at the sample element position per unit current flowing through the coil and r is the resistance of the coil.[22]

Because the numerator is proportional to the number of coil turns for a constant geometry, while the denominator is proportional to their square root, the performance will improve with additional turns. Low frequency NMR, such as Earth's field NMR, can take full advantage of this relation until the wire length becomes a non-trivial fraction of the wavelength, which is approximately 150 km! Additional improvements are possible by reducing the resistance of the coil wire, for example, by increasing the diameter of the wire, an avenue made easier in this situation simply by the large size of the coil. S/N was increased further by the use of adiabatic fast half-passage sweeps in order to cause B_1-independent nutation of magnetization into the transverse plane.

This arrangement has recently been scaled up to a coil that is 6 m across for detecting liquids 1 to 2 m away in less than a minute without pre-polarization. In order to keep the required transmitter voltage low with the increased current required to reach the sample element that is farther away, S/N was maximized not by maximizing B_1/i but by minimizing r. Therefore, the finished coil has only an 8-wire array in the center that is 1 m wide but each "wire" consists of 210 11-gauge aluminum magnet wires so that the total resistance with various connections is in the range of tens of milliohms. With the total impedance of ~4 ohms, it is possible to have an effective Q of ~100, which represents a bandwidth of 20–25 Hz. Although not portable in the usual sense, this coil can be moved around by a helicopter.

An early commercial compact NMR, aimed at industrial uses, was made by Tri-Valley Research using a Halbach magnet and was primarily aimed at examining rock cores. This is an ideal application for a compact NMR device without stringent field homogeneity requirements and has evolved into a viable field with companies such as Magritek marketing rock core analyzers, which are described in Chapter 10.

An exciting development is that of a bulk superconductor as a compact magnet that features field strength that is comparable to existing superconducting (wire) magnets and whose homogeneity is improving. The bulk superconductor is placed in a uniform field of another superconducting magnet and its temperature is lowered past the superconducting transition to trap the field. The strongest field strength attained so far is 4.74 T and the magnet operates at around 50 K, which requires only a standard cryogenic refrigerator.[23] Additional details are given in Chapter 11, but it is easy to conceive that this magnet could be configured as a desktop unit. It is also easy to imagine, because of its light weight, small size, and the relatively high operating temperature, that this magnet will be easy to transport while energized. This promising technology will join others involving permanent magnets to open the door to compact high resolution NMR. Other technological developments that may have beneficial effects on mobile and compact NMR/MRI include SQUID and atomic NMR, which are described in Chapter 7. Once again, the suitability of these techniques for mobile NMR is rooted in the lack of necessity of a large magnet although these methods must overcome some other barriers for them to be truly mobile. Specifically, the need for cryogens for SQUIDs and high temperature for atomic magnetometry plus significant shielding requirements must be overcome.

All in all, the field of mobile NMR and MRI is an exciting one with many potential applications, and significant additional progress should be forthcoming in the near future. The following chapters in this monograph offer ample evidence that the phrase "shrinking NMR" has become a distinct possibility.

Acknowledgements

The author is grateful to Armando De Los Santos of Southwest Research Institute and Daiki Tamada of University of Tsukuba for providing unpublished drawings and photographs. Discussions with colleagues Stephen Altobelli and Lana Chavez as well as editorial help from Janet Bench are acknowledged. Finally, the Earth's field NMR at ABQMR, briefly mentioned, was supported by Upstream Research Company of Exxon Mobile.

References

1. J. Mitchell, L. F. Gladden, T. C. Chandrasekera and E. J. Fordham, *Prog. Nucl. Magn. Reson. Spectrosc.*, 2014, **76**, 1–60.
2. G. S. Waters, *Nature*, 1955, **176**, 691.
3. C. M. Lerici, in *Expedition Magazine*, Spring, 1962, p. 4.
4. D. F. Brown, in *Expedition Magazine*, Winter, 1963, p. 40.
5. C. L. Strong, *Sci. Am.*, 1968, **218**, 124.
6. N. J. Wadsworth, *J. Sci. Instrum.*, 1967, **44**, 552.
7. S. Hollos and R. Hollos, in *Signals from the Subatomic World: How to Build a Proton Precession Magnetometer*, Abrazol Publishing, 2008.
8. J. Geller, *EDN*, 2012, August 12, 2012, available online at http://www.edn.com/design/analog/4392193/Geomagnetic-observatory.
9. J. A. R. Koehler, 2012, available online at http://members.shaw.ca/jark/ProtonPrecessionMagnetometers.pdf.
10. A. Mohorič and J. Stepišnik, *Prog. Nucl. Magn. Reson. Spectrosc.*, 2009, **54**, 166–182.
11. M. E. Halse and P. T. Callaghan, *eMagRes*, 2009, DOI: 10.1002/9780470034590.emrstm1025.
12. S. Appelt, F. W. Häsing, H. Kühn, U. Sieling and B. Blümich, *Chem. Phys. Lett.*, 2007, **440**, 308–312.
13. S. Appelt, S. Glöggler, F. W. Häsing, U. Sieling, A. G. Nejad and B. Blümich, *Chem. Phys. Lett.*, 2010, **485**, 217–220.
14. B. Blümich, *Sci. Am.*, 2008, **299**, 92–98.
15. Y. Geya, T. Kimura, H. Fujisaki, Y. Terada, K. Kose, T. Haishi, H. Gemma and Y. Sekozawa, *J. Magn. Reson.*, 2013, **226**, 45–51.
16. D. Capitani, F. Brilli, L. Mannina, N. Proietti and F. Loreto, *Plant Physiol.*, 2009, **149**, 1638–1647.
17. H. Van As, T. Scheenen and F. J. Vergeldt, *Photosynth. Res.*, 2009, **102**, 213–222.

18. T. Umebayashi, K. Fukuda, T. Haishi, R. Sotooka, S. Zuhair and K. Otsuki, *Plant Physiol.*, 2011, **156**, 943–951.
19. J. A. Jackson, L. J. Burnett and J. F. Harmon, *J. Magn. Reson.*, 1980, **41**, 411–421.
20. R. J. S. Brown, R. Chandler, J. A. Jackson, R. L. Kleinberg, M. N. Miller, Z. Paltiel and M. G. Prammer, *Concepts Magn. Reson.*, 2001, **13**, 335–413.
21. L. Chavez, S. Altobelli, E. Fukushima, T. Zhang, T. Nedwed, D. Palandro, L. Srnka and H. Thomann, *Near Surf. Geophys.*, 2015, **13**, DOI: 10.3997/1873.0604.2015023.
22. D. I. Hoult and R. E. Richards, *J. Magn. Reson.*, 1976, **24**, 71–85.
23. D. Tamada, T. Nakamura, Y. Itoh and K. Kose, *Phys. C*, 2013, **492**, 174–177.

CHAPTER 2

NMR Well Logging

MARTIN D. HÜRLIMANN*[a] AND NICHOLAS J. HEATON[b]

[a]Schlumberger-Doll Research, One Hampshire Street, Cambridge, MA 02139, USA; [b]Schlumberger HFE Center, 110 Schlumberger Drive MD-4, Sugar Land, TX 77478, USA
*E-mail: hurlimann@slb.com

2.1 Introduction

Well logging is the general technique of measuring physical properties of the subsurface using specialized sensors that are lowered into boreholes. NMR well logging is based on low field NMR measurements using NMR sensors. The measurements respond directly to the fluids in the subsurface and provide a unique tool for their characterization. Recent advances in the technique of NMR well logging have greatly enhanced the versatility and robustness of this measurement. Well logging provides a valuable tool in the exploration of hydrocarbon wells and has become a significant commercial application of NMR. The analysis of NMR well logging data enables the quantification of the fluids occupying the pore space and the prediction of the fluid flow properties through the reservoir. These are critical steps in the evaluation of hydrocarbon reservoirs and in the planning of production strategies. These tasks are not only important for the petroleum industry, but they lie at the heart of other applications based on the Earth Sciences. For example, hydrology focuses on the distribution, transport, and contamination of water in aquifers, and carbon sequestration deals with the deposition of carbon dioxide into earth formations. For these purposes, it is essential to

New Developments in NMR No. 5
Mobile NMR and MRI: Developments and Applications
Edited by Michael L. Johns, Einar O. Fridjonsson, Sarah J. Vogt, and Agnes Haber

Published by the Royal Society of Chemistry, www.rsc.org

understand the connectivity of the pore space and monitor the fluids inside the earth formations.

NMR measurements are of particular interest because they are able to directly detect, distinguish, and characterize both aqueous and hydrocarbon fluids *in situ*. There is no need to extract the fluids first from the formation, which is an expensive and time-consuming process. The *in situ* fluid characterization also avoids complications due to possible phase transitions induced by the process of fluid extraction. The unique value provided by NMR well logging for the characterization of subsurface earth formations can be best appreciated by comparing it to other available logging or surface measurements. The most widely used surface technique is seismic prospecting.[1] This provides large-scale information on the geometry of the geological structures deep into the earth, but yields limited information on the fluids contained in the pore space or on the flow properties of the formation. A number of surface-based measurements, including ground-penetrating radar, gravimetric measurements, electrical conductivity measurements, and the technique of surface NMR, can provide information on the fluids. While the typical resolution and range of depth accessible with these techniques is adequate for the investigation of shallow aquifers, it is clearly insufficient for the exploration of hydrocarbon reservoirs, which are often located 1 to 3 km below the surface, but can be as deep as 10 km.

Both the techniques of surface NMR and NMR well logging are based on the general concepts of nuclear magnetic resonance, but their implementations and operations are very distinct. The technique of surface NMR (see ref. 2 as well as Chapter 9 of this monograph) is based on earth field measurements using a large coil with a typical diameter as large as 100 m. The proton spins in the underlying earth formation are excited by passing an oscillating current at the relevant Larmor frequency through the large coil. The Larmor frequency is controlled by the strength of the local Earth's magnetic field and is therefore in the audio frequency range between 1 kHz and 2.8 kHz. The same coil is then used to detect the resulting free induction decay. From measurements with excitation currents of different strengths or durations, it is possible to infer a coarse one-dimensional image of the fluid distribution. The size of the sensitive region scales with the radius of the coil, which limits this technique to the exploration of the near surface region. To extend NMR measurements to greater depth, it is necessary to perform well logging measurements.

2.1.1 General Technique of Well Logging

The general method of well logging is a highly developed technique with many companies providing commercial services. In 2010, the logging industry generated total annual revenues in the range of tens of billions of dollars. Logging measurements are available with sensors based on many different measurement principles, including numerous electromagnetic, nuclear, acoustic, ultrasonic and NMR measurements. An overview of the various measurement principles can be found elsewhere, such as in the book by Ellis and Singer.[3] Here we focus on the technique of NMR well logging that allows the quantitative extraction of the relaxation and diffusion properties of the fluids. These measured properties can then be related to a wide range of other physical properties that are of direct interest to geoscientists and engineers, including porosity and fluid saturations, fluid compositions and viscosities, and estimates of the distribution of pore sizes and of flow properties.

There are a number of distinct modes of well logging. The two most important modes are *wireline logging* and *logging while drilling*, followed by the technique of *mud logging*. Wireline logging has traditionally been the dominant logging technique and has the longest history. In this mode, different sensors are connected into a so-called tool string that is lowered into the borehole at the end of a long armored cable. The cable acts simultaneously as a mechanical support, as a conduit of power, and as a link of communication to control the sensors and to retrieve the data to the surface. The assembly of sensors is typically first lowered to the bottom of the borehole. As the sensors are pulled back up to the surface at a controlled rate in the range of 2 cm s^{-1} to 30 cm s^{-1}, they are instructed to continuously perform measurements on the surrounding earth formation. The resulting data is referred to as log. Under special circumstances, it is also possible to perform so-called station measurements for a limited amount of time. Under typical borehole conditions, the duration of NMR stations has to be kept to less than 30 minutes to minimize the danger of getting stuck in the borehole and losing the sensor. While the tool string is stationary, it also becomes possible to use specialized sampling instruments that withdraw, collect, and analyze fluids directly from the formation.

The employment of the logging while drilling (LWD) method has grown rapidly in recent years and has now become of comparable importance to wireline logging in the oil and gas industry. In LWD operations, the sensors are directly integrated into the drill string and measurements are performed during the drilling operation. This has the advantage that the information from the sensor can be used to guide the direction of the drilling. This approach of directional drilling with LWD has enabled much higher accuracy in well placement at the desired geological location within the earth formation. Note that compared to wireline measurements, LWD measurements are less affected by fluid invasion (*i.e.* the replacement of formation fluid by borehole fluid) because the LWD measurements are performed shortly after the borehole is created.

Sensors for all forms of well logging have to be built to withstand the extreme conditions that can occur in a reservoir. The downhole temperatures and pressures can reach 175 °C and 140 MPa, respectively. The sensors have to be slim enough to fit into the boreholes with typical diameters in the range of 12 cm to 35 cm. Furthermore, LWD sensors have to be able to withstand high shocks and vibrations encountered during the drilling operation.

2.1.2 Overview of NMR Well Logging and its Challenges

Commercial services of NMR well logging are available for both wireline and LWD applications. There has been much progress in the development of this technique during the last two decades and advances continue at a rapid pace. Nevertheless, it is fair to say that NMR well logging is not yet as mature as other major logging techniques. A key reason is that NMR logging is not simply an adaption of a well-established laboratory measurement to downhole conditions. Important aspects of the NMR logging technique differ substantially from the traditional forms of laboratory NMR measurements, such as NMR spectroscopy and MR imaging. These differences include the configuration of the hardware, the dominant term in the underlying spin dynamics, data inversion, and data interpretation.

A fundamental difference between laboratory NMR and NMR well logging is that in the latter technique, the sample is located outside the sensor. Clearly, the earth formation cannot be placed inside the detection coil in a region of magnet field of high strength and homogeneity. NMR well logging tools therefore require an inside-out measurement scheme. The geometry of the magnet and radio frequency (rf) coil have to be adapted to this uncommon situation. As a consequence of the inside-out geometry, the applied magnetic fields across the sample are necessarily much weaker and less uniform than in standard NMR measurements. This leads to typical Larmor frequencies below a few MHz and inhomogeneities $\Delta B_0/B_0$ of order unity. This has to be compared to state-of-the-art high field laboratory NMR spectrometers where Larmor frequencies approach 1 GHz and $\Delta B_0/B_0 < 10^{-8}$. As a consequence, off-resonant effects play a key role in the spin dynamics for NMR well logging and they completely dominate chemical shift terms. This precludes the implementation of conventional NMR spectroscopy. Instead, NMR well logging relies on relaxation and diffusion measurements. The design of the sensor must also take into account that the apparatus is generally moving with respect of the sample during the measurement.

A number of different sensor configurations have emerged as solutions to these challenges. An overview of the various approaches is given in Section 2.2. In Section 2.3, we review the challenges posed by performing the NMR measurements in weak and grossly inhomogeneous fields and discuss the pulse sequences developed for well logging. Strategies to develop robust calibration procedures over the entire range of parameters are outlined in Section 2.4. There are often drastic changes in the environmental conditions (*e.g.* temperature and borehole salinity) during the logging operation that pose additional difficulties for NMR well logging. Issues related to data processing are covered in Section 2.5. The analysis of standard NMR measurements relies mainly on the technique of Fourier transformation, a robust and well-understood operation. In contrast, well logging deals with relaxation and diffusion data that require an inverse Laplace transformation. This is an ill-conditioned problem and poses many additional challenges. In Section 2.6, we give a brief overview of the relevant physics that control the relationships between the measured relaxation and diffusion properties, and the properties of the porous media and the complex fluids within. Finally, in Section 2.7, we present some representative applications of NMR logging.

2.2 Sensors for NMR Logging

2.2.1 Early Earth Field NMR Technique

Efforts to develop NMR well logging started early and began only a short time after the discovery of NMR by Bloch and Purcell in 1946. A historical overview of the activities up to the end of the 20th century can be found in.[4,5] The initial approach was based on an ingenious scheme using the earth magnetic field.[6] The logging apparatus consisted of a large solenoid with approximately 1000 turns. A 2 kW power supply was used to energize the coil with a large current that produced the strongest possible static magnetic field in the formation around the borehole and increased the magnetization of the nearby nuclear spins. To generate the maximum polarization, the field was applied for a time long compared to the longitudinal relaxation time (*i.e.* at least a few seconds). To detect the NMR signal, the current in the coil was then abruptly turned off in such a manner to result in a non-adiabatic removal of the applied field. This generated a free induction decay in the earth magnetic field that was detected by connecting the large coil to a sensitive preamplifier. The signal frequency corresponded to the Larmor frequency in the local earth magnetic field, *i.e.* 1.0 kHz to 2.8 kHz depending on the geographical location.

Despite the low detection frequency, the earth field approach was able to generate a signal with an adequate signal-to-noise ratio (SNR) as it was able to excite transverse magnetization in a relatively large volume. Nevertheless, the earth field method has a number of intrinsic shortcomings that limited its impact on formation evaluation. First, it was essential to eliminate any NMR signal that originates from the fluid inside the borehole. This was achieved by doping the borehole fluid with a sufficiently high concentration of paramagnetic salts. However, this procedure complicates the drilling procedure and increases the operation time, adding to the overall cost. Furthermore, it introduced a significant uncertainty in the calibration of the amplitude of the NMR signal in terms of the formation porosity because it requires the knowledge of the exact shape of the borehole and of the invasion profile of the paramagnetic ions into the formation, information that is generally not available. An additional challenge was posed by the sizeable dead-time of the receiver before the free induction decay could be recorded. Given the large initial current, the large inductance of the coil and therefore long intrinsic decay time, it took typically 25 ms before the current decayed to a sufficiently low level in the coil to detect the weak NMR signal. Signal components with relaxation times shorter than this dead-time could therefore not be detected. Such fast decaying components are often a significant fraction of the overall magnetization. As a consequence, it becomes impossible to relate the initial amplitude of the detected signal with this earth field NMR technique directly to the total porosity signal of the surrounding formation.

2.2.2 Modern NMR Well Logging Technique: Pulsed NMR in an Applied Magnetic Field

To circumvent the shortcomings of the Earth field approach, modern NMR well logging tools use externally applied B_0 fields and operate at a higher Larmor frequency. Conceptually, the overall architecture of modern NMR logging devices resembles those of standard laboratory NMR systems. They consist of a magnet system that generates the static magnetic field, and an rf coil that can be switched between an rf transmitter and a sensitive receiver. The rf coil is used to transmit strong rf pulses into the formation to generate net precessing magnetization, which is then detected with the same rf coil. Given the challenging environment for NMR logging, the static magnetic fields are not generated by superconducting magnets as in conventional NMR systems, but instead by permanent magnets. They are typically based on modern rare earth magnetic materials with a high Curie temperature such as SmCo. Pulsed NMR well logging using externally applied magnetic fields can overcome all the key limitations of the earth field based technique:

(i) Externally applied magnetic fields results in a strong spatial localization of the NMR signal with a well-defined volume of investigation. A signal is only generated in regions where the local Larmor frequency (proportional to the local field strength) is close to the rf frequency of the rf pulses used. This localization eliminates the need to dope the drilling fluid with paramagnetic salts. It also avoids the calibration uncertainties associated with the details of the borehole size and invasion profiles of the paramagnetic ions that were encountered in the earth field technique. However, the challenge now becomes designing a sensor with a sufficiently large sensitive volume that can generate signals with adequate signal to noise ratio.

(ii) The detection frequency is up to three orders of magnitude higher than in earth field measurements. In current logging instruments based on modern permanent magnets, the operating frequencies range from about 150 kHz to 2.3 MHz. As a consequence, the electronic recovery times are much shorter compared to earth field operation and it becomes feasible to detect fast relaxing components and to determine the total porosity.

(iii) The pulsed NMR technique allows flexibility in the type of measurements to be performed. The pulse sequence and pulse parameters can be tailored to the specific application of interest. As discussed below, measurement schemes are available to measure not only porosity, but also relaxation and diffusion properties quantitatively in grossly inhomogeneous fields. In addition, it is possible to vary the spatial location of the volume of investigation by changing the carrier frequency of the rf pulses, similar to the technique of slice selection in MRI.

2.2.3 General Considerations Governing the Tool Design

A fundamental challenge in designing a sensor for pulsed NMR is to find a configuration of magnets and rf coils that creates a sufficiently large resonant volume that leads to a resulting NMR signal with an adequate signal-to-noise ratio. All the possible designs are controlled by a number of general considerations. Given that the sample is outside the NMR apparatus, it is not possible to apply uniform B_0 and B_1 fields to the sample. This leads to a large distribution of the local Larmor frequency $\omega_L(\vec{r}) = \gamma B_0(\vec{r})$ and nutation frequency $\omega_L(\vec{r}) = \gamma B_{1,\perp}(\vec{r})/2$ across the entire sample. In inhomogeneous fields, the signal S generated by the normalized magnetization $m_\perp(\vec{r})$ is given by:[7,8]

$$S = \phi \frac{2\chi}{\mu_0} \int d\vec{r}\; B_0^2(\vec{r}) \frac{\omega_1(\vec{r})}{I} D(\omega_0(\vec{r})) \left[m_x(\vec{r}) + i\, m_y(\vec{r}) \right] \tag{2.1}$$

Here ϕ is the porosity, χ is the MKS susceptibility of the relevant nuclei, which is typically hydrogen, I is the current in the antenna, and $D(\omega_0)$ is the frequency response of the detector, including the tuned coil. We use the standard notation $\omega_0(\vec{r}) \equiv \omega_L(\vec{r}) - \omega_{RF}$ to indicate the offset of the local Larmor frequency from the applied rf frequency, ω_{rf}, in the rotating frame. The factor of ω_1/I is the detection efficiency of the antenna according to the reciprocity theorem.

Inspection of eqn (2.1) reveals a number of required design principles for the generation of a large NMR signal. The equation contains the term B_0^2, which is a combination of the Boltzmann factor, which describes the polarization effect, and the detection efficiency based on Faraday's law. This strong dependence on the magnetic field strength implies that magnet assemblies should be used that generate the highest possible static field B_0 in the formation. It is also important to generate transverse magnetization over the largest possible volume. The exact results depend on details of the field profiles and pulse sequences, but to first order the resonant volume can be approximated by the region in space where the local Larmor frequency deviates from the nominal Larmor frequency (*i.e.* the applied rf frequency ω_{RF}) by less than the nutation frequency, ω_1. In other words, the resonant region is approximately the region of space where $|\omega_0(\vec{r})| < \omega_1$. This implies that the static field generated by the magnet assembly in the earth formation should not only be as large as possible, but also be as uniform as possible. Earnshaw's theorem states that it is impossible to generate field profiles with a maximum outside the magnets.[9] Nevertheless, it is possible to generate fields that are sufficiently uniform and to create sensitive regions that extend either along a line or a plane. The first category includes magnet configurations that create field profiles with saddle points where the local gradient vanishes along a line, while the second category consists of magnet configurations that create gradient fields.

The resonant volume depends on the rf power. An increase of the instantaneous rf power results in larger values of ω_1, which increases the acceptable deviations of the static magnetic field from its target value. For this reason, an effort is made to maximize the peak rf power in commercial NMR well logging tools. Typical values are of the order of kW.

There is a long list of additional requirements for the successful design of an NMR logging tool. It is essential that the resonant region is completely localized within the earth formation surrounding the borehole and that it does not intersect the borehole. In addition, it is desirable that the resonant region is as far away from the borehole wall as possible to minimize the impact of possible formation damage induced by the drilling operation on the measurement. A large depth of investigation also reduces the fraction of the sensitive region that is contaminated with invaded borehole fluid. There is obviously a trade-off between the design criteria for a large depth of investigation and the goal of maximizing the signal-to-noise ratio of the measurement. Current commercial logging tools are able to achieve a depth of investigation of up to 15 cm from the borehole.

The sensor design also has to take into account the motion of the apparatus during the measurements. The resonant region has to extend along the direction of the motion. For wireline applications, the resonant region should extend along the length of the apparatus, whereas for LWD applications, the resonant region should have azimuthal symmetry to account for the rotational motion of the drill-string.

2.2.4 Configurations of NMR Logging Tools

The first practical solution of this optimization problem was described in 1980 by J. Jackson and collaborators at Los Alamos National Laboratory.[10–12] This group also presented the first experimental demonstration of the feasibility of a pulsed NMR logging tool based on permanent magnets. Since this pioneering work, a number of other suitable configurations have been identified and developed into commercial logging tools. The different configurations of the magnet and rf coils can be generally classified according to two overall properties:

(1) The azimuthal symmetry of the resonant volume: azimuthally symmetric *versus* one-sided.
(2) The radial profile of the magnetic field across the sensitive region: saddle-point *versus* gradient.

Logging tools with azimuthally symmetric resonant regions probe the formation all around the borehole. Such devices have to be operated while being centralized in the borehole. They are designed for a given borehole diameter. Logging tools with one-sided sensitivity patterns are pushed against the borehole wall during the logging operation. In this case, the depth of investigation is independent of the diameter of the borehole and the same tool can be used in a range of borehole sizes. This is an attractive feature that lowers the overall asset cost and simplifies service maintenance.

The radial profile of the magnetic field across the sensitive region controls another set of tradeoffs. A saddle point profile generates a relatively wide resonant region that makes the measurements less sensitive to lateral motion than in a tool based on a gradient design. Measurements in a gradient tool are generally more affected by diffusion effects. Depending on the application, this can be considered either an advantage or a disadvantage. In a gradient tool, it is straightforward to change the depth of investigation by changing the rf frequency. In a saddle point design, quantitative measurements require the setting of the rf frequency precisely to the Larmor frequency at the saddle point. In such tools, it is important to track and compensate any drifts in the Larmor frequency due to possible changes in the magnet temperature or accumulation of magnetic debris on the magnets.

Commercial logging tools have been developed with all four possible combinations of azimuthal and radial field profiles. We now present illustrative examples for each possibility.

2.2.4.1 *Centralized Logging Tool with Saddle-Point Magnetic Field Profile*

The magnet configuration introduced in the pioneering work by Jackson *et al.*[10–12] consists of two cylindrical magnets, polarized in opposite directions along the axis of symmetry, that are spatially separated along the axis. In the plane of symmetry perpendicular to the axes of the magnets, this magnet array produces a purely radial field with azimuthal symmetry. In the radial direction, the field strength starts at zero in the center, rises to a maximum, and then falls off at larger distance. By adjusting the magnet spacing, it is possible to achieve a relatively uniform saddle-point-like field profile in a toroidal region. An efficient way to generate an rf field perpendicular to the static field in the resonant region is to place a tuned solenoid coaxially between the magnets. The resulting resonant region has the form of a toroid with azimuthal symmetry, but without a large extent along the sensor axis. Therefore, this configuration is suited for LWD applications where the dominant sensor motion is circular, but not suited for wireline logging where the sensor moves along the borehole. Several commercial logging-while-drilling devices[13–15] are based on variations of this basic tool configuration. A schematic drawing of the device described in[13] is shown in Figure 2.1.

2.2.4.2 *Centralized Logging Tool with Gradient Magnetic Field Profile*

Taicher and Shtrikman[16] developed an alternative tool configuration that is based on two-dimensional dipole profiles for both the static and rf magnetic fields. The static field is created by a long cylindrical magnet that is magnetized perpendicular to its axis. Neglecting end effects, the resulting magnetic field lies in the transverse plane and has a magnitude that depends

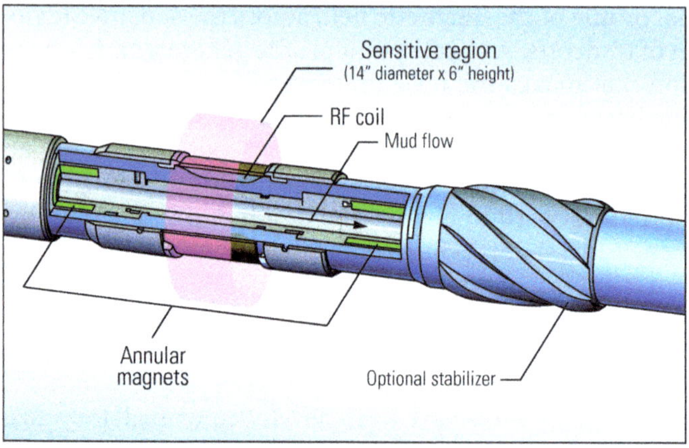

Figure 2.1 Schematic view of the NMR logging-while-drilling device with an azimuthally symmetric sensitive region described in ref. 13. The two permanent magnets, shown in green, are magnetized along the tool axis in opposite directions and generate a magnetic field that is radial in the sensitive region. Illustration copyright Schlumberger. Used with permission.

only on the distance from the tool. The direction of the field varies along the azimuth. This static field is matched by an rf field generated by a coil that is wound around the magnet and oriented in such a way that it produces a two-dimensional dipole rf field that is everywhere perpendicular to the static field. The magnets are built from non-conducting ferrite magnets to make them transparent to the rf fields. This elegant design results in an azimuthally symmetric resonant region that consists of a thin cylindrical shell extending along the tool. The diameter of the shell is controlled by the operating rf frequency. In practical implementations, the magnets extend beyond the rf coils to attenuate the impact of the end effects on the field profiles in the sensitive zone and to help polarize the spins before they enter the measurement zone. The symmetry of this design (see Figure 2.2) makes it suitable for both wireline and logging-while-drilling applications, and commercial devices have been developed for both applications.[17,18] In these devices, the operating frequencies are in the range of 500 kHz to 750 kHz, and the corresponding radii of the resonant regions are in the range of 17 to 21.5 cm. Within the resonant region, the field profile is characterized by a well-defined gradient that falls in the range of 14 G cm^{-1} to 21 G cm^{-1}. This leads to thin shells with a typical thickness of about a millimeter.

2.2.4.3 Single-Sided Logging Tool with Saddle-Point Magnetic Field Profile

Kleinberg and Sezginer[19] introduced a single-sided design with a magnetic field profile characterized by a saddle point that extends along the tool direction. This field profile is achieved with the use of two SmCo magnets that

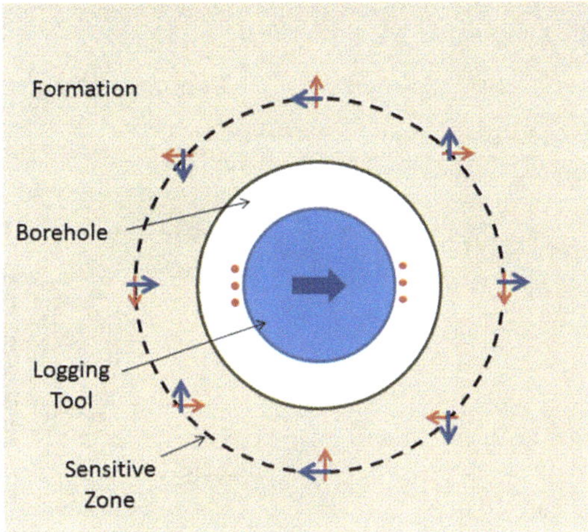

Figure 2.2 Centralized logging tool with a gradient field profile. Schematic cross section of a wireline NMR logging tool[17] based on the design by Taicher and Shtrikman.[16] The tool consists of a long cylindrical permanent magnet, shown in blue, magnetized perpendicular to its axis, as indicated by the arrow, and an rf coil shown in red. The sensitive region consists of thin cylindrical shells (indicated by the dashed line). By symmetry, in these shells the static magnetic field (shown as blue arrows) and the rf magnetic field (shown as thin red arrows) have a constant magnitude and are everywhere perpendicular to each other.

are both magnetized transverse with respect to the sonde axis. The resulting magnetic field is predominantly radial into the formation and its profile has a saddle point about 2.5 cm inside the formation. The field strength in the resonant region is 55 mT, corresponding to the Larmor frequency of 2.3 MHz. The rf coil occupies a semicircular cylindrical cavity on the face of the sonde. It can be thought of as half a coaxial cable 15 cm long that irradiates the formation with a field transverse to the static field. It is loaded with ferrite to improve the reception of weak signals from the nuclear spin system. This device, shown in Figure 2.3, has been commercialized for wireline applications.

2.2.4.4 *Single-Sided Logging Tool with Gradient Magnetic Field Profile*

When a single-sided logging tool with a saddle point is operated at an rf frequency that does not correspond to the saddle point, the sensor becomes a one-sided logging tool with a field profile characterized by a gradient. More generally, any sensor that has an rf coil located on one side operates as a single-sided gradient tool, unless special care is taken to design it as a saddle point tool. This configuration has the advantage that the sensitive region can be easily moved to a different depth away from the borehole by changing

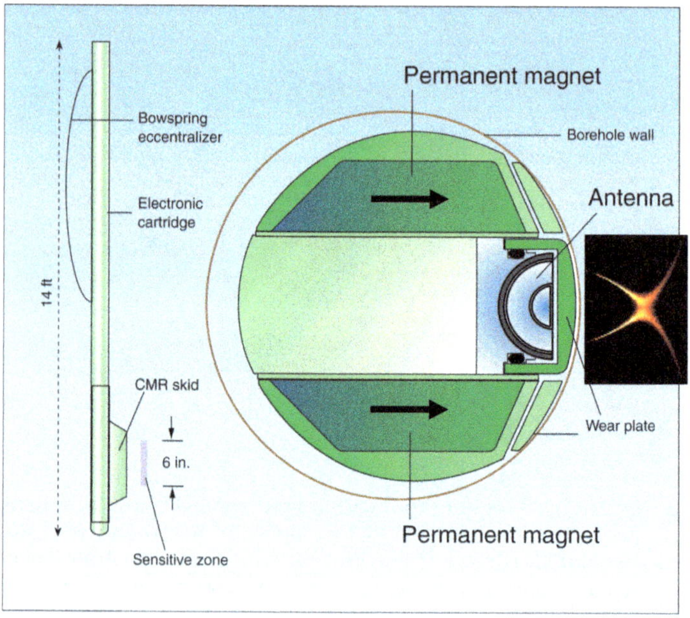

Figure 2.3 Single-sided logging tool with a saddle point field profile. Schematic view of the NMR logging tool designed by Kleinberg *et al.*[19] The permanent magnets are magnetized in the direction of the arrows. The sonde is pressed against the borehole wall. The magnitude of the magnetic field forms a saddle point about 25 mm inside the formation. The resulting sensitive region is shown in the insert. The color intensity indicates the local signal sensitivity. Illustration copyright Schlumberger. Used with permission.

the rf carrier frequency of the pulses. Examples of such sensors used in commercial wireline operations are described in.[20,21] As an illustration, Figure 2.4 shows the wireline NMR logging tool of[20] that can be operated at different rf frequencies between 1000 kHz and 500 kHz, corresponding to depths of investigations ranging from 2.5 to 10 cm. In each case, the field profile is well characterized by a uniform gradient in the range between 38 G cm^{-1} and 12 G cm^{-1}.

2.3 NMR Measurement Techniques

2.3.1 Measurements in Grossly Inhomogeneous Fields

It is a general feature of inside-out measurements on extended samples, such as those encountered in NMR well logging, that the range of Larmor frequencies across the entire sample is much larger than the amplitude of the rf field. Therefore, all pulses act as slice selective pulses. As a consequence, the $T_2{*}$ dephasing time of the free induction decay is very short and

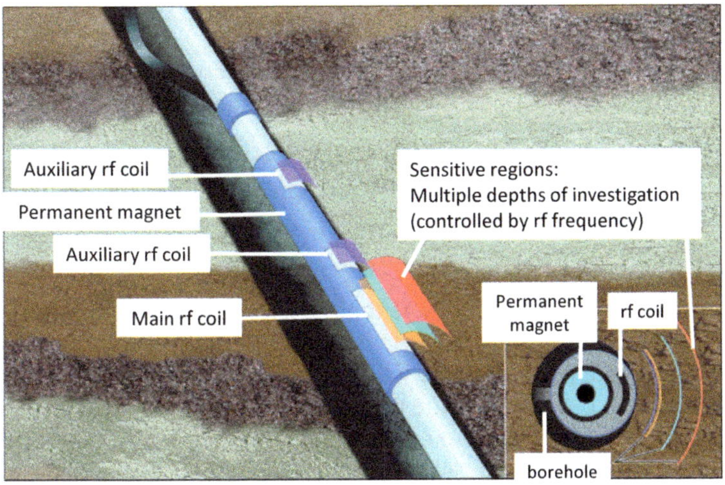

Figure 2.4 Single-sided logging tool with a gradient field profile. Schematic view of the NMR wireline logging device described in.[20] The magnet (shown in blue) generates a gradient field in the formation. The main rf coil can transmit and receive at different rf frequencies, which correspond to different sensitive regions as shown by the differently colored surfaces. In addition, there are two auxiliary rf coils that are operated at a higher rf frequency and are associated with a shorter sensitive region. The sensor is pushed against the wall of the borehole by the bow spring placed on the opposite side of the tool, indicated on the top left of the figure. Illustration copyright Schlumberger. Used with permission.

only of the order of the pulse duration. This makes it difficult to acquire the signal of the free induction decay, especially at these low Larmor frequencies where in all practical cases T_2* is shorter than the dead-time of the electronics. It is therefore essential to use multi-pulse sequences to generate echoes that can be more readily detected. However, the detailed NMR response of such multi-pulse sequences and the extraction of quantitative information is complicated by the grossly inhomogeneous fields of logging tools as all pulses act as slice selective pulses. In particular, it is critical to use sequences that are robust against the rapid accumulation of the large resonance offset effects.

The prototypical multi-pulse sequence for moderately inhomogeneous fields widely used in conventional NMR applications is the Carr–Purcell–Meiboom–Gill (CPMG) sequence[22,23] shown in Figure 2.5. It consists of an initial 90° excitation pulse, followed by a long series of 180° refocusing pulses that are phase shifted by 90° with respect to the excitation pulse. Echoes appear at the mid-points between successive refocusing pulses. In the practical implementation, a simple two-step phase cycling scheme of the first pulse is used to eliminate baseline offsets and ringing effects. This basic sequence and modifications thereof form the basis of all pulsed NMR logging measurements. It allows the determination of the overall amplitude and the

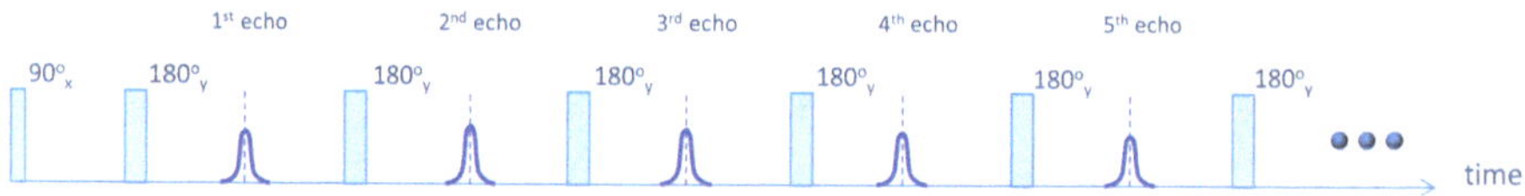

Figure 2.5 Basic Carr–Purcell–Meiboom–Gill (CPMG) sequence used to measure amplitude and transverse relaxation time. The dashed lines indicate the nominal echo times.

decay time of the magnetization. As long as the echo spacing is shorter than any intrinsic relaxation time, the measured amplitude of the early echoes is proportional to the initial longitudinal nuclear magnetization of the sample, which in turn can be related to the porosity, as is further discussed in Section 2.4. The decay time of the echo amplitudes, T_2, is related to the pore size or viscosity of the fluid, as discussed in Section 2.6.

In conventional applications, the inhomogeneities in the applied B_0 field are typically much smaller than the strength of the rf field, B_1. In this case, the refocusing pulses in the CPMG sequence are able to compensate the dephasing caused by the field inhomogeneities and the magnetization forms echoes that are all in-phase, as shown in the top row of Figure 2.6.

When the inhomogeneities in B_0 approach or exceed B_1, dephasing during the application of the pulses becomes significant. As a consequence, the phase of the magnetization generated by the CPMG sequence becomes highly dependent on the offset frequency and the echo number, as illustrated in Figure 2.6. The simulation demonstrates that the resulting magnetization at the time of the nominal echo can lie anywhere on the Bloch sphere. At first sight, this appears to rule out the CPMG sequence as a suitable sequence for quantitative measurements in NMR logging.

Fortunately, the situation is not as dire as suggested by the results shown in Figure 2.6. In fact, the CPMG is surprisingly well suited for robust measurements in grossly inhomogeneous fields.[8] This is evident from the experimental results shown in Figure 2.7. The top panel shows the traces of the first ten CPMG echoes on an extended sample of doped water in a gradient field, while the lower panel shows the decay of the amplitudes of all the echoes. After the first few refocusing pulses, the echoes quickly approach a constant shape that is in phase with the refocusing pulses.

To reconcile the apparent contradictions between Figures 2.6 and 2.7, we have to consider that the detected echo signal is the sum of the contributions from all positions in the inhomogeneous field. Even though the local magnetization can point in any direction on the Bloch sphere, there is a net excess of magnetization along the axis of the refocusing pulses. The experimental results show that this excess magnetization is a quantity that is robust from echo to echo and gives rise to the characteristic echo shape. After the first few echoes, the overall amplitudes of these echoes decay exponentially. The CPMG sequence therefore allows the robust extraction of relaxation times even in grossly inhomogeneous fields.

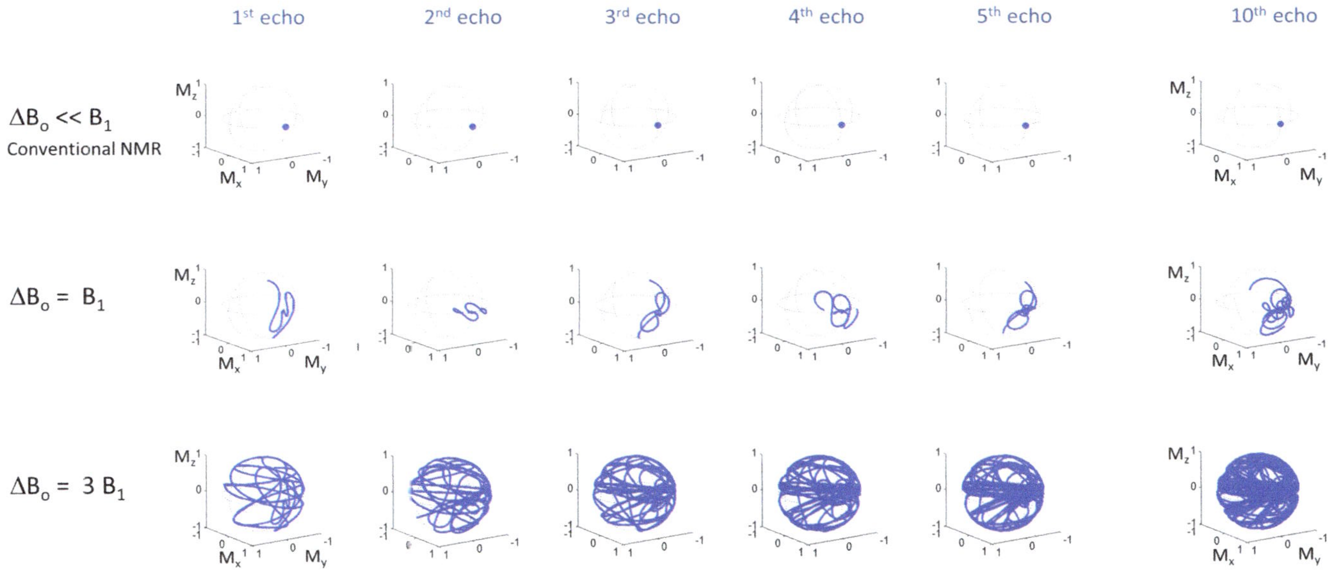

Figure 2.6 Spin dynamics of the CPMG sequence in inhomogeneous fields: Simulation of the magnetization on the Bloch sphere at the nominal echo times of echoes 1 to 5 and echo 10 for three different B_0 field inhomogeneities. The results show that in fields as inhomogeneous as in well logging ($\Delta B_0 \gg B_1$), the magnetization cannot be fully refocused and can lie anywhere on the Bloch sphere.

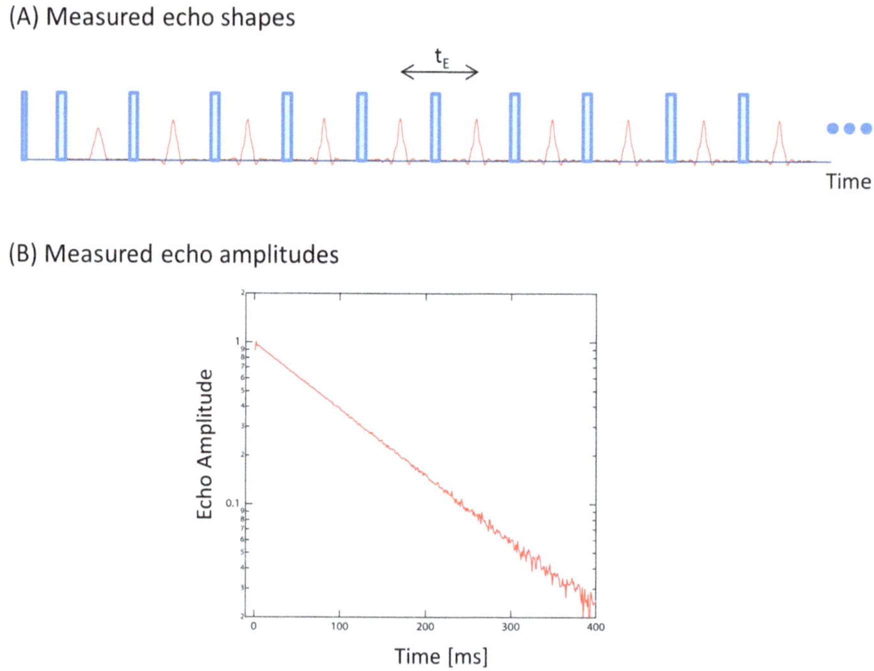

Figure 2.7 (A): CPMG pulse sequence (blue) and measured echo signal (red) for the first ten echoes of a sample of doped water in a gradient field. After the second echo, the shapes of the echoes are invariant from echo to echo. (B): The measured decay of the echo amplitudes *versus* time is governed by the intrinsic relaxation time of the sample, $T_{2,\mathrm{eff}}$.

This simple behavior of the detected signal of the echoes can be understood by analyzing the underlying spin dynamics in terms of the refocusing cycle.[8] The key step is to divide the magnetization for any offset frequency into a CPMG component that falls along the axis of the refocusing cycle and a CP component perpendicular to it. The CPMG components are refocused from echo to echo and decay with a time constant $T_{2,\mathrm{eff}}$ that reflects the intrinsic relaxation properties of the sample. In contrast, the CP components oscillate rapidly and average out after only one or two echoes. The observed signal, a superposition of the two components, is dominated by the CPMG component that is coherent from echo to echo. The CP component is only noticeable for the first two echoes where it gives rise to a characteristic transient in the measured echo amplitudes. Applying a multiplicative correction factor to the amplitudes of the first few echoes can easily compensate for this spin dynamics induced transient. In the representation in Figure 2.6, the CPMG components appear as a surplus of magnetization along the x-axis on the Bloch sphere. It is easier to see the CP components that are spread across the whole Bloch sphere, but they do not contribute significantly to the detected echoes.

2.3.2 Relaxation Measurements

After applying the correction factor that accounts for the transient effect to the amplitudes of the first few echoes, the decay of the l-th echo amplitude with echo spacing t_E is described by the following kernel:

$$k_{T_2}(lt_E) = \exp\left\{-\frac{lt_E}{T_2}\right\} \tag{2.2}$$

The observed relaxation rate $1/T_{2,\text{eff}}$ of the CPMG sequence in inhomogeneous fields is determined by the sample, but it can deviate from $1/T_2$. The analysis in[8] shows that $1/T_{2,\text{eff}}$ is in general a weighted sum of the $1/T_2$ and $1/T_1$ relaxation rates:

$$1/T_{2,\text{eff}} = (1-\varepsilon)1/T_2 + \varepsilon/T_1 \tag{2.3}$$

The parameter ε is a tool-specific parameter, but it is typically quite small in the range of 0.09 to 0.12. Therefore, for most practical purposes in NMR well logging, the deviation of $T_{2,\text{eff}}$ from T_2 is ignored and the approximation $T_{2,\text{eff}} \approx T_2$ is implicitly made.

For practical applications, it is critical to keep the echo spacing t_E as short as possible. In practice, the minimum value of t_E is controlled by the electronics design and the efficiency of quenching the antenna ring-down, as well as by constraints on power and heat dissipation. In current NMR logging tools, values of t_E as short as 200 μs are achieved. Such a short echo spacing has several benefits: First, it allows the detection of the short T_2 components that are encountered in zones with significant bound water or heavy oil, or in shales. Second, it reduces the effect of diffusion on the measured relaxation times, as discussed further below. Third, it effectively increases the signal to noise ratio of the measurements, as adjacent echo amplitudes can be averaged. At the same time, it is necessary to acquire a large number of echoes in order to characterize the longest relaxation times. It is common to acquire a few thousands echoes in a CPMG train.

2.3.3 Diffusion Measurements

In the presence of large field inhomogeneities and the associated field gradients, diffusion effects shorten the measured relaxation time. To a first approximation, diffusion adds a rate given by:[22]

$$1/T_{2,\text{diff}} \approx 1/12\gamma^2 g^2 D t_E^2 \tag{2.4}$$

Here γ is the gyromagnetic ratio, g is gradient in the static field, D is the diffusion coefficient, and t_E is the echo spacing. By monitoring the relaxation time as a function of echo spacing, eqn (2.4) enables the extraction of the diffusion coefficient from the NMR measurements. This approach was first demonstrated in an NMR well logging application by Akkurt *et al.*[24] and used to identify a gas zone.

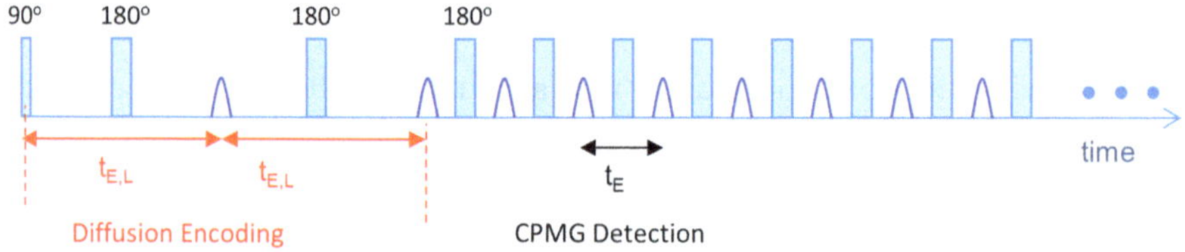

Figure 2.8 Diffusion editing sequence used to measure diffusion in the static gradient of a logging tool.

Eqn (2.4) is exact on resonance where the refocusing pulses act as proper 180° pulses. However, it is only an approximation for measurements in grossly inhomogeneous fields.[25] In well logging applications, the observed diffusion induced relaxation rate is found to be about 30% larger than indicated by eqn (2.4) as off-resonance contributions with higher coherence pathways and higher diffusion sensitivity contribute to the signal. The exact behavior depends on experimental details, including the filtering of the echo.

A more quantitative approach for diffusion measurement is achieved with the so-called diffusion editing sequence shown in Figure 2.8.[26,27] It is a simple modification of the basic CPMG sequence where the first two echo spacings, indicated by $t_{E,L}$, are systematically increased between different measurements, while all subsequent echo spacings are kept at the minimal value t_E. In this case, the diffusion response is dominated by diffusion during the two-echo encoding time and it leads to an attenuation of the subsequent echoes by:

$$k_D(t_{E,L}) = \alpha \exp\left\{-\frac{1}{6}\gamma^2 g^2 D\, t_{E,L}^2\right\} + \beta \exp\left\{-\frac{1}{3}\gamma^2 g^2 D\, t_{E,L}^2\right\} \tag{2.5}$$

Here α and β are coefficients that are tool dependent and indicate the relative contributions from the direct and stimulated echo coherence pathways, respectively.[28]

2.3.4 T_1 Measurements

The longitudinal relaxation time, T_1, can be measured by adapting a saturation-recovery measurement technique. A typical implementation consists of a series of CPMG sequences that are run consecutively with a variable delay time T_w. At the end of the CPMG sequence, the magnetization is saturated and the longitudinal magnetization recovers towards the thermal equilibrium with the time constant T_1. When the recovery time T_w is not much longer than T_1, the amplitude of the subsequent CPMG echoes are reduced by the factor:

$$k_{T_1}(T_w) \approx 1 - \exp\{-T_w/T_1\} \tag{2.6}$$

This kernel is exact on-resonance, but again only an approximation for operations in inhomogeneous fields. The CPMG is able to fully saturate the magnetization only if it is close to resonance. It is possible to account for off-resonance effects as discussed in ref. 25.

2.3.5 Distribution Functions

2.3.5.1 One-Dimensional Distribution Function—T_2 Distribution

Given the heterogeneous nature of the fluid-saturated rock samples of interest here, the relaxation and diffusion properties can generally not be described by a single relaxation time T_1 or T_2, or by a single diffusion coefficient D. Instead, it is necessary to introduce distribution functions for these quantities, such as $f(T_1)$, $f(T_2)$, or $f(D)$, to characterize their NMR behavior.[29] The decay of the echoes of a CPMG sequence measured on such samples is therefore in general not single-exponential. The amplitude of the l-th echo is then described by:

$$S_l = \int \mathrm{d}T_2 \; f(T_2) \; k_{T_2}(lt_\mathrm{E}) \tag{2.7}$$

Since the kernel k_{T_2}, given in eqn (2.2), has an exponential form, the signal is essentially the Laplace transform of the distribution function $f(T_2)$, which is commonly referred to as T_2 distribution. The inversion of eqn (2.7), *i.e.* the extraction of the T_2 distribution, $f(T_2)$, from the data S_l is discussed in Section 2.5.

2.3.5.2 Multi-Dimensional Distribution Function

In addition to the T_2 distribution function, it has proven particularly powerful to consider multi-dimensional distribution functions, such as $f(T_1,T_2)$,[30] $f(D,T_2)$,[26] and $f(D,T_1,T_2)$.[31] In particular, distribution functions involving diffusion and relaxation have become essential tools for fluid typing, as discussed in more detail in Sections 2.6 and 2.7.

Such multi-dimensional distribution functions can be measured with repeated measurements where at least one parameter is systematically varied. Figure 2.9 shows pulse sequences that are suitable for the determination of the two-dimensional distributions functions $f(D,T_2)$ and $f(T_1,T_2)$. In the case of D-T_2 distribution functions, the sequence consists of a series of diffusion editing sequences (see Figure 2.8) with a range of different initial echo spacings $\{t_\mathrm{EL}^{(j)}\}$. For the j-th diffusion editing sequence, the amplitude of the l-th echo is given by:

$$S_{j,l} = \iint \mathrm{d}D \, \mathrm{d}T_2 \; f(D, T_2) \; k_\mathrm{D}(t_\mathrm{EL}^{(j)}) \; k_{T_2}(lt_\mathrm{E}) \tag{2.8}$$

For the measurement of T_1–T_2 distribution functions, the overall sequence consists of multiple CPMG sequences with a range of different wait

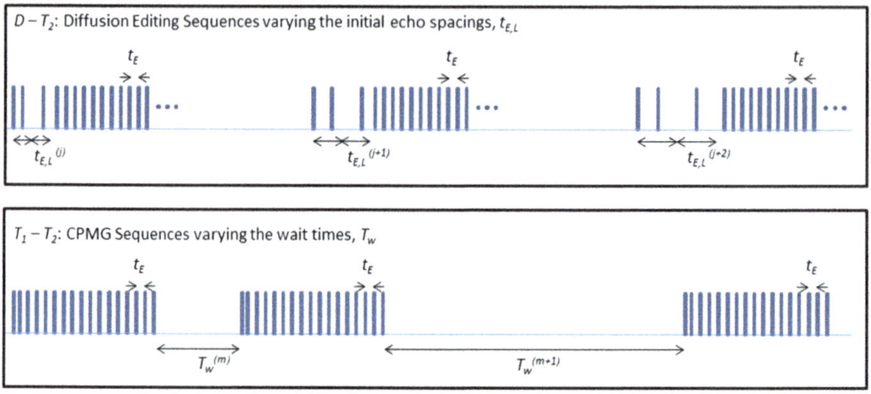

Figure 2.9 Two representative pulse sequences to determine multi-dimensional distribution functions: $f(D,T_2)$ on the top, and $f(T_1,T_2)$ on the bottom.

times $\{T_W^{(m)}\}$. The resulting amplitude of the l-th echo for the m-th wait time is given by:

$$S_{m,l} = \iint \mathrm{d}T_1\ \mathrm{d}T_2\ f(T_1, T_2)\ k_{T_1}(T_W^{(m)})\ k_{T_2}(lt_E) \tag{2.9}$$

In current logging operations, the most advanced sequences combine the measurements of T_1, T_2, and diffusion properties at multiple depths of investigations, r. The depth profiling is achieved by varying the rf frequency of the pulses, in a way that is analogous to multi-slice measurements in standard MRI. In NMR well logging, it enables spatial profiling of the NMR properties away from the borehole.

Such 4-dimensional measurements[31] allow the determination of $f_r(D,T_1,T_2)$, the distribution function of D, T_1, and T_2 at a radius r, and are based on a suite of CPMG and diffusion editing sequences where the values of $t_{E,L}$, T_W, and rf frequency are systematically varied. The signal can be described by:

$$\begin{aligned} &S_{j,m,l}^{(r)}\left(t_{E,L}^{(j)}, T_W^{(m)}, lt_E, \omega_{rf}(r)\right) \\ &= \iiint \mathrm{d}D\mathrm{d}T_1\ \mathrm{d}T_2\ f_r(D, T_1, T_2) k_D\left(t_{E,L}^{(j)}\right) k_{T_1}\left(T_W^{(m)}\right) k_{T_2}\left(lt_E\right) \end{aligned} \tag{2.10}$$

The inversion procedure required to extract these multi-dimensional distribution functions from the measured echo amplitudes is discussed in Section 2.5.

2.3.6 Tool Motion Effects

The kernels given above apply strictly speaking only when the tool is stationary with respect to the formation. Fast tool motion can noticeably modify the measured response. It leads to extra signal decay that affects in particular the determination of long relaxation times. For wireline logging, the rate of the extra signal decay scales to first order as v/L, where v is the logging speed and L is the length of the sensitive region. With higher logging speeds and longer echo spacings, the form of the extra decay becomes more complicated and additional odd-even echo modulation can appear in the signal decay. In LWD operations, lateral motion is often the dominant source of the extra motion-induced signal decay. In this case, the extra decay scales to first order with v/w_r, where w_r is the radial width of the sensitive region. However, the form of the extra decay can be rather complicated and does not even need to be monotonic, as documented by Speier *et al.*[32]

In addition to the extra signal decay, tool motion can also affect the initial amplitude measured by the NMR tool. Depending on the situation, motion can increase or decrease the amplitude. If the tool moves between subsequent CPMG measurements a distance that is significant compared to the length of the rf antenna, magnetization from regions not affected by the previous rf pulses enters the sensitive region and gives rise to a larger signal than anticipated by eqn (2.6). It is possible to take advantage of this effect to speed up the measurement process, as first discussed in ref. 33. If the tool moves fast enough so that the subsequent CPMG sequences probe non-overlapping regions of the formation, the initial amplitude of the signal becomes insensitive to T_w/T_1. This approach requires that there is a long enough section of magnets above the rf section so that the magnetization entering the sensitive zone is fully polarized.

Using this technique of non-overlapping measurements with a fast-moving tool that has a long prepolarization region, it is therefore possible to generate CPMG measurements that show the full initial amplitude, even if T_w is not long compared to T_1. While the tool motion will leave the initial amplitudes of the measurements unaffected, and therefore allow a robust measurement of porosity, it will however corrupt the measured T_2 relaxation time.

At even higher logging speed, components with long T_1 will not be exposed long enough to the prepolarization region and will fail to reach the thermal equilibrium magnetization. In this case, the measured NMR porosity will underestimate the true formation porosity and it is necessary to reduce the logging speed to obtain reliable data.

2.4 NMR Log Calibration

2.4.1 Calibration Requirements and Challenges

The calibration process is a critical part of NMR well logging and poses more challenges than are typically encountered in other NMR applications. It is essential that the measured signal amplitude can be quantitatively related to the density of protons in the formation, which then allows the determination of the formation porosity. Each commercial NMR logging service has its own specifications for accuracy and precision of the measured quantities. In most cases, porosity errors of up to 5% are considered acceptable, including both statistical (precision) and systematic (accuracy) errors. To meet this demanding requirement over the complete range of operating parameters, it is essential that the sensor is first rigorously characterized during the engineering phase, that the established validation and verification procedures are then strictly followed during the manufacturing process, and that the tool is periodically recalibrated while in operation.[34] Detailed procedures are developed for each sensor. Key responses are periodically checked and tracked for every tool throughout its deployment history. If deviations of any of these responses outside their pre-defined tolerance limits are detected, the tool will be withdrawn from operation and then subjected to detailed inspection and maintenance.

As indicated, the main calibration challenge in NMR well logging concerns the calibration of the amplitude of the measured NMR signal. The proportionality constant between the amplitude and the porosity is controlled not only by the tool design, including the configuration and strengths of the static and rf magnetic fields and the electronics, but also by the operating parameters of the pulse sequence used during the measurement, including rf frequency and pulse durations. The calibration process is further complicated by its dependence on a range of environmental parameters, such as borehole salinity and temperature, which can change drastically during the logging operation.

In comparison to the amplitude calibration, the issues related to the calibration of relaxation and diffusion behaviors are less daunting. A reliable determination of relaxation times requires the absence of ringing artifacts and a calibration of the transient effect in the first few echo amplitudes, as discussed in Section 2.3. During the sensor calibration procedure, these spin dynamics effects are determined through the analysis of experimental data and compared to the modeling predictions. Similarly, diffusion measurements are performed on a standard sample to calibrate the strength and distribution of the B_0 gradient in the sensitive zone and to determine the corresponding values of the parameters α and β of the diffusion kernel, given in eqn (2.5).

2.4.2 Sensor Characterization/Engineering

As outlined in Section 2.2, a number of commercial NMR logging tools are currently in operation that are based on radically different designs and operate at frequencies ranging from ~150 kHz up to ~2.3 MHz. Given these design differences, the response of the NMR tools to changing environmental conditions is instrument specific, not universal. A comprehensive description of the sensor response models is beyond the scope of this article. However, there are some general properties shared by all NMR tools. The initial signal amplitude, S, extracted from a CPMG sequence after properly phasing the signal and correcting for the transient effect may be written as:

$$S = \phi \; HI \; M_0 \; F_{\text{pol}} \int I(\omega) \; D(\omega) \; \zeta(\omega) \, \mathrm{d}\omega \tag{2.11}$$

The different quantities in this equation are described in Table 2.1.

Table 2.1 Key parameters determining the signal amplitude in an NMR logging tool.

Parameter	Description	Factors controlling this parameter	Parameter and condition during calibration
ϕ	Porosity	Property of formation	1.0 (For calibration bottle)
HI	Hydrogen Index	Property of formation fluid	1.0 (For water)
$M_0 \sim \frac{B_0}{T}$	Magnetization density at thermal equilibrium	Formation temperature; static field strength in the sensitive volume	Ambient temperature; default value of static field strength in sensitive volume
F_{pol}	Polarization factor	T_1, logging speed, pulse sequence, B_0 field profile of tool	1.0 (Stationary measurement; long polarization time)
$\zeta(\omega)$	NMR response; spectrum of emf induced in the receiver coil due to transverse magnetization	Sensor design including magnet and rf field configuration; design of transmitter electronics and power level; pulse sequence and operating parameters	Standard calibration sequence with default parameters; rf frequency tuned to magnetic field in sensitive volume
$D(\omega)$	Response of receiver	Receiver electronics, including gain and filters; quality factor and tuning of receiver coil at relevant borehole temperature and conductivity	Standard gain and filter setting; quality factor and tuning at controlled/ambient conditions
$I(\omega)$	Acquisition and detection filter	Design of digital filters and integration windows	Set to standard values

The polarization factor F_{pol} describes the degree of polarization relative to the thermal equilibrium magnetization M_0 that the spins have acquired at the start of a CPMG sequence. For stationary tools, this factor is identical to the kernel $k_{T1}(T_w)$ given in eqn (2.6) and depends only on the ratio of the wait time T_w and the longitudinal relaxation time T_1. When the tool is moving quickly, the initial magnetization can deviate significantly from M_0 even when $T_w \gg T_1$ because the spins have not been exposed sufficiently long to the magnetic field generated by the tool. At the limit of $T_w \gg T_1$, the polarization factor, F_{pol}, is controlled by the spatial field profile, $B_0(\vec{r})$, of the magnet and the motion of the tool relative to the formation, $\vec{r}(t)$:

$$F_{pol} = \frac{1}{T_1}\int_0^{\infty} \frac{B_0(\vec{r}(t_0 - t))}{B_0(\vec{r}(t_0))} \exp\left\{-\frac{t}{T_1}\right\} dt \tag{2.12}$$

Here t_0 indicates the starting time of the CPMG.

A change in the environmental conditions, such as temperature or borehole salinity, affects the NMR response, $\zeta(\omega)$, and the receiver response, $D(\omega)$, in non-trivial ways. The key objective of the sensor characterization is to determine these dependences quantitatively and parameterize the sensitivities in terms of properties, P_i, that can be readily measured during logging. The auxiliary measurements P_i typically include temperature, tuning frequency, antenna gain and Q factor, Larmor frequency at the sensitive zone, and some measure of the rf field strength, B_1 (*i.e.* rf pulse amplitude).[5] For gradient logging tools, the Larmor frequency at the sensitive zone is set by the rf carrier frequency of the pulses. For saddle point logging tools, the Larmor frequency at the sweet spot can be determined from an analysis of the echo shapes.[35]

Essentially, it is necessary to determine the sensitivity function F that describes the dependence of the integral in eqn (2.11) on the auxiliary measurements:

$$\int I(\omega)D(\omega)\zeta(\omega)\mathrm{d}\omega \equiv F(P_1, \ldots, P_n) \tag{2.13}$$

The form of eqn (2.13) is quite general and applies to virtually any NMR measurement. However, well logging tools must operate in an extremely wide range of environments, including at temperatures up to 175 °C and in salt saturated boreholes with electrical conductivities up to 100 S m^{-1}. Consequently, the range of the sensitivity function F can be large. A generic dependence of the sensitivity function on borehole conductivity is plotted in Figure 2.10. Also shown is a corresponding curve for the variation of antenna Q. The sensitivity function will also depend on temperature, requiring at least a second dimension to characterize F. Note that in the hypothetical example shown, the sensitivity function decays rapidly at high conductivity. This

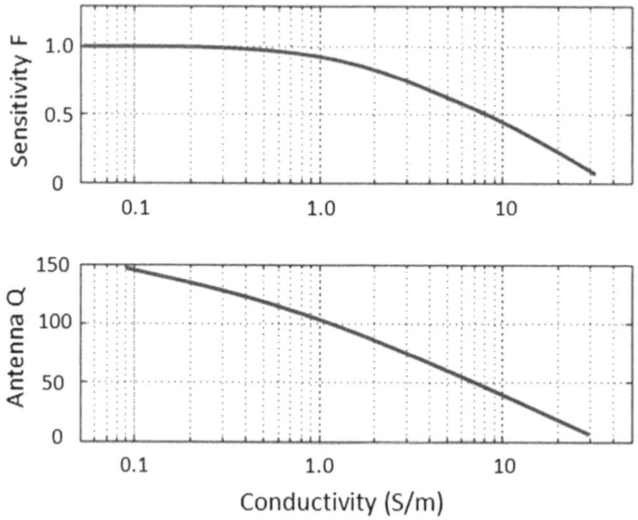

Figure 2.10 Schematic dependence of calibration function, F, and of antenna quality factor, Q, on borehole conductivity.

effect, as well as increased power requirements, limit the highest conductivity in which the NMR logging tool can operate.

2.4.3 Validation and Verification/Manufacturing

After a tool of a given design has been thoroughly tested and characterized, the manufacturing of additional tools begins. Each individual tool is subject to a validation and verification (V&V) procedure prior to shipping to a field location. This procedure tests the tool's calibration and overall performance. It ensures that each tool meets the established specifications, including the ratings on temperature, pressure, and shock tolerance. Key performance indicators, such as signal-to-noise ratio and level of antenna ringing, are also tracked.

2.4.4 Periodic Calibration/Operations

Once a tool is deployed to the field, it is subject to periodic tests and calibrations throughout the course of its operational life. Depending on the tool type and company-specific procedures, different levels of calibration tasks may be established. The most commonly performed calibration for NMR tools involves the measurement of the signal amplitude for a known standard, typically a bottle of water. In this case, a purpose-built container filled with water is placed on the tool such that the measurement volume resides completely inside the container. Figure 2.11 shows a schematic of a typical calibration setup for an azimuthally symmetric logging tool. A practical aspect of surface calibration that warrants mention is the need for a Faraday shield. This is a necessity in typical oilfield facilities to avoid interference with external rf noise. Fortunately, this is no longer necessary once the tool is in a borehole. In this situation, the brine-filled formation surrounding the borehole acts as a large Faraday shield. To confirm the amplitude calibration, an NMR acquisition sequence is run with the calibration setup and the signal amplitude, S(cal), is measured. This reference value for HIϕ = 1 at ambient

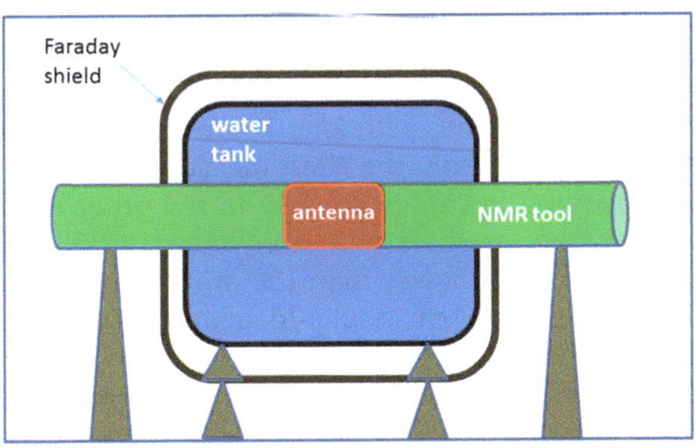

Figure 2.11 Schematic of calibration setup for NMR logging tool.

condition is stored with the tool, along with the characterization parameters, $\{P_1(\mathrm{cal}),..., P_n(\mathrm{cal})\}$, until the next calibration is performed.

2.4.5 Application of Calibration While Logging

During logging, the measured signal $S(t)$ is compared to the calibration signal $S(\mathrm{cal})$ and the most recent calibration parameters, $\{P_1(t),..., P_n(t)\}$, are used to normalize the signal and express it in terms of porosity units

$$[\phi \mathrm{HI}](t) = \frac{S(t)}{S(\mathrm{cal})} \times \frac{T(t)}{T(\mathrm{cal})} \times \frac{B_0(\mathrm{cal})}{B_0(t)} \times \frac{F(P_1(\mathrm{cal}),\ldots,P_n(\mathrm{cal}))}{F(P_1(t),\ldots,P_n(t))} \tag{2.14}$$

The "NMR porosity", $[\mathrm{HI}\phi](t)$, is generally plotted as a function of depth, rather than time, so that it may be combined with other log measurements for interpretation purposes. It should be stressed that the "NMR porosity" reflects both the matrix porosity, ϕ, and the average fluid hydrogen index, HI. While it is often a reasonable approximation to assume HI = 1 and equate NMR porosity with formation porosity for a "quicklook" interpretation, more rigorous log analysis is required to derive an accurate formation porosity.[36,37]

It is also important to detect and flag any occurrence of extreme conditions where the data quality might not be able to meet the specifications. We illustrate this point with two examples. Excessive magnetic debris in the mud (*e.g.* from shavings of the casing) can accumulate around the magnets,[35] distort the magnetic field and detune the antenna, resulting in a loss of NMR signal that is difficult to predict accurately. As a second example, excessive lateral motion during LWD operations can lead to additional signal loss.[32] In favorable cases, some degree of correction can be applied to the data based on additional auxiliary measurements. In severe cases, it is only possible to flag the corrupted data.

It is essential to avoid measurement errors due to the use of an inappropriate acquisition sequence for the particular logging environment. For example, acquisition sequences with short polarization times may be employed for logging environments where no long T_1 fluids are anticipated, such as sandstone formations containing heavy oil. In this case, $F_{\mathrm{pol}} = 1$. However, if these same sequences are used to log formations containing light hydrocarbons, it is likely that these fluids with long T_1 are not fully polarized and $F_{\mathrm{pol}} < 1$. The porosities inferred from the NMR measurements will then exhibit a deficit with respect to the true porosities. In the other limit, formations containing tar, bitumen, or shale may likely contain fluids with intrinsic transverse relaxation times shorter than the minimum echo spacing. This corresponds to HI < 1. This again results in an apparent porosity deficit. In such cases, the initial NMR amplitude does not directly correspond to the porosity and the NMR data must be interpreted in conjunction with other log measurements in order to derive accurate formation properties.[38] This is further discussed in Section 2.7.

2.5 Data Processing and Analysis

2.5.1 Inversion Challenges

After the first generation pulsed NMR logging services were introduced in the early 1990s,[17,39] it was quickly recognized that data processing posed a major challenge. In these early tools, the data consisted of single CPMG decays. The echo amplitudes, S_l, after proper calibration outlined in the previous section, are related to T_2 distribution functions, $f(T_2)$, through eqn (2.7). The extraction of the distribution function from the measured echo amplitudes essentially consists of an inverse Laplace transform (ILT) with a non-negativity constraint. In practice, the ILT is generally framed as a least squares minimization. For example, in the case of a single CPMG measurement the cost function to be minimized is:

$$\Psi = \sum_{l=1}^{N_E} \left(\left[\sum_{j=1}^{N_j} k_{T_{2,j}} (l t_E) \ f(T_{2,j}) \right] - S_l \right)^2 \tag{2.15}$$

where S_l is the l-th echo amplitude, $k_{T_{2,j}}(lt_E)$ is the measurement kernel given in eqn (2.2), and $f(T_{2,j})$ is the discrete T_2 distribution function to be estimated.

This inversion task is considerably more difficult than the processing encountered in standard NMR for three distinct reasons. First, the NMR well logging data is generally much noisier, with typical signal to noise ratio (SNR) of less than 10 for a single echo. Second, unlike the inverse Fourier transform, the ILT is an ill-conditioned problem.[40,41] This implies that even for data with very high SNR, many different distribution functions exist that fit the data within the experimental uncertainties. Third, practical considerations limit the number of measurements and the amount of data that can be acquired during each measurement cycle or "scan". Constraints are imposed by various factors, such as the minimum logging speed (the rate at which tools are conveyed along the borehole), the required SNR and vertical resolution, the telemetry bandwidth, the power drain, or the available downhole processing resources.

The non-uniqueness of the inversion is illustrated in Figure 2.12. Panel (A) shows the CPMG signal of a high SNR laboratory measurement with two fits superposed to the data. The two fits correspond to the two T_2 distribution functions shown in panel (B). While these distribution functions show significant differences, they generate two almost identical fits to the data that cannot be easily distinguished in panel (A). In fact, the fits differ by less than ±0.02 pu over the entire range, as shown in panel (C). This is over an order of magnitude smaller than the uncertainties in the experimental data, indicated by the dashed lines in (C). For all practical purposes, it is therefore impossible to use experimental relaxation measurements to distinguish between the two T_2 distributions. From a mathematical point of view, both T_2 distributions describe the relaxation data well. This non-uniqueness calls for caution when interpreting details of the distribution functions. As an

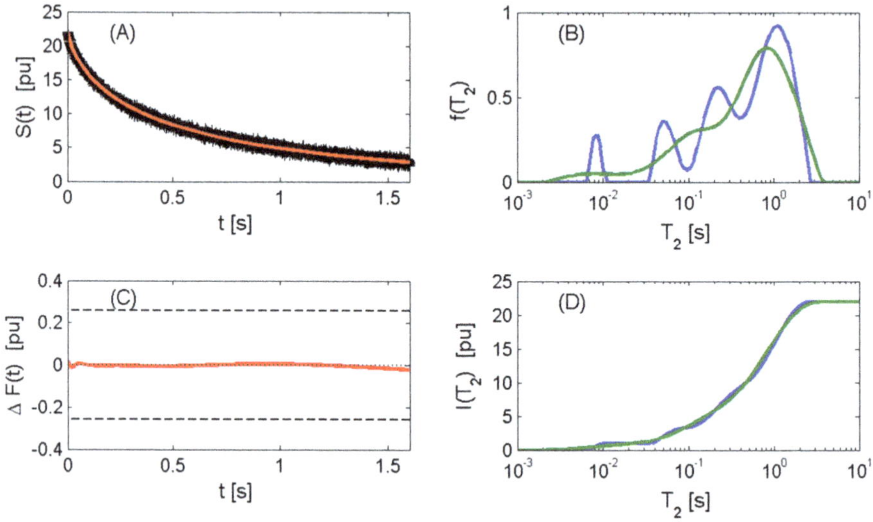

Figure 2.12 Non-unique inversion of relaxation data. Panel (A) shows experimental CPMG data in black, and two fits to the data in red. The two fits correspond to the two different distribution functions shown in (B). Panel (C) shows that the difference of the two fits (red line) is less than ±0.02 pu over the entire range, as compared to the noise in the experimental amplitudes of ±0.26 pu (dashed line). Panel (D) compares $I(T_2)$, the porosity associated with relaxation times less or equal than T_2, for the two solutions.

example, it is in general difficult to distinguish a distribution consisting of multiple discrete components from a more continuous distribution.

Fortunately, many important quantities extracted from the distribution functions are intrinsically more robust than the distribution functions themselves. As discussed in Section 2.6, quantities such as porosity, bound fluid volume, saturations, permeability, or viscosity are derived from integral functions of the distribution function. Panel (D) of Figure 2.12 compares the integral function $I(T_2) \equiv \int_{T_{2,\min}}^{T_2} \mathrm{d}T_2' \, f(T_2')$ for the two distribution functions. $I(T_2)$ corresponds to the porosity associated with relaxation times less than or equal to T_2. In contrast to the distribution function, this integral function shows little variation between the two different solutions. This demonstrates that despite the ill-conditioned nature of the ILT and the non-uniqueness of the extracted distribution function, many quantities derived from the distribution functions are generally robust.

2.5.2 Inversion with Regularization

It is impossible to directly assign an error bar to the distribution function.[42] However, it is feasible to analyze thousands of the possible solutions of the ill-conditioned ILT inversion problem of eqn (2.15) and derive probability distributions for derived quantities using a Monte Carlo technique.[43] The most common approach is to find a 'representative' solution. The 'representative' solution is typically chosen to be the distribution function with the least structure among all possible solutions, *i.e.* the smoothest distribution function that still fits the data. Finding such a solution can be accomplished by including a regularization term in the cost function of the minimization. Most inversion methods adopt Tikhonov regularization,[44] which uses the L2 norm as the regularizing term and the cost function is then:

$$\Psi = \sum_{l=1}^{N_E}\left(\left[\sum_{j=1}^{N_j} k_{T_{2,j}}(lt_E)\, f(T_{2,j})\right] - S_l\right)^2 + \chi\sum_{j=1}^{N_j}\left[f(T_{2,j})\right]^2 \tag{2.16}$$

with the additional non-negativity constraint $f(T_{2,j}) \geq 0$. Here χ is the regularization parameter that has to be chosen appropriately. The resolution of the distribution functions is limited both by the intrinsic properties of the ILT[45] and the value of χ, and by the acquisition parameters and number of measurements (*e.g.* number of echoes, N_E). Various implementations for solving eqn (2.16) have been described, differing primarily in the method used to reduce the data or solution space and to select the regularization parameter χ.[46–48]

As the measurements have become increasingly more complex and now often include multi-dimensional acquisition schemes, the complexity of the corresponding data processing has also increased. This has required the development of suitable inversion algorithms to analyze data sets with multiple sets of parameters, such as timings (polarization time, echo spacing, diffusion encoding times) or rf frequencies.[31,49,50] Job planning has also become a more critical process. It is essential to select acquisition parameters that ensure sensitivity over the entire range of anticipated properties while satisfying the logging requirements and constraints.

2.5.3 SNR and Spatial Resolution

NMR well logging suffers in general from a low signal to noise ratio (SNR). Noise levels for NMR logging tools are typically reported in terms of "porosity units" (pu) of noise per echo for one phase cycle, or scan of the CPMG sequence. By definition, 1 pu represents 1% of the signal measured for a water bottle (*i.e.* 100% porosity). In practice, values for tool noise range from less than 2 pu to more than 15 pu, depending on tool design, operating frequency, depth of investigation (DOI), and the environmental conditions (in particular temperature and borehole conductivity). Considering that formation porosities rarely exceed 35 pu, SNR of less than 10 are therefore routine for NMR log data.

The effect of low SNR can be ameliorated by decreasing the echo spacings, t_E. A short value of t_E not only enables the detection of short T_2 components, but it is also beneficial for the detection of longer T_2 components. While the reduction of t_E does not increase the SNR for a single echo, it results in a larger number of echoes during the T_2 decay and thereby increases the effective SNR per unit time. This benefit comes with the drawbacks of higher data volumes and increased rf power requirements, both scaling linearly with echo number.

Another common approach to increase the SNR is to average data over several successive "scans". However, data averaging results in a degradation of spatial resolution along the borehole because well logging is generally performed with tools moving upwards (wireline) or downwards (LWD) along the borehole. Figure 2.13 illustrates the effects of sequential averaging for an NMR tool moving at constant velocity along a borehole. The + and − signs refer to the relative phase of each CPMG train. Most NMR logging tools employ simple two-step phase cycling in which the phase of the 90° excitation pulse is alternated between successive echo trains. For n-level averaging, the resulting decrease in the effective noise per echo, $\sigma(n)/\sigma(n = 1)$, and the degradation in vertical resolution, $\Delta(n)/\Delta(n = 1)$, are given in Figure 2.13 for the case of non-overlapping measurements with an antenna of length L, and sampling interval δ. Evidently, the degree of averaging applied to the data represents a compromise between vertical resolution and measurement repeatability (precision), since both are desirable log attributes. In practice, "depth averaging" is ultimately determined by the logging objectives. It is possible to invert the data using a smoothing constraint along the depth axis.[51] In situations where high axial resolution is required to delineate boundaries in thinly bedded formations, it may be preferred not to perform any vertical averaging, with the understanding that the precision (repeatability) of the inferred NMR properties will be generally degraded. In this case, processing techniques have been proposed to extract some specific properties with maximal axial log resolution at a minimal cost in precision.[52]

For NMR logging tools with multi-frequency capability, there exists an analogous trade-off between SNR and spatial resolution in the radial direction.

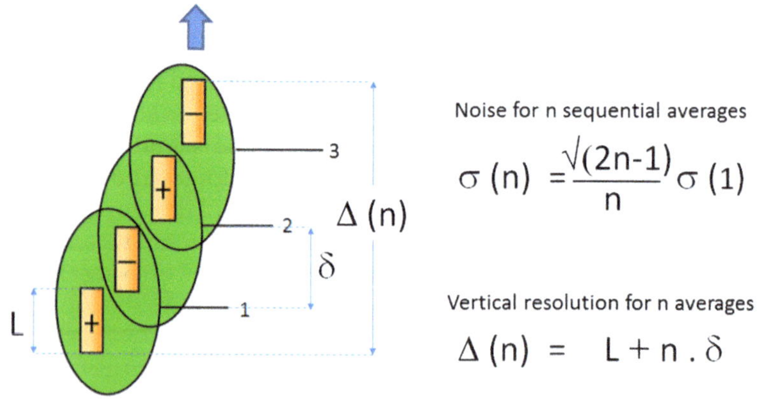

Figure 2.13 Schematic illustrating sequential data averaging for an NMR logging tool.

Such logging tools enable radial profiling applications whereby, similar to multi-slice acquisition schemes in MRI, NMR data is acquired at different radial depths in the formation. The data may be combined to improve the overall SNR, but at the price of losing information on the radial variation of formation and fluid properties. Although the range of depths of investigation (DOI) typically spans just a few cm, there are often systematic changes in fluid saturations over this range, caused by the invasion of drilling fluid or mud fines from the borehole into the near-wellbore region of the formation.

Figure 2.14 illustrates an early example of a multi-DOI log of T_2 distributions in a well with a plastic casing that shows radial variations.[53] Data were acquired for different DOIs between about 2 cm and 7 cm and separately inverted. The variation in the amplitude of the long T_2 components indicates that the different DOIs correspond to different physical regions. The deeper measurements (lower frequency) probe at least part of the geological formation of interest that comprises some clean sands (hence the long T_2 components), whereas the shallower measurements sample the region between the plastic well casing and the formation that is filled with cement and clay.

These examples highlight an important point regarding the combinability of data acquired at different locations. Because fluid saturations may vary in the region close to the borehole, it is important that any comprehensive inversion model properly accounts for such spatial variation of the measured distributions. In the case that the SNR is insufficient for a full inversion of the

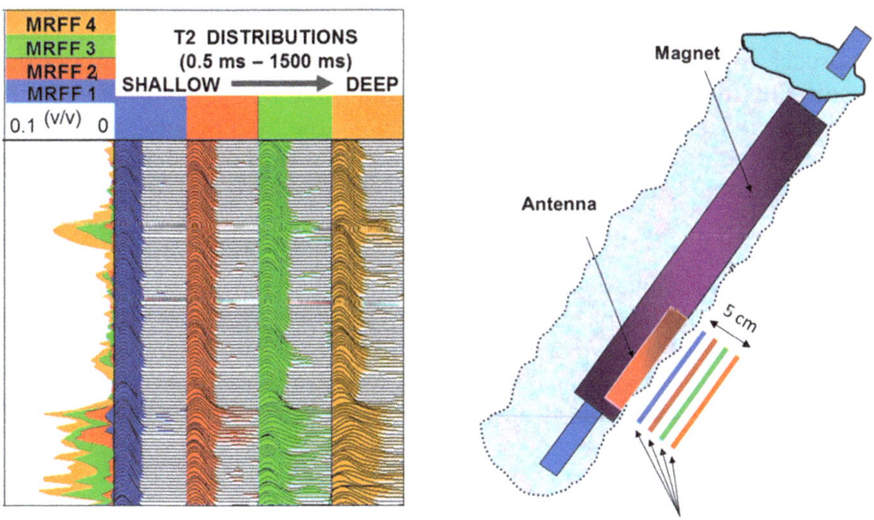

Figure 2.14 Multifrequency NMR log data acquired in a test well, showing the variation in T_2 distributions over a range from about 2 cm to 7 cm into the formation. Adapted from Heaton *et al.*, 2002.[53] © 2002 Society of Petroleum Engineers Inc. Reproduced with permission of SPE. Further reproduction prohibited without permission.

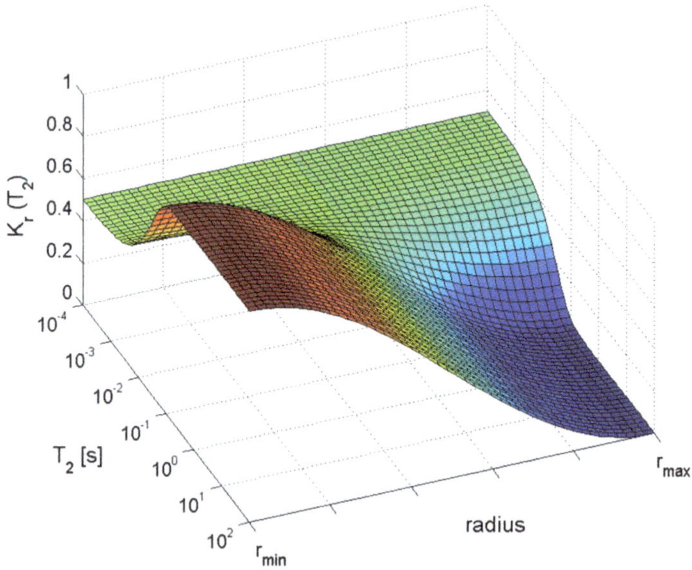

Figure 2.15 Radial function $K_r(T_2)$ used in the inversion of multi-frequency NMR log data. It imposes radial invariance of the bound fluid response at short T_2 while allowing radial variation of the mobile parts of the distribution at long T_2. Here r_s and r_d are the shallowest and deepest depths of investigation, respectively (adapted from ref. 31).

data, it is useful to perform constrained inversions. An example of such an approach is described in,[31] where it is imposed that the components of the T_2 distributions that correspond to non-moveable fluids (typically short T_2 components) are independent of DOI, whereas the long T_2 components representing movable fluids are unconstrained. This was implemented by using the following ansatz for the radial dependence of the T_2 distribution function:

$$f_r(T_2) = K_r(T_2)f_{r,\min}(T_2) + [1 - K_r(T_2)]f_{r,\max}(T_2) \tag{2.17}$$

where the radial function $K_r(T_2)$ is assumed to be known and has a typical form as shown in Figure 2.15, and $f_{r,\min}(T_2)$ and $f_{r,\max}(T_2)$ are two distribution functions to be optimized. The ability to combine multi-frequency data in a simultaneous inversion has resulted in a remarkable improvement of fluid characterization, particularly at the deeper DOI where the data suffers from reduced SNR.

2.5.4 Robustness of Inversion

2.5.4.1 Porosity Estimate

Given the challenges of the inversion, it is critical to evaluate the robustness of any inversion result. We illustrate this here on the example of the extracted porosity. Clearly, it is more difficult to extract reliable porosities from short

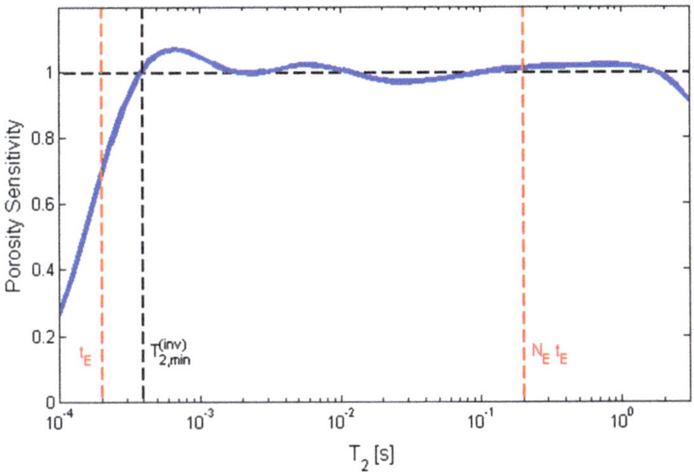

Figure 2.16 Porosity sensitivity *versus* T_2 for an acquisition scheme with an echo spacing of 200 μs, a main CPMG with 1000 echoes and wait time of 10 s, and a burst with 50 repetitions, 30 echoes, and a wait time of 20 ms (adapted from ref. 55). The echo spacing, t_E, the minimum T_2 limit used for the inversion, $T_{2,min}^{(inv)}$ and the duration of the CPMG, N_E t_E, are indicated with dashed lines.

T_2 formations than from long T_2 formations with fixed acquisition parameters, reflecting the small number of echoes contributing to the fast-relaxing components. It was recognized by the late 1990s that significant improvements in the final answer quality can be obtained by adding "bursts" of short echo trains with short polarization times in the acquisition sequence.[33,54] Repeating the short echo trains multiple times effectively increases the SNR for the fast relaxing components and substantially reduces the uncertainty in the amplitude determination of short T_2 components. However, the kernel now depends not only on the T_2 relaxation times, but also on the T_1 relaxation times.

The quality of the inversion and acquisition scheme can be assessed by performing simulations for different model distributions and different SNR.[43] An insight into porosity (*i.e.* amplitude) accuracy is provided by sensitivity plots, which indicate the porosity response of the inversion as a function of relaxation time. Figure 2.16 shows an example of such a sensitivity plot for an acquisition sequence that consists of a main CPMG with 1000 echoes and a wait time of 10 s, and 10 repetitions of a burst with 30 echoes and a wait time of 20 ms. In all cases, the echo spacing is 200 μs.

At short T_2 values, the sensitivity diminishes as the limit for the minimum T_2 considered in this inversion was set at $T_{2,min}^{(inv)} = 2t_E$. If this limit is decreased, the porosity determination becomes unstable and highly sensitive to noise. This reflects the low contribution of these fast-relaxing components to the measured echo data. For relaxation times longer than $T_{2,min}^{(inv)}$, the inversion is able to determine the porosity robustly even for values of T_2 that are significantly longer than the duration of the main CPMG, $N_E\ t_E = 0.2$ s. For components with such long relaxation times, the extracted amplitude is accurate but T_2 cannot be well determined. At the longest T_2 values considered in Figure 2.16, the sensitivity starts to decrease because the magnetization does not get fully polarized during the wait time between the measurements.

This effect of incomplete polarization is further emphasized in the examples of Figure 2.17 that show the porosity sensitivities as a function of T_1 for a series of 6 CPMG acquisitions with wait times T_w that are approximately geometrically distributed between a minimum and maximum wait time, $T_w^{(min)}$ and $T_w^{(max)}$. In these examples, we set $T_w^{(min)} = 10$ ms and varied $T_w^{(max)}$ between 4 s and 15 s. The plot on the left compares the porosity sensitivities for the three different values of $T_w^{(max)}$, all other parameters being identical.

When T_1 approaches or exceeds $T_w^{(max)}$, the initial magnetization for none of the 6 CPMG acquisitions is fully polarized. This incomplete polarization is quantitatively described by the kernel in eqn (2.6). When this kernel is included in the inversion, it should be in principle possible to compensate for incomplete polarization. However, the results in Figure 2.17 demonstrate that this approach is only partially successful. The observed oscillations indicate that the inversion is not able to accurately estimate the T_1 values from only the 6 measurements. For some values of the relaxation time, T_1 and the degree of under-polarization are overestimated, which leads to an overestimation of the porosity, while for other values of the relaxation time, the opposite is true. The plot on the right shows results of the porosity sensitivities for the same input data, generated with an inversion kernel that omitted the T_1 kernel of eqn (2.6). In this case, no polarization correction is applied. The oscillatory behavior has now disappeared from the porosity sensitivity functions, but they now underestimate the porosity more significantly at longer T_1. In some cases, it might be preferable to have the more predictable but somewhat more biased response on the right over the oscillatory behavior presented by the unconstrained inversion depicted on the left.

The last example demonstrates the more general point that sensitivity plots can be adjusted for a given application by modifying the inversion (*e.g.* regularization or optimization method) or by changing the acquisition sequence. Recent examples of this approach include applications in unconventional shale reservoirs where much emphasis falls on short relaxation times.[55,57]

2.5.4.2 *Estimation of Relaxation Time*

The sensitivity plots discussed above dealt with the porosity response for a given acquisition and inversion scheme as a function of T_2 or T_1. However, these results do not inform us about the accuracy and resolution of the extracted relaxation times. A CPMG sequence of N_E echoes with echo spacing t_E can resolve relaxation times T_2 between roughly $T_{2,min} \approx t_E$ and $T_{2,max} \approx N_E\ t_E$. Components with relaxation times shorter than $T_{2,min}$ do not generate an appreciable signal. Consequently, the data has little sensitivity

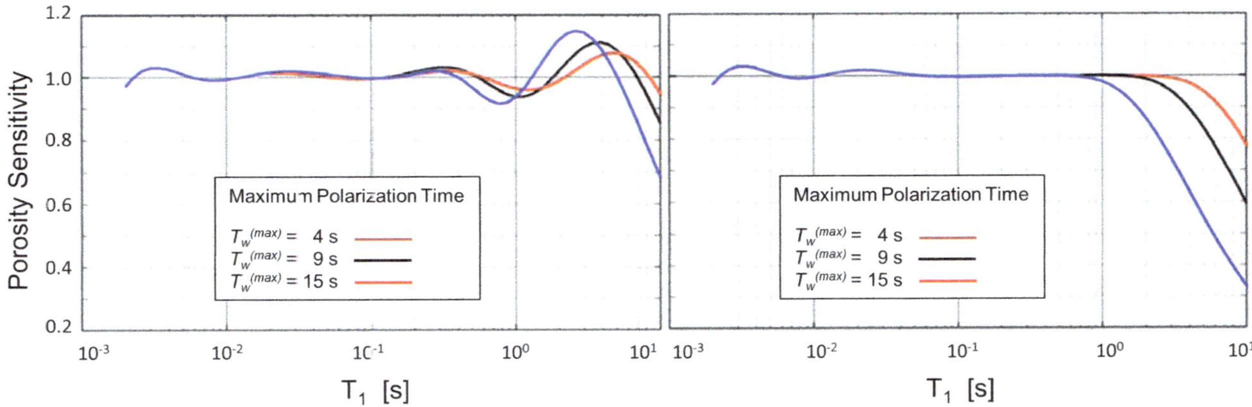

Figure 2.17 Porosity sensitivity *versus* T_1. The left hand plot shows T_1 sensitivity for an acquisition of a series of CPMG echo trains with 6 polarization times logarithmically spaced between $T_w^{(min)}$ and $T_w^{(max)}$, where $T_w^{(min)} = 10$ ms and $T_w^{(max)} = 4$ s (blue), $T_w^{(max)} = 9$ s (black), and $T_w^{(max)} = 15$ s (red). The right hand plot shows the inversion results of the same data when the T_1 kernel is omitted in the inversion. Adapted from Bachman *et al.*, 2007.[56] © 2007 Society of Petroleum Engineers Inc. Reproduced with permission of SPE. Further reproduction prohibited without permission.

to these components and inversion should be avoided. Attempts to estimate such short T_2 components will lead to noise amplification and result in large uncertainties. Components longer than $T_{2,\mathrm{max}}$ contribute to the signal provided that they are sufficiently polarized, *i.e.* the polarization times are longer than the corresponding T_1 values. However, the extracted values of T_2 are not reliable in this regime. Regularization and the intrinsic sensitivity of the ILT will broaden the distributions beyond $N_\mathrm{E}\, t_\mathrm{E}$ to $T_{2,\mathrm{max}}^{(\mathrm{inv})}$, the maximum T_2 value considered in the inversion.

The question of relaxation time resolution has been explored by several authors using modeling and analytical approaches.[45,58,59] While the average T_2 rate is generally a robust parameter to extract as it is directly related to the initial slope of the echo decay, it is more difficult to resolve individual components and determine their relaxation times. Not surprisingly, the capability of resolution increases with SNR and depends on the measurement sequence used to acquire the data. Interestingly, for typical SNR values encountered in NMR well-logging (*e.g.* SNR ~10), it has been demonstrated that a difference in relaxation rate of almost a factor of 4 is required to resolve two signals of similar amplitude.[45]

The discussion for T_1 relaxation times follows the same general principles as for T_2. In practice, T_1 and T_2 measurements are often acquired and analyzed simultaneously. Since these quantities are typically strongly correlated, constraints are typically placed on the range of T_1/T_2 ratios in the inversion. The resolution in T_1 ultimately depends on the number and the range of wait times used in the acquisition, and the corresponding SNR.

2.5.4.3 *Diffusion Sensitivity*

The question of sensitivity is equally important for fluid characterization applications that use diffusion contrast to distinguish between different fluids. Resolution in diffusion depends on the range of echo spacings t_EL chosen in the measurements. Using diffusion editing sequences with a range of echo spacings between $t_{\mathrm{EL,min}}$ and $t_{\mathrm{EL,max}}$, the measurements will have a resolution for diffusion coefficients between $D_\mathrm{min} \sim 6/\gamma^2 g^2 t^3{}_{\mathrm{EL,max}}$ and $D_\mathrm{max} \sim 6/\gamma^2 g^2 t^3{}_{\mathrm{EL,min}}$, provided that the associated relaxation times are sufficiently long. Here γ is the gyromagnetic ratio and g is the tool gradient. For diffusion coefficients outside this range, only a lower or upper limit of D can be established. When the relaxation times are too short, the signal decays before the diffusion information can be encoded.[60] In this case, no information on diffusion can be extracted from the signal. This occurs when $T_2 < \epsilon(\gamma^2 g^2 D)^{-1/3}$, with $\epsilon \approx 3$ and only weakly dependent on SNR. This limitation in diffusion resolution is illustrated in Figure 2.18, which displays a series of inversion results of model data for a single component with a diffusion coefficient of 7×10^{-9} m^2 s^{-1} (corresponding to water at an elevated temperature), and a T_2 relaxation time that varies between 1 ms and 100 ms within the series. The color-maps in the top row represent the extracted D-T_2 distribution functions with diffusion coefficients on the vertical axis and T_2 on the horizontal axis. The yellow

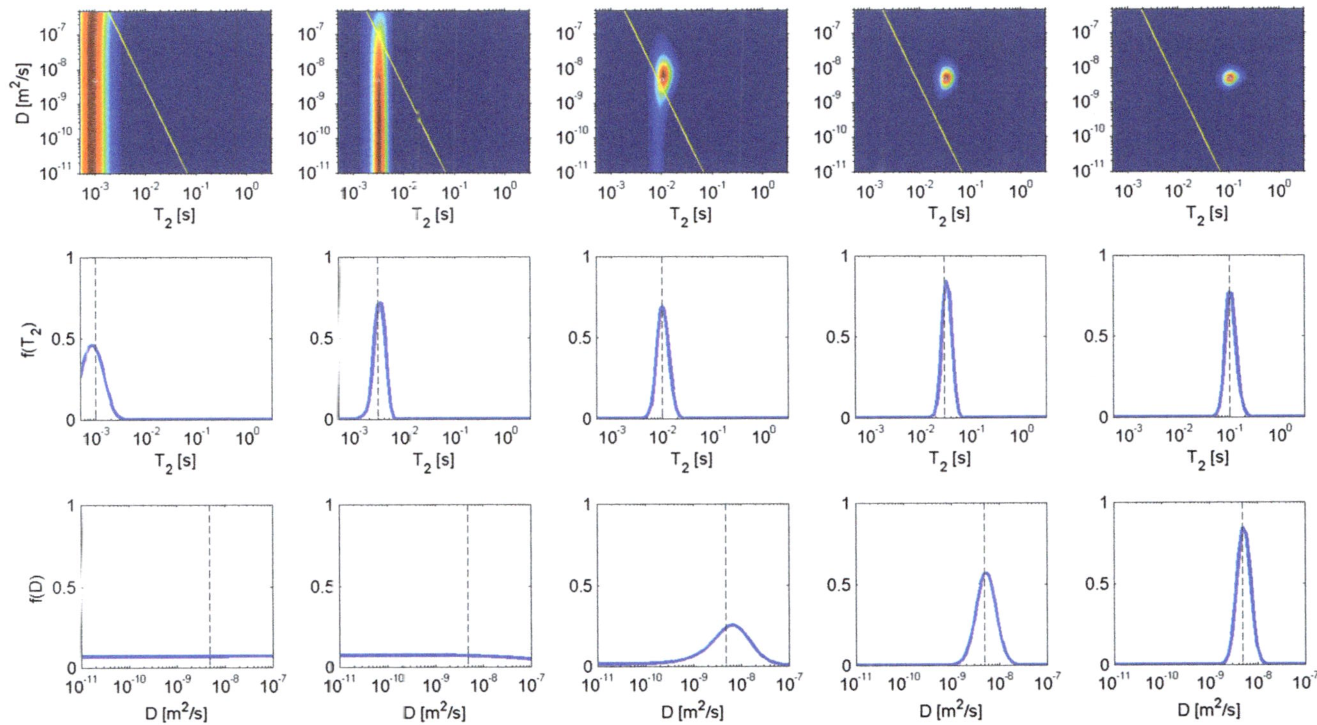

Figure 2.18 Diffusion sensitivity: D-T_2 maps (top row), T_2 (middle row), and D distributions (bottom row) generated from simulated data. All data sets were generated assuming a single fluid with a diffusion rate of $D = 7 \times 10^{-9}$ m^2 s^{-1} but with different T_2 values ranging from 1 ms to 100 ms. For the T_2 and D distributions, the blue curves are obtained from the projection of the D-T_2 maps, while the black broken lines correspond to the input values. The yellow lines on the D-T_2 maps represent the diffusion sensitivity limit $T_2 = \epsilon(\gamma^2 g^2 D)^{-1/3}$.

lines indicate the diffusion sensitivity limit defined by $T_2 = \epsilon(\gamma^2 g^2 D)^{-1/3}$. The middle and lower panels show the resulting T_2 and diffusion distributions, respectively. They are obtained by projecting the two-dimensional distribution functions $f(D,T_2)$ onto the corresponding single dimension. When the transverse relaxation time drops below the critical limit (~14 ms in this case for water), it is evident that the measurement loses sensitivity to diffusion, but retains sensitivity to T_2. In this regime with the inversion scheme that incorporates regularization, $f(D)$ becomes evenly distributed between the minimum and maximum diffusion coefficient considered in the inversion and contains no information on the fluids anymore. It is important to consider this intrinsic limitation of diffusion measurements when interpreting the results to distinguish the signals of hydrocarbon and water. This limitation is particularly significant for heavy oil evaluation and separating its response from that of capillary- and clay-bound water.

2.5.5 Alternative Inversion Approaches

2.5.5.1 *Maximum Entropy*

An alternative to the method of least squares minimization with regularization is based on the maximum entropy formulation of the distribution function.[61] For a simple T_2 distribution, this can be expressed as:

$$f(T_{2,j}) = \frac{\phi}{Z}\exp\left\{\sum_{n=1}^{N_c} c_n V_{n,j}\right\} \tag{2.18}$$

where ϕ is the NMR porosity and Z is a normalization factor that ensures that the area of the distribution is equal to ϕ. The basis vectors $V_{n,j}$ are typically chosen to be the singular vectors derived by singular value decomposition (SVD) of the kernel function. In effect, the selection of the number of singular vectors, N_c, replaces the estimation of χ in the Tikhonov regularization algorithms. In practice, it is often possible to set N_c at a fixed value based on the anticipated range of the SNR. Provided that the SNR does not substantially exceed the expected range, this procedure generally yields stable inversion results. Although not widely used for conventional T_2 inversion applications, the maximum entropy approach has been applied to multi-dimensional inversion where multi-dimensional distribution functions of T_2, T_1, and diffusion are estimated from multi-measurement datasets.[50]

2.5.5.2 *Monte Carlo*

Interesting alternatives to conventional Laplace inversion based on Monte Carlo methods have also been reported for NMR relaxation analysis.[43,62] A Monte Carlo-based inversion scheme incorporating NMR, dielectric, and nuclear log measurements was recently described with application to heavy oil evaluation.[63] Such statistical approaches to NMR inversion have not yet been widely adopted in the industry, partly because of their increased computational demands. However, as high-performance computing resources become more available, the more detailed uncertainty information provided by statistical inversions may render these methods attractive for NMR log data processing and interpretation in the future.

2.5.5.3 Model-Based Inversion

The introduction of diffusion measurements to NMR logging services spawned a range of new approaches for fluid characterization. Beyond the inversion approach for a general D-T_2 distribution function,[26] alternative inversion methods were developed which assumed specific NMR responses for water, oil, and gas.[64–66] These methods all impose constraints on the D-T_2 distribution function based on prior knowledge or assumptions concerning the NMR responses. For example the "Constituent Viscosity Model" (CVM) proposed by Freedman and co-workers[65] imposes a relationship between the T_2 (or T_1) distribution and the distribution of diffusion rates for crude oils. For each component, it is assumed that $D = \lambda f(\mathrm{GOR})T_2$. Here λ is a constant that describes the alkane line indicated in Figure 2.23. The factor $f(\mathrm{GOR})$ represents a correction for live crude oils that accounts for additional relaxation mechanisms introduced when the gas : oil ratio (GOR) becomes significant.

By reducing the degrees of freedom in the inversion, these methods can provide more robust mathematical solutions than model independent inversion schemes. However, their use has been somewhat limited by the uncertain validity of the input assumptions over the vast range of fluids and reservoir conditions encountered in well logging. As an example, deviations from the linear dependence of diffusion coefficient on relaxation time, assumed in the CVM model, are observed in crude oils containing asphaltene, in biodegraded oils[67] (see Figure 2.24), or in oils occupying small pores and affected by restricted diffusion (see Figure 2.25).

Model-based analysis methods have also been applied to the inversion and interpretation of NMR log data in carbonate reservoirs.[68] In this case, the T_2 distribution function was assumed to be a superposition of a small number of components, each with a Gaussian distribution with respect to $\log(T_2/T_{2,\mathrm{peak},j})$. It was shown that the components obtained from such an analysis correlated well with the different pore types found in the formation.

2.5.5.4 Direct Estimation of Petrophysical Properties from Magnetization Data

The approaches of data analysis discussed so far all involved the extraction of the appropriate distribution function as a first step. The answer products, such as porosity, bound fluid volume, permeability, and viscosity, are then extracted from these distribution functions by different integral transforms. An alternative approach is to estimate a given parameter directly from the measured magnetization data using a linear transform without the intermediate step of first estimating the distribution of relaxation times. An early example of such an approach is the extraction of a high-resolution permeability log from the sum of the first few echo amplitudes.[69] This technique can be generalized to other applications involving linear functionals of the distribution function by selecting appropriate kernels in the integral transforms of the measured echo amplitudes. Examples include the Mellin transform or the exponential Haar transform.[70,71] This approach avoids the complications associated with ILT, *i.e.* the inversion of an ill-conditioned, non-linear problem. Provided that the relationship between the NMR measurement and the parameter of interest is well understood, it results in more accurate estimates for fluid and formation parameters with well-defined uncertainties that are directly related to the SNR of the measured data.

2.5.5.5 Data-Based Inversion

In practice, every inversion constitutes an individual step in a larger workflow whose ultimate aim is to derive useful reservoir and petrophysical properties from the measured data. In the conventional approach, the inversion routines incorporate a description of the sensor response and a physical model relating the measured quantities to the parameters of interest. Given the complexity of geological formations, these models inevitably include assumptions or approximations that might not be universally valid. To circumvent this shortcoming, new inversion methods have been proposed that do not attempt to describe the underlying physics in detail, but instead rely on a comprehensive database of laboratory measurements on samples with known properties. In this approach, measurements are compared with all the lab measurements of the database and the properties are then inferred by interpolation between the known properties of the closest lab measurements. Using radial basis functions to measure the distance between measurements,[72] this technique has been successfully applied to evaluate crude oil properties from NMR laboratory measurements[73] and permeability from NMR logs.[74] This is a fairly general approach, but its main challenge is the requirement of a high-quality data base that covers the entire range of possible acquisition parameter and sample properties.

2.6 Data Interpretation: Physics of Relaxation and Diffusion

NMR measurements are intrinsically very versatile and provide information on a wide range of properties. As discussed in Section 2.3, NMR logging sensors can measure amplitudes, relaxation times, diffusion coefficients and various distribution functions with some spatial resolution along the borehole and into the formation. To extract the maximum value from these measured NMR properties, it is critical to relate them to petrophysical properties and key reservoir attributes. A robust interpretation must be based on the underlying physics that governs relaxation and diffusion in porous media and in complex fluids. Over the last two decades, much progress has been made and we can provide only a brief overview of the main developments here. It is also important to keep in mind that the NMR measurements are generally not analyzed in isolation, but combined with information obtained from other types of sensors.

We first give a short overview of key reservoir properties of interest and then discuss how the different measured NMR properties (such as initial amplitude, relaxation times, diffusion coefficients, and various distribution functions) are related to these reservoir properties.

2.6.1 Reservoir Properties

2.6.1.1 Conventional Reservoirs

Conventional hydrocarbon reservoirs consist of porous sedimentary rocks that are capped by an impermeable layer that acts as a seal and traps the hydrocarbons in the pore space of the reservoir rock. Key reservoir parameters include the porosity and the permeability. The porosity of a reservoir rock typically falls in the range from a few percent to over 30%. In exceptional cases such as diatomite formations, it can even exceed 60%. The permeability is controlled by the pore sizes and the connectivity of the pores, and typically falls in the range of 1 mD to several Darcies. In conventional reservoirs, typical pore sizes fall in the range of 0.1 μm to 500 μm. The geometry of the pore space in a typical reservoir rock can be quite varied, as illustrated by Figure 2.19. The picture on the left shows an example of a sandstone, a rock that consists predominately of quartz grains and can contain various amounts and types of clays. The picture on the right shows a limestone, which is an example of a carbonate rock. The pore space of carbonates is typically more complex than that of sandstones.

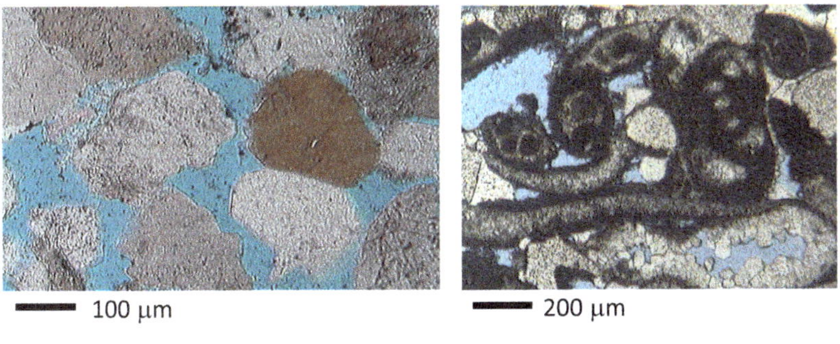

Figure 2.19 Pore geometry in conventional reservoirs. Examples of the geometry of the pore space in sedimentary rocks. The thin section on the left shows a sandstone, on the right a limestone. The pore space appears in blue.

The fluid occupying the pore space of the reservoir rock is typically a complicated mixture of brine and hydrocarbons. The hydrocarbons can possibly be further separated into distinct phases, *e.g.* an oil-rich and a gas-rich phase. The saturations of the different phases (*i.e.* the fraction of the pore space that is filled with a given fluid phase) and the viscosity of the hydrocarbon phases are further key reservoir properties. The mobility of the oil is determined by the oil viscosity and the formation permeability. The detailed composition of the crude oil is also considered critical information. This includes the concentration of dissolved gases, such as methane, carbon dioxide, nitrogen, or hydrogen sulfide, and the possible presence of dissolved asphaltenes and waxes that may become unstable, damage the formation, or cause flow assurance problems. Near the borehole, the reservoir fluids can be disturbed by invasion of fluid from the borehole into the reservoir. As a result of this process, the fluids probed by the NMR sensors are generally a spatially varying mixture of the native reservoir fluids and the filtrate.

The producibility of a reservoir is another key parameter. It is not only determined by the properties of the rock, in particular the geometry of its pore space, and by the quantities and properties of the fluids, but also by the rock–fluid interactions. This interaction is commonly described by the wettability parameter. The affinity of the rock surfaces for the different fluids controls the microscopic configuration of the phases within the pores, which in turn has a strong impact on the relative permeabilities and the residual saturations of oil and water. After the initial deposition, sandstone and carbonate formations generally originate as strongly water-wet. After migration of hydrocarbons into the reservoir, surface-active components in the crude oil (*e.g.* asphaltene and resin molecules) can turn reservoir rocks oil-wet or mixed-wet. The wettability is often variable across a reservoir.

2.6.1.2 Unconventional Resources

The past decade has seen remarkable growth in the development of so-called unconventional resources. In particular, shale-oil and shale-gas reservoirs account for an increased proportion of new hydrocarbon reserves and production. These formations generally contain large amounts of organic matter that range in composition from methane to bitumen and kerogen. The porosities are typically low and dominated by nanometer-scale pores in the organic kerogen. This leads to extremely small intrinsic permeabilities in the nano- to pico-darcy range. As a consequence, the economic production of the resources requires hydraulic fracturing that creates wider flow channels.

The properties of shale reservoirs differ significantly from those of conventional sandstone and carbonate formations, and present new challenges for the interpretation of well logs. A new key parameter is the concentration of "Total Organic Carbon" (TOC), which represents all carbon contributions from organic sources in the formation, including gas, oil, bitumen, and kerogen. In addition, it is critical to distinguish between the different types of organic matter. Even the characterization of gas requires a new approach that includes distinction between absorbed and free gas. In nano-pores, a significant fraction of the gas occurs in the adsorbed phase on the pore surface with a higher density than the free gas in the interior of the pores.[75]

The complex shale reservoirs provide particularly challenging environments for NMR logging because much of the NMR signal is characterized by low amplitudes and very fast relaxation times. The optimization of the signal-to-noise ratio and the minimization of the echo spacings are therefore vital requirements for the data acquisition.[57] Interpretation is also complicated by the different types of fluid–matrix interactions occurring in shales.[76] They reduce the hydrocarbon relaxation times[77] and often lead to high T_1/T_2 ratios.[78,79] Diffusion measurements are strongly affected by the high degree of restrictions in the pore space.[75,80] While the study of the NMR properties of these reservoirs has only just started[76–79] and is currently still incomplete, the early results indicate that NMR logging will play a key role in the further characterization of such unconventional reservoirs and the definition of the optimal production strategies.

2.6.2 Amplitude

2.6.2.1 Amplitude-Derived NMR Porosity

The initial amplitude of the CPMG measurement is proportional to the number of spins in the sensitivity volume of the NMR sensor. This quantity is in turn proportional to the product of porosity ϕ and the spin density. It is customary to express the spin density in terms of the hydrogen index, HI. The hydrogen index is dimensionless and is defined as the proton density of the fluid normalized to that of water at ambient conditions. For alkanes and oils of moderate viscosity, the hydrogen index is close to 1 and the deviation from unity can often be ignored. However, the HI is significantly smaller than 1 for heavy oils, brine with high salt content, or gas.[36] After proper tool calibration, as discussed in Section 2.4, the NMR amplitude can therefore be converted to the 'NMR porosity', ϕ HI.

In order to extract an accurate porosity from the amplitude of CPMG measurements during NMR logging, it is important that the longitudinal magnetization is at thermal equilibrium before the sequence is applied. This requirement can be challenging to fulfill in formations filled with light hydrocarbons that can have T_1 relaxation times of many seconds. In this case, it is necessary to use long wait times between subsequent CPMG acquisitions and limit the logging speed.

2.6.2.2 Comparison of NMR Porosity and Density Porosity

Density logging is another type of logging that is routinely used to determine the formation porosity. Useful information can be obtained from a comparison of NMR- and density-derived porosities. The density logging technique relies on the attenuation of gamma rays and measures the effective density, ρ_{eff}, of the formation. This quantity is related to the porosity by:

$$\rho_{\text{eff}} = (1 - \phi)\rho_{\text{grain}} + \phi\rho_{\text{fluid}}, \tag{2.19}$$

where ρ_{grain} and ρ_{fluid} are the densities of the rock grains and the pore fluid, respectively. Since the grain densities vary significantly between rocks of different lithology, the application of this technique requires a good *a priori* knowledge of the mineralogy of the reservoir rocks. In contrast, the NMR measurement of porosity is insensitive to the lithology of the formation. In addition, the density technique also shows some sensitivity to the type of fluid occupying the pore space. Similar to NMR, the deviation of the derived porosity from the true porosity is most pronounced when gas is present. Interestingly, the deviations for the NMR- and density-based determinations of porosity have opposite signs: in the presence of gas, density measurements overestimate ϕ, while NMR measurements underestimate ϕ. Systematic deviations between density- and NMR-derived porosities are therefore good indicators for gas, as was first demonstrated by Freedman *et al.*[81] The comparison of NMR and density porosities is also a useful tool to monitor increases in the viscosity of heavy oils and to detect the presence of tar mats.[38] As shown in Figure 2.22, the apparent HI of heavy oils decreases from unity with increasing viscosity. This leads to an increasing deficit of the NMR porosity relative to the density porosity.

2.6.3 Relaxation

Spin relaxation is controlled by the details of the molecular dynamics of reorientation of individual fluid molecules,[82] which in turn is controlled by the composition of the fluids and the geometry of the pore space. The measured values of NMR relaxation times in NMR logging often span several decades and can extend from less than 1 ms to several seconds. The relaxation rate for fluids in the pore space of rocks can be separated into distinct contributions of surface, bulk and diffusion induced relaxation:[83]

$$\frac{1}{T_2} = \frac{1}{T_{2,\mathrm{surf}}} + \frac{1}{T_{2,\mathrm{bulk}}} + \frac{1}{T_{2,\mathrm{diff}}} \tag{2.20}$$

$$\frac{1}{T_1} = \frac{1}{T_{1,\mathrm{surf}}} + \frac{1}{T_{1,\mathrm{bulk}}} \tag{2.21}$$

Surface relaxation is caused by interaction between the fluid and the grain surfaces, while bulk relaxation is an intrinsic property of the fluid. The transverse magnetization can also get dephased due to diffusion in field inhomogeneities, as already pointed out in eqn (2.4).

2.6.3.1 Relaxation of Wetting Phase: Surface Relaxation

The relaxation rates of the fluid that is the wetting phase are generally dominated by surface relaxation. The fluid molecules inside the pore space undergo Brownian motion and collide with the grain surfaces. The resulting spin relaxation is caused by interactions that occur while the spins are in the immediate vicinity of the pore surface. Such processes include intermolecular dipolar coupling with electron spins associated with paramagnetic impurities on the grain surfaces, and enhanced intramolecular dipolar coupling caused by a reduced mobility while the molecule is in contact with the grain surfaces. While efforts have been made to develop microscopic theories for this effect,[84–86] such detailed descriptions are fortunately not necessary for practical applications. Instead, it is sufficient to use the two phenomenological parameters of surface relaxivities ρ_1 and ρ_2 to characterize the relaxation strength for the longitudinal and transverse magnetization, respectively. Following the analysis of Brownstein and Tarr,[87] the magnetization decay follows the diffusion equation in the pore space with the appropriate boundary conditions at the grain surfaces based on the values of ρ_1 and ρ_2. In the important limit of weak relaxation when the spins undergo on average many collisions with the walls before they relax, only the lowest eigenmode contributes to the solution of the problem. This is generally referred to as the fast diffusion regime. In this regime, the transverse magnetization decays uniformly within a given pore with the relaxation rate:

$$\frac{1}{T_{2,\text{surf}}} = \rho_2 \frac{S}{V_\text{p}} \tag{2.22}$$

Here S is the surface and V_p is volume of the pore. An analogous expression holds for the longitudinal relaxation rate.

2.6.3.1.1 Relaxation-Derived Pore Size Distribution. The ratio V_p/S is a geometrical length that scales to first order with the size of the pore. The result of eqn (2.22) therefore indicates that the distribution of the measured relaxation times, $f(T_2)$, is directly related to the distribution of pore sizes, as was first pointed out by Seevers,[88] and by Loren and Robinson.[89] Long relaxation times correspond to large pores, and *vice versa*. Kenyon *et al.*[29] first tested the validity of this interpretation on a number of sandstones cores by comparing the distribution function of NMR relaxation times to the image analysis of thin sections. As an example, Figure 2.20 shows a comparison for a limestone of the measured T_2 distribution and the pore size distribution inferred from mercury porosimetry.

The log of the measured T_2 distributions $f(T_2)$ is a key output of NMR logging and allows the detection of changes in the pore size distribution along the well. The total area under $f(T_2)$ corresponds to the total porosity:

$$\phi = \int_{T_{2,\text{min}}}^{T_{2,\text{max}}} f(T_2)\,\mathrm{d}T_2 \tag{2.23}$$

To relate the T_2 relaxation time quantitatively to a pore size requires knowledge of the surface relaxivity. As indicated in Figure 2.20, it is possible to determine the surface relaxivity from the comparison of T_2 measurements and mercury porosimetry. Another method is based on the comparison of relaxation and diffusion measurements.[91,92] Typical values of ρ_2 are found to

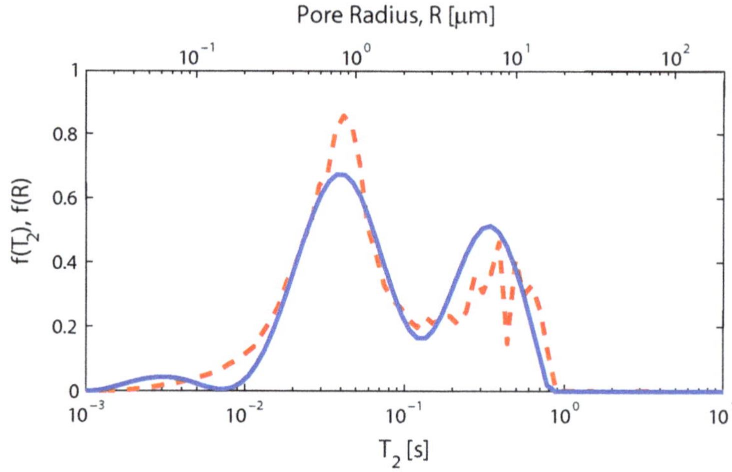

Figure 2.20 Comparison of the T_2 distribution, $f(T_2)$, shown in blue and associated with the bottom scale, and the pore size distribution, $f(R)$, derived from mercury porosimetry, shown as a dashed red line and associated with the top scale for a limestone core. This graph corresponds to a surface relaxivity $\rho = 6.6\ \mu m\ s^{-1}$. Adapted with permission from Venkataramanan *et al.*, 2014.[90]

be in the range of 5–20 μm s.[92] It is found that ρ_2 tends to be larger in sandstones than in carbonates. This is not surprising given that the clay content and the concentration of paramagnetic impurities is typically much higher in sandstones than in carbonates. It was found that the relaxivities show generally only a weak temperature dependence[93] and that the ratios of ρ_2/ρ_1 for brine-saturated samples cluster around 1.6.[94]

2.6.3.1.2 Relaxation-Derived Bound and Free Fluid Volume. There are two other important applications that are based on the measured distribution $f(T_2)$: the estimate of the so-called bound fluid volume (BFV) and the estimate of the permeability. In sufficiently small pores, the capillary pressure required to mobilize water by immiscible displacement exceeds a few atmospheres. For all practical purposes, water in such pores is effectively immovable and is therefore referred to as bound fluid. Figure 2.21 demonstrates that the bound fluid is associated with relaxation times below a certain cutoff value, $T_{2,\text{cutoff}}$.

The bound fluid volume is obtained from the T_2 distribution $f(T_2)$ by:

$$\text{BFV} = \int_{T_{2,\text{min}}}^{T_{2,\text{cutoff}}} f(T_2)\, \mathrm{d}T_2 \tag{2.24}$$

The cutoff relaxation time $T_{2,\text{cutoff}}$ depends somewhat on the surface relaxivity, but it is often adequate to use the empirical default values of 33 ms for sandstones and 90 ms for carbonates. The free fluid volume (FFV) is then

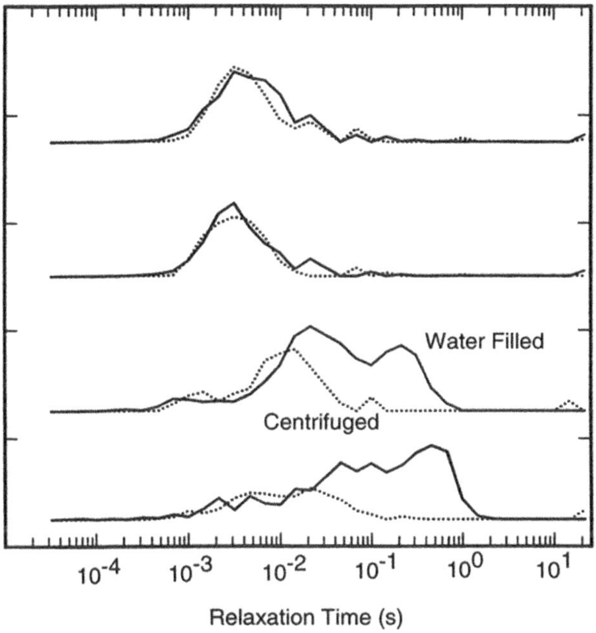

Figure 2.21 T_2 distributions and bound fluid volume: The solid lines show the T_2 distributions for four different cores that are fully water saturated, while the dotted line show the distributions after spinning the cores in the centrifuge at a rate that corresponds to a drawdown pressure of around 100 psi. Adapted with permission from Straley *et al.*, 1995.[95]

given by FFV = ϕ − BFV and is an important parameter to predict the water cut in the produced fluid.

2.6.3.1.3 Relaxation-Derived Permeability Estimate. The hydraulic permeability is controlled by the geometry of the pore space, in particular by the sizes of the pores and their connectivity. It is therefore reasonable to expect that the permeability can be estimated from the measured relaxation time distribution $f(T_2)$.[88,96,97] Two forms of the permeability estimator[29,98] have found widespread applications:

$$k_{\text{SDR}} = a\langle T_2\rangle^2\phi^4 \tag{2.25}$$

$$k_{\text{TC}} = b\,\phi^4\left(\frac{\text{FFV}}{\text{BFV}}\right)^2 \tag{2.26}$$

Given that for a given porosity and pore network topology, the permeability scales as the square of the pore size, it is understandable that the permeability estimator k_{SDR} in eqn (2.25) depends on the square of the average NMR relaxation time. For best results, the coefficients a and b have to be locally calibrated as they depend somewhat on the surface relaxivity.[92]

The NMR-based permeability prediction is generally more reliable in sandstone formations than in carbonates, where the pore space is often much more heterogeneous (see Figure 2.19) and also frequently contains micropores. In such systems, the T_2 distribution cannot be directly compared to pore size distributions obtained from mercury porosimetry, as diffusional coupling between the different pores becomes important[99] and averages out the small pore features in the T_2 distribution. A number of research groups are actively engaged in developing new NMR techniques based on two-dimensional relaxation exchange measurements[100,101] to probe directly the connectivity between different pores.

2.6.3.2 *Relaxation of the Non-Wetting Phase: Bulk Relaxation*

For non-wetting fluids, surface relaxation becomes ineffective if a layer even as thin as a few monolayers of the wetting fluid covers the rock grains. In this case, the observed relaxation of the non-wetting phase is dominated by the intrinsic bulk relaxation of the fluid. In the general case of mixed wettability, both surface and bulk relaxation can give sizeable contributions. It is this sensitivity to the microscopic configuration of the fluid arrangement that makes relaxation measurements sensitive to wettability.[102,103]

2.6.3.2.1 Relaxation-Derived Oil Viscosity. The bulk relaxation behavior of crude oils is controlled by their composition. To first order, the average relaxation time $T_{2,\mathrm{lm}}$ is inversely proportional to the viscosity of the oil, η.[36,104–106] This is illustrated in Figure 2.22. The panel on the top left displays T_2 distributions for a large set of crude oils, ordered according to their viscosities ranging from 20 cP on top to 10^6 cP at the bottom. On the bottom, values of $T_{2,\mathrm{lm}}$, the average log mean T_2, are plotted *versus* viscosity. For viscosities below 1000 cP, the simple inverse relationship $T_{2,\mathrm{lm}} \propto \eta^{-1}$ holds. As the oils become more viscous and the relaxation times decrease, some of the components relax with relaxation times below the detection limit of the spectrometer, $T_{2,\mathrm{min}}$. In this regime, $T_{2,\mathrm{lm}}$ levels out at $T_{2,\mathrm{min}}$ and loses sensitivity to viscosity. However, the overall amplitude becomes sensitive to viscosity with an apparent hydrogen index, $\mathrm{HI}^{\mathrm{app}}$, that decreases with increasing viscosity.[107,108] By combining the information from the measured values of $T_{2,\mathrm{lm}}$ and $\mathrm{HI}^{\mathrm{app}}$, it is possible to predict the oil viscosity over a range of over 6 decades in viscosity.[109] The expression relating $T_{2,\mathrm{lm}}$ and $\mathrm{HI}^{\mathrm{app}}$ to the viscosity has the form:

$$\eta \approx \frac{\alpha}{T_{2,\mathrm{lm}}} e^{\beta(1-\mathrm{HI}^{\mathrm{app}})} \tag{2.27}$$

The coefficient β depends on the lower T_2 detection limit of the tool and the acquisition sequence. To apply this expression to log data, it is essential to determine $\mathrm{HI}^{\mathrm{app}}$ as the apparent hydrogen index of the oil fraction only and exclude any contribution from the brine signal.

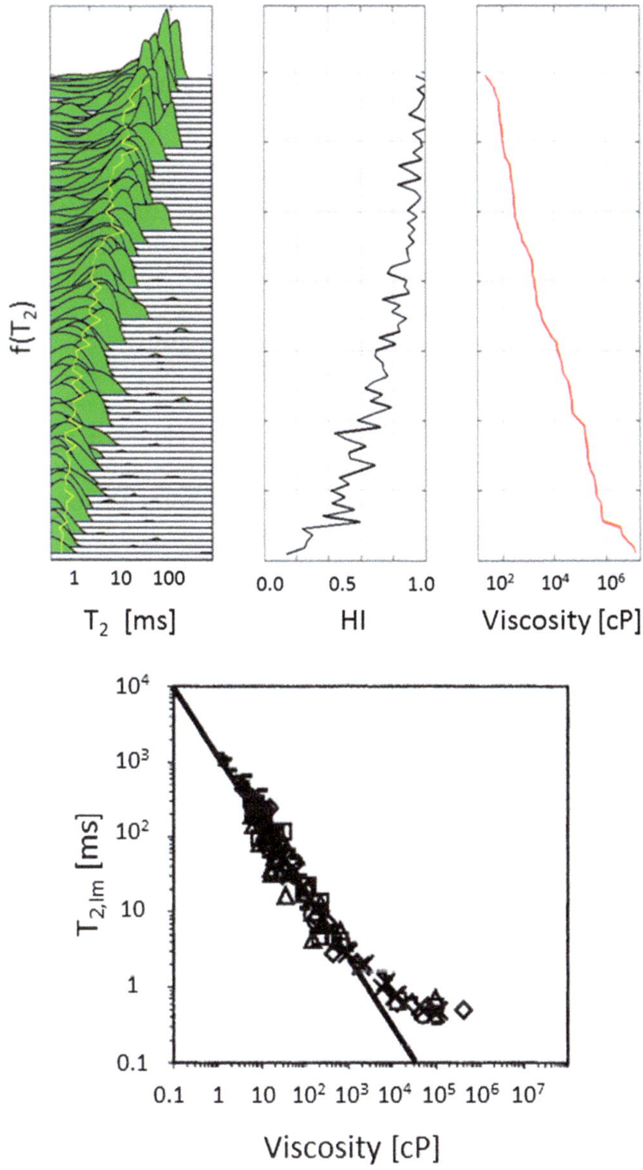

Figure 2.22 Relationship between T_2 relaxation times and viscosity of crude oils. Figure on the top left shows T_2 distributions of oils displayed according to their viscosity, adapted from ref. 63. The yellow line indicates the log mean value of T_2. The second and third columns show the corresponding hydrogen index and viscosities of the oils, spanning a range from 20 cP (top) to over 10^6 cP (bottom). The figure on the bottom shows the relationship between the log mean average T_2 and the viscosity, adapted with permission from Zhang *et al.*, 2002.[105]

This NMR-based viscosity relationship even applies for emulsions. In this case, the NMR measurement gives an indication of the intrinsic viscosity of the constituent oil phase, which in general will be very different from the overall viscosity of the emulsion determined by rheological measurements.

2.6.3.2.2 Relaxation-Derived Oil Composition. The magnetization decay of bulk oil samples is often distinctly non-exponential, leading to a distribution of relaxation times. This is especially pronounced in oils with a wide distribution of molecular sizes. The width in the T_2 distribution gives an indication of the distribution of molecular sizes in the oil. Small molecules undergo faster tumbling than bigger molecules and therefore better average the local dipole interaction. Accordingly, this leads to longer relaxation times.[82] Freed[110] has analyzed in detail the relaxation process for crude oils and has derived expressions that quantitatively relate the measured relaxation time distribution to the distribution of molecular size. She showed that for oils without asphaltene or dissolved methane, the molecules with a measured relaxation time $T_{2,i}$ have a size corresponding to approximately N_i carbon atoms, where:[111]

$$N_i = B^{\frac{1}{\gamma+\kappa}} \langle T_2^{\frac{1}{\kappa}} \rangle^{\frac{\gamma}{\gamma+\kappa}} T_{2,i}^{-\frac{1}{\kappa}} \tag{2.28}$$

Here the exponents $\gamma = 1.25$ and $\kappa = 1.3$. This expression implies that it is possible to derive from an NMR relaxation measurement the equivalent of a low-resolution gas chromatogram of a crude oil (containing no asphaltene). Interestingly, this can be achieved without making physical contact with the sample. As discussed in,[111] methane deviates somewhat from this relationship as it is also affected by the relaxation mechanism of spin rotation.[105]

2.6.3.2.3 Effect of Asphaltene on Oil Relaxation. Dissolved asphaltene molecules act as relaxation contrast agents[67] and shorten the relaxation time of all other components by an overall multiplicative factor ξ_1 or ξ_2 for T_1 or T_2, respectively.[112] Oil molecules get intermittently entangled with asphaltene aggregates and this process leads to a temporary slowdown of the molecular motion and reduction of relaxation time. The presence of such slow components in the molecular motion can be detected through the resulting property $T_1/T_2 > 1$. For any oil with at least 1% asphaltene, it is observed that $T_1/T_2 > 1$,[67,105] while light oils without asphaltene are characterized by $T_1/T_2 = 1$. The presence of asphaltene also leads to a dependence of T_1 on the Larmor frequency.[113] Remarkably, almost no frequency dependence is observed for T_2, even for oils rich in asphaltenes.[114] Slow molecular motions are also responsible for the large ratios of T_1/T_2 observed in kerogen and bitumen. It has been proposed to use this effect to separate the bound water signal from the viscous hydrocarbon signal in shale reservoirs.[79]

2.6.3.3 Relaxation Due to Diffusion in Gradient

In inhomogeneous fields, diffusion leads directly to extra T_2 relaxation even when the pores are large and surface relaxation is ineffective. The relevant expression depends on the strength of the local gradient, the echo spacing, and the diffusion coefficient, and was already given in eqn (2.4). The source of the local gradient can be external to the sample and created by the magnets of the logging tool or internal, generated by the contrast in magnetic susceptibility between the fluid and the rock grain.

2.6.3.3.1 Relaxation-Based Gas Detection. In wireline logging applications, the relevant gradient in eqn (2.4) is usually dominated by the tool gradient. This relaxation effect is minimized by using short echo spacings, but it typically remains an important contribution for the measured T_2 relaxation of methane. Since there is no corresponding contribution to T_1 relaxation, the resulting signal for methane is characterized by long T_1 and $T_1/T_2 > 1$. This characteristic has been used to identify methane in NMR logging.[115]

2.6.3.3.2 Relaxation Due to Internal Gradient. Local susceptibility-induced gradients are typically negligible in carbonates, but can become comparable to tool gradients in sandstones, especially if they contain large amounts of clay or magnetic minerals such as hematite.[116] The presence of large internal gradients will further shorten the measured relaxation times of the gas signal, but will have otherwise only a minor impact on relaxation measurements provided they are performed with the minimum echo spacing. Internal gradients generally have a larger impact on diffusion measurements than on relaxation measurements.[117]

2.6.4 Diffusion

When the pore space is filled with multiple fluid phases, relaxation measurements are generally insufficient to unambiguously identify and separate the corresponding signals. The distributions of relaxation times of the different phases are often at least partially overlapping. In this situation, diffusion measurements can provide crucial additional information. There is typically a significant contrast in the diffusion properties of the different fluid phases, such as brine, oil, and gas. For example, the diffusion coefficient of gas is under common conditions at least an order of magnitude higher than that of brine.

2.6.4.1 Diffusion of Bulk Fluids

The diffusion coefficient D of a simple fluid is determined by the size of its constituent molecule and is inversely proportional to the viscosity, η, *i.e.* $D \sim \eta^{-1}$. More complex fluids, such as crude oils, that consist of a mixture of molecules, require a distribution of diffusion coefficients, $f(D)$, to adequately describe the diffusion properties. The width of $f(D)$ is related to the range of molecular sizes of the oil and the viscosity can be inferred from a weighted sum of the inverse diffusion coefficient.[118] As demonstrated by Freed *et al.*,[118] the shape of $f(D)$ is directly related to the distribution of molecular sizes in the oil. A fluid component with a diffusion coefficient D_i corresponds to a molecule with approximately N_i carbon atoms, where:

$$N_i = A^{\frac{1}{\beta+\upsilon}} \langle D^{\frac{1}{\upsilon}} \rangle^{\frac{\beta}{\beta+\upsilon}} D_i^{-\frac{1}{\upsilon}} \tag{2.29}$$

Here the exponents are $\beta = 1.73$ and $\nu = 0.7$.

Given the more direct relationship between diffusion and molecular size, the inference of composition from diffusion measurement given in eqn (2.29) is generally more robust than the relaxation-based estimates given in eqn (2.28). Diffusion-based estimates are valid even in the presence of dissolved methane and asphaltene. As discussed in the next section, Donaldson *et al.*[119] have successfully used this approach to determine the mean molecular size of the oil and gas phase in a reservoir at different DOIs. In heavier oils, short relaxation times decrease the diffusion sensitivity, as shown in Figure 2.18. This prevents a full characterization of the diffusion properties of viscous oils. For this reason, the diffusion-based composition analysis works well for oils with viscosities below about 40 cP.

2.6.4.2 Restricted Diffusion

While the diffusion properties of fluids filling the pore space or rocks are primarily determined by the chemical composition of the fluids, they are also affected by the confining geometry of the pore space. When the displacement of spins undergoing Brownian motion approaches the pore size during the diffusion encoding time, the motion becomes restricted and the effective diffusion coefficient decreases.[120] This effect of restricted diffusion can in turn be used for a direct determination of the pore size that is independent of surface relaxivity. Mitra *et al.*[121] showed that the reduction of the effective diffusion coefficient $D(T_d)$ from the bulk diffusion coefficient D_o at a diffusion time of T_d is to first order proportional to S/V_p with a known proportionality factor:

$$\frac{D(T_d)}{D_o} = 1 - \frac{4}{9\sqrt{\pi}} \frac{S}{V_p} \sqrt{D_o T_d} + \mathcal{O}(T_d) \tag{2.30}$$

This expression allows the direct extraction of S/V_p from diffusion measurements. Furthermore, based on eqn 2.22 and 2.30, it is possible to determine the surface relaxivity ρ_2 by combining diffusion measurements with relaxation measurements and analyzing the T_2 dependence of the measured diffusion coefficient:[91,92]

$$\frac{D(T_2)}{D_o} = 1 - \frac{4}{9\sqrt{\pi}} \frac{1}{\rho_2} \frac{\sqrt{D_o T_d}}{T_2} + \ldots \tag{2.31}$$

2.6.5 Diffusion–Relaxation Distribution Functions

Multi-dimensional distribution functions contain intrinsically more information than the corresponding one-dimensional diffusion or relaxation distribution functions. After the first demonstration of D-T_2 measurements in 2002[26,28] it has become evident that these measurements are particularly useful for fluid typing applications and the characterization of complex samples containing multiple fluid phases.

2.6.5.1 D-T_2 Based Fluid Typing and Saturation Monitoring

Contrasts in the relaxation and diffusion properties between the brine and the hydrocarbon phases give rise to separable contributions in the corresponding D-T_2 distribution functions. This allows the identification and quantification of the different fluid phases, including the determination of the relative saturations. To first order, the gas signal can be identified by its high diffusion coefficient, the water signal is distributed along an approximately horizontal line near the molecular diffusion coefficient of water, $D \approx D_{water}$, and the oil signal appears near the diagonal 'alkane line', characterized by $D = \lambda T_2$, with $\lambda \approx 5 \times 10^{-10}$ m^2 s^{-2}. The total area under all the peaks is proportional to the porosity, while the relative area of a given peak allows the determination of the saturation of this phase.

An illustration of the wealth of information contained in such multi-dimensional distribution functions is given in Figure 2.23. It shows D-T_2 laboratory measurements on a sandstone in three different saturation stages: first, fully brine saturated, followed by drainage with a crude oil, and finally after forced imbibition with water.[122] The signals from the oil and water phases are easily identified and separable. The location of the water signal along the relaxation axis gives an indication of the relative pore size occupied by the water phase, while the peak location of the oil signal along the alkane line gives a good measure of its viscosity and the width of the oil signal indicates the relatively wide distribution of molecular sizes in this particular crude oil.

2.6.5.2 D-T_2 Based Oil Composition and SARA Analysis

The oil contribution is generally expected to fall close to the alkane line. To first order, both the diffusion coefficient and the relaxation time for each oil molecule are inversely proportional to its radius of gyration. This gives rise to the strong correlation between D and T_2 along the alkane line evident in Figure 2.23. Further measurements on a larger set of oils have shown that the exact correlation between D and T_2 is also affected by the chemical composition of the oil,[67] as shown in Figure 2.24. In particular, the correlation depends not only on the size, but also on the shape of the oil molecules. The D-T_2 correlations for oils rich in sheet-like aromatic molecules are systematically different from those of oils rich in rod-like saturate molecules. Furthermore, the presence of asphaltene in the oil moves the contribution off the alkane line to shorter relaxation times. The shape of the D-T_2 contribution

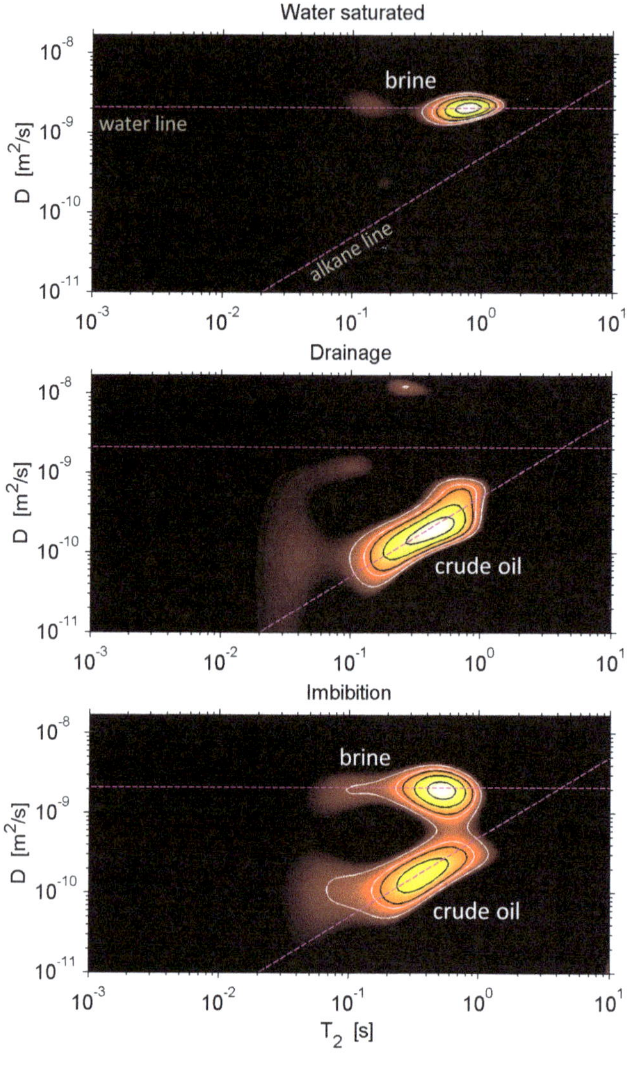

Figure 2.23 Laboratory D-T_2 measurements on a sandstone core in three different saturation stages. The horizontal dashed line is the water line, while the diagonal line is the alkane line. Figure adopted from ref. 122.

of the oil together with the T_1/T_2 ratio can therefore be used to estimate the SARA (saturate, aromatic, resin, asphaltene) composition of the oil.[67,123]

2.6.5.3 *D-T_2 Derived Information on Fluid Configuration and Wettability*

The flow properties of oil and water through a reservoir rock are not only determined by the geometry of the pore space and the fluid saturations, but also by the geometrical configuration of the fluid phases within the pore space. These geometrical configurations are sensitive to the wettability and

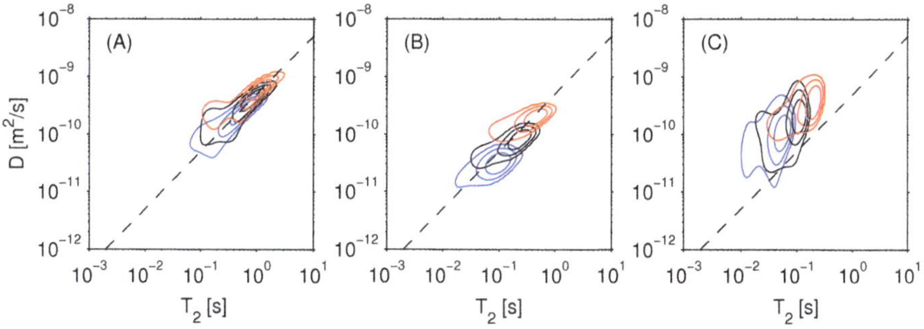

Figure 2.24 D-T_2 distribution functions for three representative crude oils with contrasting composition at the three temperatures 10 °C (blue), 30 °C (black), and 50 °C (red). The crude oil on the left is rich in saturates, the oil in the middle is biodegraded and rich in aromatics, whereas the oil on the right contains asphaltene. The dashed line indicates the alkane line. Adapted with permission from Mutina and Hürlimann, 2008.[67] © 2008 American Chemical Society.

can be probed with D-T_2 measurements. This is accomplished by comparing the measured distribution function of a given fluid inside and outside the core. Confinement of the fluid within the pore space reduces the measured relaxation time by an amount that depends on the degree of surface wetting, since only fluid–grain interfaces lead to significant surface relaxation. The impact of wettability on D-T_2 measurements is illustrated in Figure 2.25 with an example of laboratory measurements on a set of three cores obtained from a single well. The cores were fine grained and had nearly identical pore size distributions (as measured by Hg porosimetry). They were initially brine saturated and then underwent a drainage cycle with dodecane. From the measurements in Figure 2.25, it is apparent that the diffusion properties show in all cases a noticeable reduction from the bulk value of the oil (shown as white circles) by about the same amount. This comparable degree of restricted diffusion indicates that the oil has a similar geometrical configuration with characteristic dimensions smaller than 10 μm in each core. However, the relaxation properties of the three cores show large differences. In the core from the water zone, the measured relaxation time is barely reduced from its bulk value, whereas in the core from the oil zone, there is a much more pronounced reduction. This indicates that in the core from the oil zone, the oil molecules are in good contact with the surfaces and thus affected by surface relaxation. It can be concluded that this core is at least partially oil wet. In contrast, the oil molecules in the core from the water zone are not able to get in close contact with the grain surfaces. In this case, the grain surfaces must be wetted by a thin layer of water and the core is therefore mainly water wet. This contrasting wetting behavior of the three cores from the different sections of the reservoir reflects the different exposures to crude oil over geological times, which resulted in a wettability transition from an original water-wet condition to a more oil-wet condition within the oil reservoir.

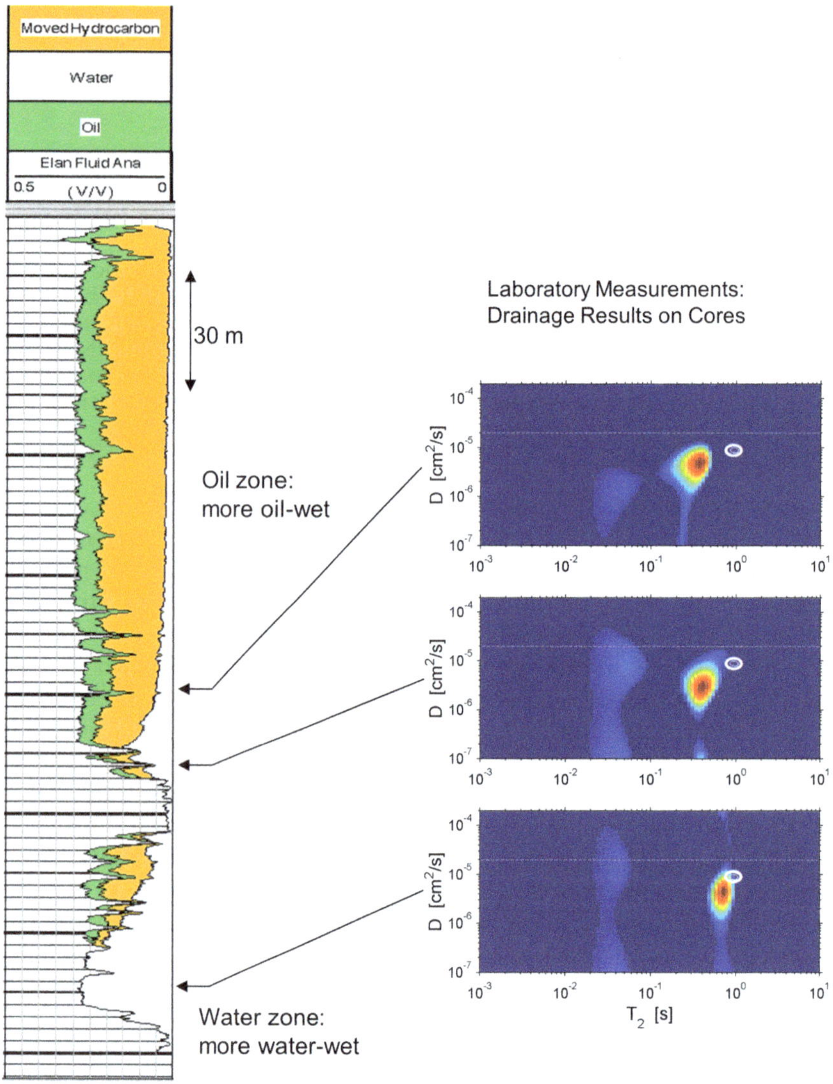

Figure 2.25 The impact of wettability on D-T_2 measurements is demonstrated on three cores taken from different depths of a carbonate well at the locations indicated on the left. The cores were initially water saturated and showed essentially identical D-T_2 distribution functions. After drainage with dodecane, the responses shown on the right exhibit clear differences that are related to wettability. As a comparison, the response of bulk dodecane is indicated by the white ellipses.

2.7 Applications

2.7.1 Overview

In this section, we present a number of representative NMR logging examples. They were chosen to illustrate the wide range of applications where the NMR technique provides key petrophysical or reservoir information. In typical logging operations, a number of different logging tools are combined into a "toolstring" or "bottom hole assembly" (BHA). The most common tools include resistivity sensors, various nuclear measurements (natural gamma ray emission, gamma ray absorption, neutron absorption), and acoustic measurements. As the tool moves along the borehole (upwards in wireline logging, downwards in LWD logging), the sensors continuously perform measurements. The results of these different measurements are then jointly displayed *versus* depth, *i.e.* the distance along the borehole to the surface. Measurements from all the sensors are then displayed in adjacent tracks in a display commonly referred to as a "log".

Broadly speaking, hydrocarbon reservoirs are identified as formations of high porosity that contain hydrocarbons. In conventional logging operations, porosity is generally determined from nuclear measurements and the presence of hydrocarbons is inferred from high resistivity readings. As "conventional" hydrocarbon reservoirs are becoming depleted, the focus is shifting towards more complex reservoirs that are not so easily identified. The traditional porosity–resistivity strategy is often no longer sufficient to provide a robust formation evaluation and additional logging measurements, including NMR, are required.

To interpret well-log data, it is important to account for the drilling history of the well. In most cases, wells are drilled in "overbalanced" conditions, such that the wellbore pressure is maintained at or slightly above the formation pressure. Under these conditions, some "invasion" of drilling fluids into the formation occurs. During the invasion process, solids from the drilling muds accumulate at the borehole-formation interface to form a "mudcake" which acts as a filter, allowing only fluid to enter the formation. This invading fluid, referred to as filtrate, displaces the movable fluid in the near wellbore region (see Figure 2.26). The radial extent of the invasion can range from a few millimeters to meters, depending on the time since drilling, the overbalance pressure, the formation porosity and permeability, and on the composition of the drilling mud itself. There are many different drilling muds in use today, both aqueous and oil-based, that have been designed for specific environments.

As discussed in Section 2.2, the sensitive volume for NMR measurements is determined by the resonant condition between the local applied magnetic field and the rf carrier frequency. This results in well-defined depths of investigation (DOI) for the NMR measurements that are tool specific and typically fall into the range of about 2.5 cm to 15 cm from the borehole wall. A number of NMR tools can be operated at multiple DOIs by changing the operating frequency. In wireline logging, which usually occurs at least a few

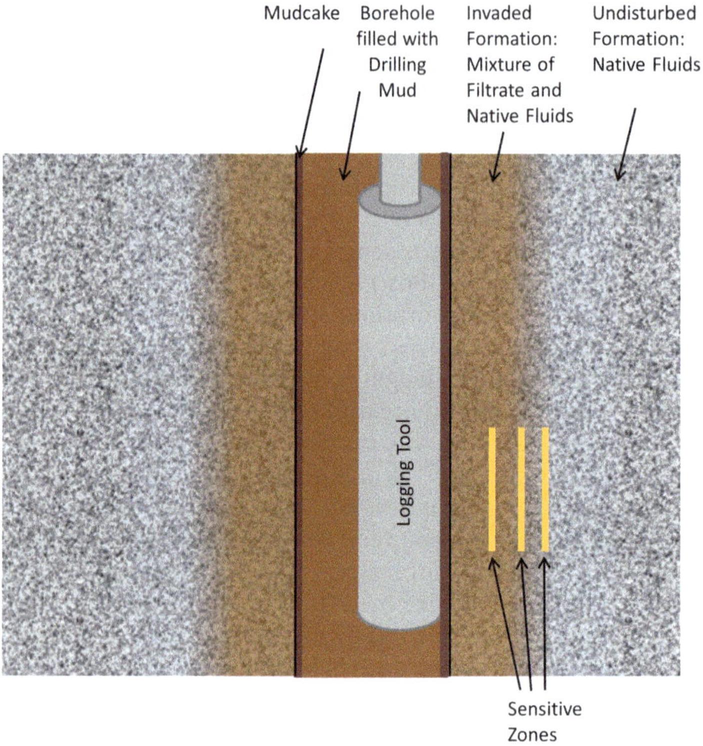

Figure 2.26 Schematic view of the borehole environment during well logging. Invasion of the borehole fluid into the formation leads to the creation of a mudcake on the walls of the borehole and the displacement of some of the native fluids by the filtrate close to the borehole. NMR measurements at different depths of investigation can probe this invasion process.

hours after drilling, the region probed by the NMR sensors generally covers the invaded part of the formation. However, in some cases, including in shales, in formations bearing heavy oils, or in porous formations drilled with oil-based muds, the invasion can be shallow. In this case, the wireline NMR measurements observe the undisturbed fluids. The situation for NMR LWD measurements also depends on the drilling rate and the placement of the NMR sensor within the BHA. The NMR LWD measurements more frequently occur prior to invasion of the probed region, in which case information on the native reservoir fluids is obtained. Understanding the degree of invasion of borehole fluids into the near well bore region of the formation is therefore a critical part of the proper interpretation of both wireline and LWD measurements. Failure to do so will introduce significant uncertainties in the log interpretation.

We next present a number of examples of NMR logs that are ordered according to the complexity of the NMR measurements. We start with examples that rely primarily on the NMR signal amplitude. This is followed by the important applications where relaxation information plays the critical role in deriving the petrophysical answers. Finally, we discuss applications that also involve diffusion information and enable a detailed characterization of the formation fluids and their immediate environment.

2.7.2 NMR Amplitude-Based Applications

2.7.2.1 Porosity in Formations with Complex Lithology

Porosity is a key petrophysical quantity. In conventional logging operations, porosity is determined with nuclear sensors, *i.e.* from the measured rate of attenuation of gamma rays or the rate of neutron absorption.[3] The interpretation of both these nuclear measurements requires knowledge of the lithology. As discussed in Section 2.6.2.2, the gamma ray measurement determines the average formation density. This quantity can be related to the formation porosity using eqn (2.19) if the density of the rock matrix is known. The neutron measurements in turn are related to the total concentration of hydrogen in the formation. In order to infer porosity from the neutron log, it is essential to correct for the presence of hydrogen in shales and clays. In contrast, NMR is unique amongst well log measurements in that the primary quantity (signal amplitude) is essentially independent of the rock matrix and depends only on the formation fluids. This is a significant benefit from the point of view of interpretation, especially in wells with complex lithology. In some cases, NMR logging is also used as the primary porosity log to avoid the use of radioactive sources, required for conventional nuclear porosity measurements.

Figure 2.27 shows an example from a carbonate formation with complex lithology[124] where the determination of porosity is straightforward from NMR measurements, but difficult from conventional nuclear measurements. The left track shows the lithology of the well, inferred mainly from nuclear measurements. In this case, the formation lithology changes from predominantly dolomitic at the top to largely calcitic in the lower section. In the right column, the NMR-derived porosities (see eqn (2.14)) are compared with two density-derived porosities based on eqn (2.19), using ρ_{grain} appropriate for calcite and dolomite matrices, respectively. Without prior detailed knowledge of the lithology, it is not possible to derive an accurate porosity across this interval purely from the density log, whereas NMR is valid regardless of lithology.

2.7.2.2 Radial Profiling of Hydrogen Index and Gas Detection

The NMR porosity is independent of lithology, but it is affected by the hydrogen index of the fluid. This effect can be used to infer the presence of gas, as demonstrated in the example[20] shown in Figure 2.28. This application takes advantage of the multi-frequency capability of modern wireline NMR logging

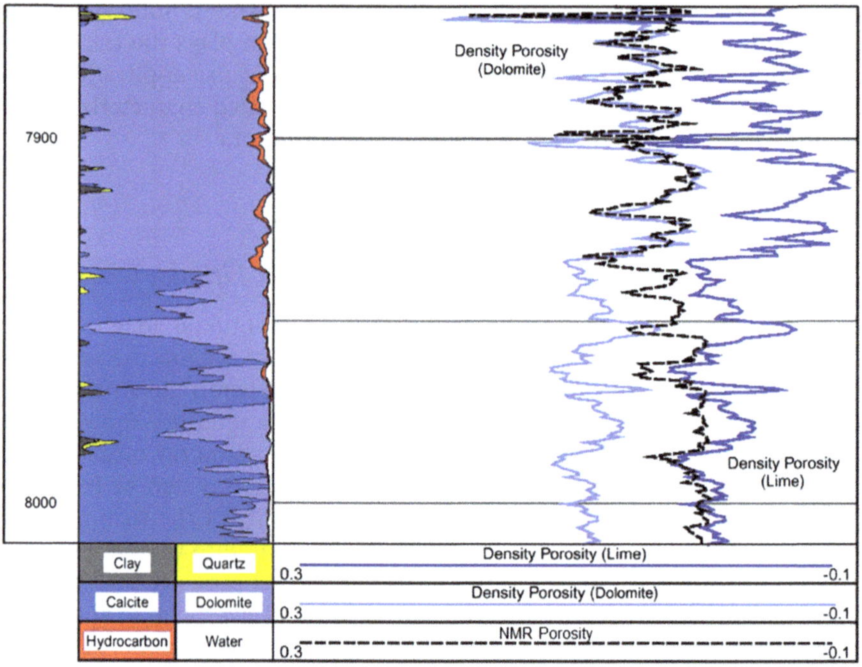

Figure 2.27 Lithology independent porosity. Example comparing NMR porosity with density porosity logs derived from nuclear measurements. Vertical axis is depth in feet and the porosity units are v/v (fractional porosity). Adapted from Boyd *et al.*, 1995.[124] © 1995 Society of Petroleum Engineers Inc. Reproduced with permission of SPE. Further reproduction prohibited without permission.

tools that allows measurements at different depths of investigations (DOI). This capability of "radial profiling" is particularly helpful for understanding log responses where invasion is shallow.

The log shows two sections from a well in the Gulf of Mexico, drilled with oil-based mud (OBM), in a formation consisting of alternating layers of sand and shale. The right-most column shows the measured NMR porosities for 4 different DOI between 1.5″ and 2.7″. Over those intervals we observe a systematic reduction in NMR porosity as the DOI increases. This is caused by partial invasion of OBM filtrate into the formation where it replaces gas. In effect, the NMR logs detect a radial variation of the fluid hydrogen index, HI. The results suggest that the radial extent of fluid invasion covers the same range as the NMR measurements (~2 inches). The shallower NMR measurement (1.5″) detects predominantly the invaded filtrate, which has a HI close to 1, whereas the deeper NMR measurement (2.7″) detects more gas, which has a low HI. The red shading (track 4) is the gas volume inferred simply from the difference in NMR "porosities" between the 1.5″ and 2.7″ measurements.

In the upper section (400 ft long), the NMR measurements indicate that the formation consists of 5 distinct layers of gas-filled sands, separated by

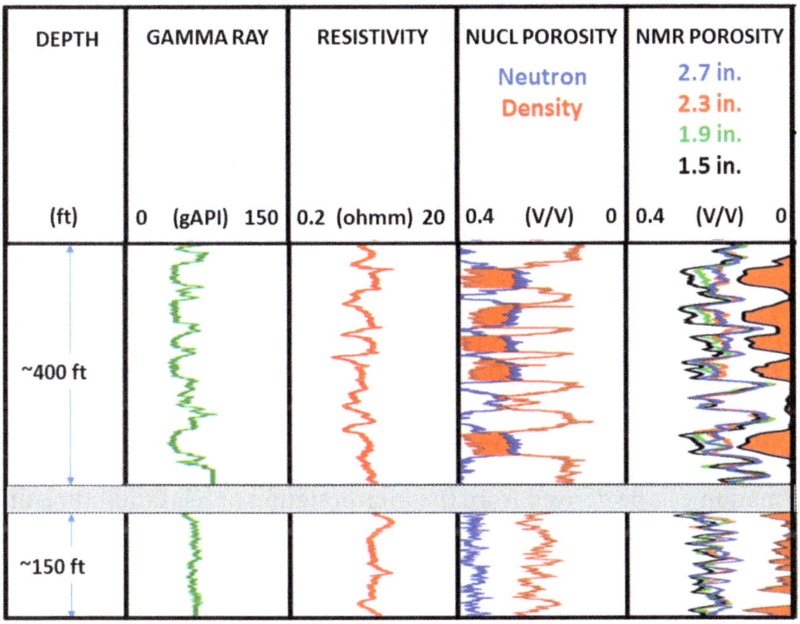

Figure 2.28 Radial profiling of hydrogen index and gas detection. Natural gamma ray log (track 1), deep resistivity (track 2), neutron and density porosity (track 3) and NMR porosities at four DOIs from 1.5″ to 2.7″ (track 4) from two intervals in a well drilled with oil-based mud in the Gulf of Mexico. Adapted from DePavia *et al.*, 2003.[20] © 2003 Society of Petroleum Engineers Inc. Reproduced with permission of SPE. Further reproduction prohibited without permission. Red shading in tracks 3 and 4 indicates the presence of gas based on a simple comparison of log responses.

sizeable layers of water-filled shales. This NMR based interpretation is supported by the interpretation of the conventional measurements: The logs of natural gamma ray in column 1 and deep resistivity in column 2 signal the presence of five separate hydrocarbon-filled sands by the low gamma ray counts and high resistivity readings. Furthermore, the crossover of neutron and density porosity logs, shown in column 3 and highlighted by red shading, is a clear sign of gas.[3]

The conventional log measurements results are more difficult to interpret in the lower interval. In this section, the gamma ray log is quite featureless and there are no crossovers of the neutron and density porosity logs. The conventional logs do not give any clear indication of hydrocarbons in this section. However, the NMR logs detect a noticeable radial reduction in HI, indicating the presence of gas, similar to that observed in the upper interval. In this case, the thickness of individual sand laminations is too small to be resolved by the logging tools. The NMR measurements provide a volumetric average across the fine laminations that is quantitative and that can be used to infer the amount of gas contained in these layers. Since conventional log responses are generally unable to detect such potential hydrocarbon-bearing reservoirs, NMR logging has become a principal method for identifying and evaluating such low resistivity laminated formations.

2.7.3 Applications Based on NMR Relaxation Measurements

2.7.3.1 Pore Size Distributions, Bound Fluid and Producibility Estimate

While porosity applications can be important drivers for NMR logging, there are often other faster logging options available that provide adequate porosity logs. The key driver that fuelled the rapid growth in NMR logging over the last two decades is the ability to provide robust estimates of producibility, including the independent measurement of movable and non-movable fluids, and the development of reliable NMR permeability transforms. This information can be derived from the interpretation of relaxation time distributions, which are now often regarded as the primary deliverables of NMR logging.

A representative example[125] is shown in Figure 2.29. It shows a suite of petrophysical logs acquired while drilling in an oil-bearing sandstone formation. The first two tracks in Figure 2.29 show resistivity logs and a lithology log derived from nuclear capture spectroscopy measurements. The third track shows the T_2 distributions, with a logarithmic T_2 scale covering 0.5 ms to 5 s. The fourth track shows the total NMR porosity and the density-derived porosity. Blue and grey shading indicates free and bound fluid volumes, respectively, derived from the NMR T_2 distributions (eqn (2.24)). The fifth column shows the NMR relaxation-based permeability logs (eqn (2.26)), which are compared with mobility estimates (black points in track 5) obtained from formation pressure measurements.

There are two sections with high resistivity readings that identify hydrocarbon-bearing zones of sands (shown as yellow shading in the lithology track 2). In these two sections, the T_2 distributions show only long relaxation times, which indicate high quality rock with large free fluid volumes and high permeabilities. Indeed, the NMR-based permeability estimates agree well with mobility estimates derived from formation pressures. The conventional log analysis can easily identify these two reservoir intervals as hydrocarbon bearing "pay zones", but it likely associates the rest of the logged section, exhibiting low resistivity (indicating high water saturation) and increased shale content (dark shading in track 2), as "non-pay" zones. However, the measured T_2 distributions show the presence of large pores with associated mobile fluid above and below the obvious pay zones in the sands. The additional free fluid identified by the NMR measurements resides in thin beds that are interspersed with shales. Conventional log analysis based on resistivity and nuclear porosity measurements often fails to identify such thinly laminated formations. However, the cumulative free fluid volume associated with hydrocarbons in the sand layers may be comparable to or even greater

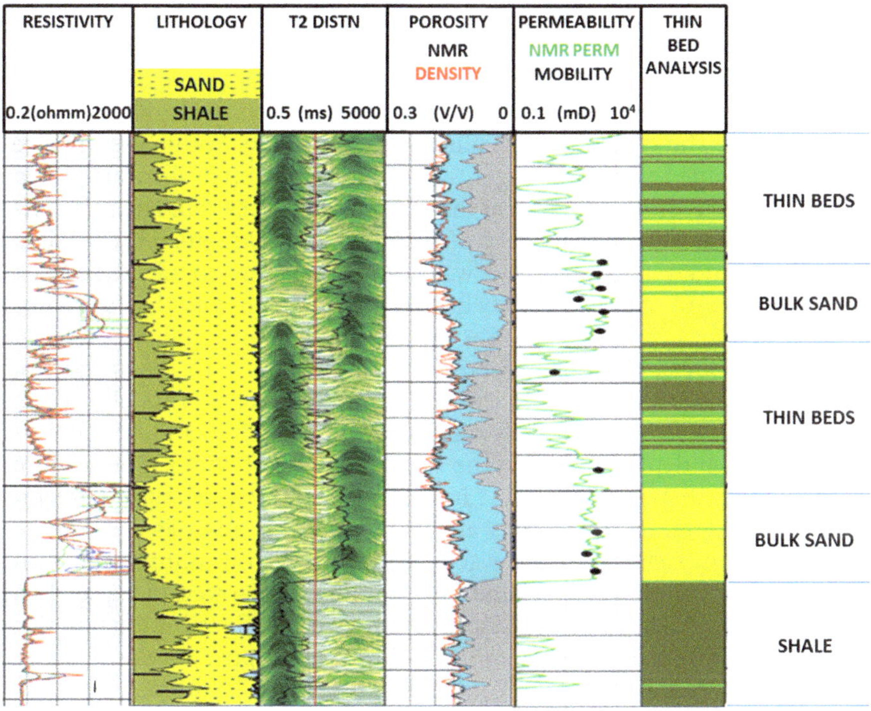

Figure 2.29 NMR and conventional logs acquired while drilling in a laminated sandstone formation. Vertical axis represents depth. The interval displayed extends approximately 600 ft. Adapted from Heaton *et al.*, 2012.[125] © 2012 Society of Petroleum Engineers Inc. Reproduced with permission of SPE. Further reproduction prohibited without permission.

than that in the main sands. A more detailed analysis of the NMR T_2 distributions[126] allows a separation of the formation into sand, thin beds, and shale (see track 6).

2.7.3.2 *In situ Viscosity Determination*

When invasion is very shallow, the native hydrocarbons are not flushed away by the filtrate and the NMR measurements can detect the hydrocarbon signal directly. In this case, the measured relaxation times can be used to infer the viscosity of the oil, using eqn (2.27). This application is illustrated in the next example[63] presented in Figure 2.30. It shows the results of an analysis of wireline data acquired in a Californian oil well drilled in a diatomite formation. While the porosity is exceptionally high across the entire reservoir, the pores are extremely small, resulting in a small permeability and an almost total absence of invasion. To assess the viability of conventional water flooding or other enhanced oil recovery techniques, it is critical to determine the viscosity of the oil. Since this parameter is known to vary across the reservoir, it is best to perform an *in situ* viscosity determination along the well.

NMR relaxation and diffusion measurements were acquired together with dielectric dispersion measurements and nuclear porosity logs. The T_2 distributions are shown in the first column of Figure 2.30. At the bottom of the well, the T_2 values extend to longer times than at the top of the well, indicating that the viscosities are lower at the bottom than at the top. The main challenge for a quantitative evaluation of viscosity is to distinguish the oil signal from the bound water signal. This was accomplished by taking advantage of the dielectric and nuclear porosity logs as constraints in a statistical analysis of the NMR logs.[63] Using a Monte Carlo approach with realistic fluid and formation models, all the logs were forward modeled with the appropriate sensor responses and compared with the measurements. This allowed the determination of the most likely oil viscosity, oil volume, and bound water volume at each depth, including the associated uncertainties. The resulting viscosities are displayed in track 2 in Figure 2.30. The width of the line provides the uncertainty in viscosity at each depth. This analysis confirms the known, but unusual, "inverted" viscosity profile. The oil volume and bound water volume including associated uncertainties are plotted in tracks 3 and 4, respectively. To the left of the depth log are two D-T_2 distribution functions for averaged intervals corresponding to lower and upper sections. Consistent with results shown in Figure 2.18, the diffusion measurements can only

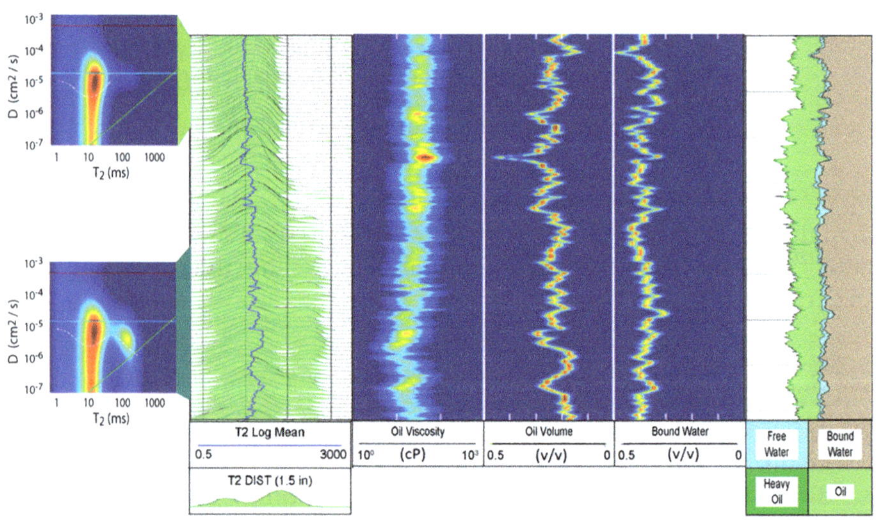

Figure 2.30 *In situ* viscosity determination. Measured T_2 distributions (track 1), oil viscosity (track 2), oil volume (track 3), and bound water logs (track 4), together with associated uncertainties from a Monte Carlo analysis of NMR, dielectric and nuclear porosity logs. The inferred fluid analysis is shown in track 5. The D-T_2 distribution functions on the left correspond to averaged intervals for the lower and upper sections. Adapted with permission from Heaton *et al.*, 2012.[125]

resolve components with $T_2 > 10$ ms. The peak appearing at ~100 ms in the lower interval is absent in the upper interval, further confirming the increase in oil viscosity from the lower to the upper section.

2.7.4 Applications Based on Diffusion Measurements

2.7.4.1 Saturation Profiling and Gas Quantification

The latest generation wireline NMR logging tools enable the acquisition of diffusion measurements at multiple depths of investigation for directly monitoring the variations of fluid saturations due to filtrate invasion (see Figure 2.26). Figure 2.31 shows excerpts from a D-T_1 log acquired in a gas reservoir at two radial depths of investigation.[127] The well was drilled through a sandstone formation with oil-based mud as the drilling fluid. As mentioned before, the high resistivity peaks (track 2) around a depth of 50 ft and the crossover of density porosity and neutron density logs in track 3 are classic indicators for the presence of gas.[3] This circumstance significantly complicates the accurate determination of porosity. Both nuclear porosity logs are sensitive to gas, but a robust porosity interpretation requires prior knowledge of the gas properties, in particular the gas density and hydrogen index, which are typically not available. This difficulty can be overcome with NMR measurements. The right hand section of Figure 2.31 displays the extracted

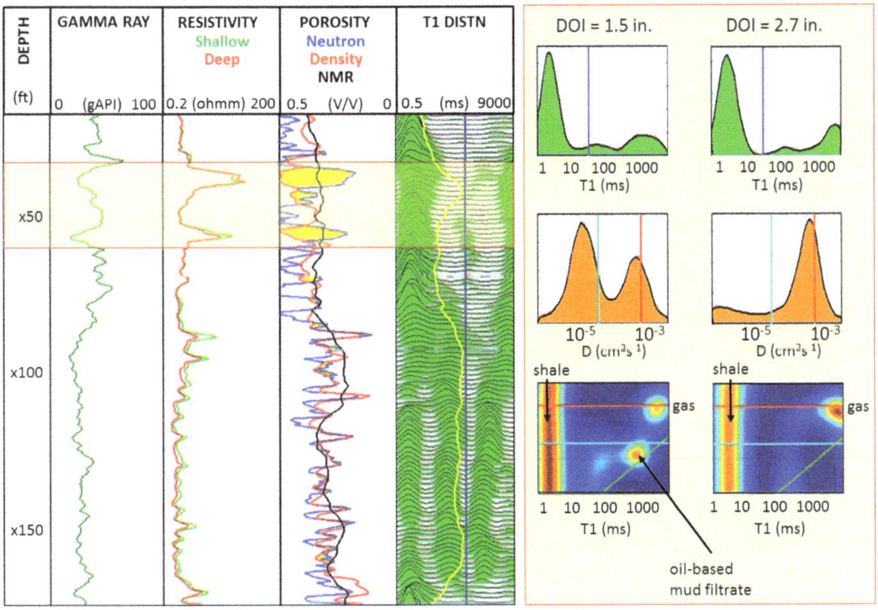

Figure 2.31 Saturation profiling and gas quantification. The diffusion-relaxation measurements allow the determination of the DOI dependent gas saturations and are used for the accurate determination of the formation porosity. Adapted with permission from Guru *et al.*, 2008.[127]

D-T_1 distribution functions together with the corresponding T_1 and diffusion distributions for the DOI of 1.5″ and 2.7″, respectively, averaged over the gas-containing interval indicated by yellow shading. For both DOIs, a gas signature is clearly visible in the top right corners of the D-T_1 maps. Also clearly visible is the broad stripe on the left of the map, which corresponds to shale, characterized by the short T_1 (and T_2) values and unresolved diffusion (see Figure 2.18). A qualitative difference between the two D-T_1 maps is the peak at around T_1 = 1 s, which appears at a DOI of 1.5″, but is absent at the DOI of 2.7″. This low diffusion peak corresponds to filtrate from the oil-based mud drilling fluid, which has invaded the near-wellbore region of the formation, displacing some of the gas. By comparing the two measurements and knowing the chemical properties (hydrogen index) of the drilling fluid filtrate, it is possible to extract the gas hydrogen index and hence determine the formation porosity.

2.7.4.2 Continuous Composition Logging

Diffusion measurements are particularly well suited for the characterization of fluids with long relaxation times, including gas and gas condensates. As discussed in Section 2.6.4.1 and eqn (2.29), the diffusion distribution function $f(D)$ can be related to the distribution of molecular sizes, $f(N)$. In practice, the signal to noise ratio of logging measurements is often too low to extract a detailed log of size distributions from the diffusion measurements, but as demonstrated in the next example, it is possible to obtain a continuous log of the average molecular size. Figure 2.32 shows a wireline example from a well in a gas condensate reservoir, drilled with oil-based mud.[119] In this case, reservoir fluids were also collected with a sampling tool and analyzed in the laboratory. This allowed a direct comparison of the downhole composition log with laboratory results.

In this application, diffusion–relaxation data were acquired for two different depths of investigation, DOI = 1.5″ and 4″. The resulting D-T_1 distribution functions were then analyzed based on eqn (2.29). From the contributions with $T_1 > 400$ ms and D between 5×10^{-11} m^2 s^{-1} and 2.8×10^{-8} m^2 s^{-1}, an average chain length for the oil component was extracted, and from the contributions with $T_1 > 400$ ms and D above 2.8×10^{-8} m^2 s^{-1}, an average chain length for the gas component was extracted. The two panels in Figure 2.32 show the results for the two DOIs. In the shallower DOI, the extracted mean chain length is close to that of the filtrate, whereas at the deeper DOI, the measurements indicate a smaller chain length that is close to that of the produced condensate, as measured in the laboratory. These results reflect the invasion process that leads to a DOI-dependent mixing of the native hydrocarbon with the filtrate. At the top of this reservoir section, the measurement at the deeper DOI also detects a section of gas that partially overlaps with the oil phase. These results illustrate that under favorable condition, it is possible to obtain a detailed profile of the hydrocarbon fluids inside the formation.

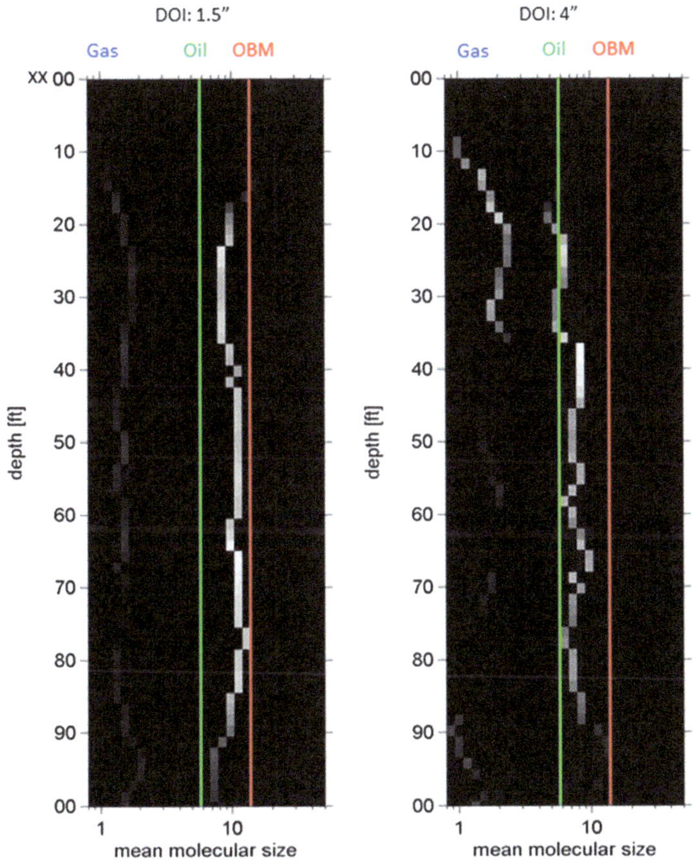

Figure 2.32 Continuous composition log based on NMR diffusion–relaxation measurements for two different depth of investigations, DOI = 1.5″ on the left and DOI = 4″ on the right. The mean chain lengths for the oil and gas phase are displayed. The shading of the mean carbon number indicates the relative saturation—higher saturation is indicated by higher intensity. The vertical green line indicates the mean carbon number of the oil-base drilling mud, whereas the red line indicates the mean carbon number of the produced oil, measured by gas chromatography in the laboratory. Adapted with permission from Donaldson *et al.*, 2013.[119]

2.7.4.3 Mobility of Heavy Oil

As was pointed out already in Section 2.5.4.3, the relaxation times of heavy oils are too short to allow proper characterization with diffusion measurements. Nevertheless, it can still be beneficial to perform diffusion–relaxation measurements in heavy oil reservoirs. In such reservoirs, it is indispensable to assess the mobility of the heavy oil and to avoid zones of water production.

An example[31,128] of this application is presented in Figure 2.33. It shows wireline measurements from a well in a heavy oil reservoir in Venezuela that

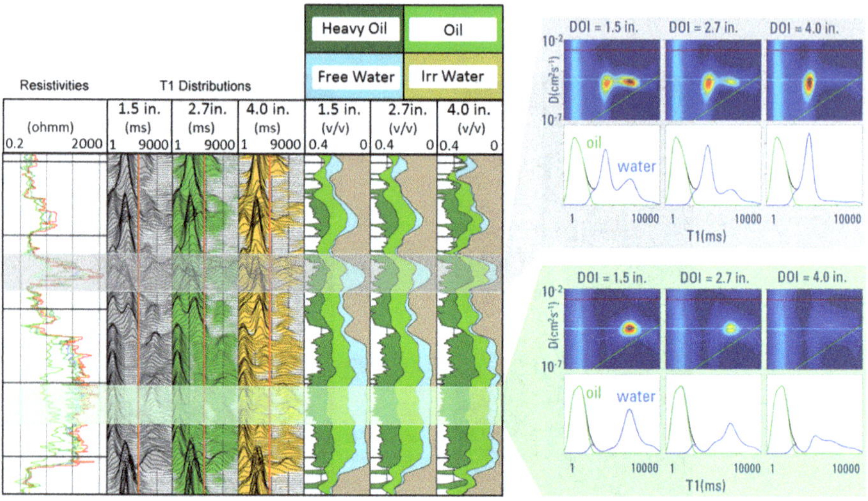

Figure 2.33 Mobility of heavy oil. Radial fluid profiling in a heavy oil reservoir using constrained 4D inversion of multi-frequency diffusion NMR measurements. Adapted with permission from Heaton *et al.*, 2008,[31] and Decoster and Carmona, 2008.[128]

was drilled with water-based mud. To assess this heavy oil reservoir, multi-frequency NMR depth logs were acquired in addition to resistivity and nuclear logs. The high resistivity values in the lower section of the log (leftmost track) are indicative of hydrocarbons, which in this case were known to be heavy (*i.e.* viscous) oil. The T_1 distributions corresponding to depths of investigation of 1.5″, 2.7″, and 4.0″ are displayed in the next three tracks. There is a reduction in the volume of movable water (components to the right of the T_1 cutoff) with increasing depth of investigation that is readily visible from the distributions. The next three tracks display the volumetric division of the T_1 distributions into bound water (brown), movable water (light blue), and oil (light green) for the three depths of investigation. The dark green shading represents the fraction of total porosity that is not detected by the NMR measurement, and is assigned to the heavier ends of the oil. These components have transverse relaxation times shorter than the minimum echo spacing of the NMR measurements. In heavy oil reservoirs such as this, the total porosity must therefore be derived from other logs, typically density and neutron absorption measurements. To the right of the depth logs, D-T_1 maps are displayed for two averaged intervals that are indicated by the horizontal shaded bars. The decrease in movable water volume with increasing radial depth is readily verified from the maps and the corresponding T_1 distributions. This observation suggests that some heavy oil was displaced by the filtrate and is therefore mobile. Furthermore, the disappearance of the free fluid volume at large DOI demonstrates that this reservoir is at “irreducible water saturation”, implying that the produced fluid will be predominantly hydrocarbon. This information is particularly important in heavy oil reservoirs, since in this case even low saturations of free water can disrupt oil production due to the high viscosity contrast between the hydrocarbon and water fractions.

2.8 Conclusion

NMR well logging is a prime example of "inside-out-NMR" where relaxation and diffusion measurements are performed on samples located outside the magnet. The technique provides a unique tool to quantitatively detect and characterize fluids in geological formations around the borehole. Consequently, NMR well logging has become a widely used commercial service for the exploration and development of hydrocarbon reservoirs. All aspects of the technique have undergone remarkable developments during the last two decades and continued progress is anticipated. New configurations of the permanent magnets and the rf coil were developed that have enabled the implementation of different modalities of NMR well logging, including logging-while-drilling applications. With novel measurement schemes, ever more sophisticated electronics, and advanced algorithms to process and analyze the data, it has become possible not only to determine quantitatively amplitudes, average relaxation times, and diffusion coefficients, but also to extract detailed multi-dimensional distribution functions of relaxation and diffusion. A critical step in this progress has been the development of a detailed understanding of the underlying spin dynamics in inhomogeneous fields and under various environmental effects, including variable temperature and salinity. This knowledge has led to new methods that mitigate the impact of these variations and improve the calibration procedures. Furthermore, with some of the logging tools it has become possible to perform these measurements spatially resolved not only along the direction of the borehole, but also along the radial direction away from the borehole. The extracted one-dimensional relaxation and multi-dimensional relaxation–diffusion distribution functions give detailed information on the quantity and identity of the fluids, their molecular composition, the size of the pores occupied by the wetting phase, and the geometrical configuration of the fluid phases within the pore space. As these relationships between the quantities measurable by NMR and the reservoir properties of interest are further clarified and discovered, new applications for NMR logging will emerge. It is also anticipated that the well logging techniques of inside-out NMR developed for the characterization of geological formations will find further applications in other fields, including medicine, process control, material and food sciences, and engineering.[129]

References

1. R. E. Sheriff and L. P. Geldart, *Exploration Seismology*, Cambridge University Press, 1995.
2. A. Legchenko and P. Valla, *J. Appl. Geophys.*, 2002, **50**, 3–19.
3. D. V. Ellis and J. M. Singer, *Well logging for earth scientists*, Springer, 2007.

4. R. J. S. Brown, R. Chandler, J. A. Jackson, J. L. Kleinberg, M. N. Miller, Z. Paltiel and M. G. Prammer, *Concepts Magn. Reson., Part A*, 2001, **13**, 335–411.
5. G. R. Coates, L. Xiao and M. G. Prammer, *NMR logging principles and applications*, Halliburton Energy Services, 1999.
6. R. J. S. Brown and B. W. Gamson, *Trans. Am. Inst. Min. Metall. Eng.*, 1960, **219**, 199–207.
7. D. I. Hoult and R. Richards, *J. Magn. Reson.*, 1976, **24**, 71–85.
8. M. Hürlimann and D. Griffin, *J. Magn. Reson.*, 2000, **143**, 120–135.
9. S. Earnshaw, *Trans. Cambridge Philos. Soc.*, 1842, 7, 97.
10. R. K. Cooper and J. A. Jackson, *J. Magn. Reson.*, 1980, **41**, 400–405.
11. L. J. Burnettt and J. A. Jackson, *J. Magn. Reson.*, 1980, **41**, 406–410.
12. J. A. Jackson, L. J. Burnett and J. F. Harmon, *J. Magn. Reson.*, 1980, **41**, 411–421.
13. J. Horkowitz, S. Crary, K. Ganesan, R. Heidler, B. Luong, J. Morley, M. Petricola, C. Prusiecki, M. Poitzsch, J. R. Scheibal and M. Hashem, in *Transactions of the SPWLA 43th Annual Logging Symposium*, Oiso, Japan 2002, p. Paper EEE.
14. M. Borghi, F. Porrera, A. Lyne, T. Kruspe, V. Krueger, H. Thern and R. Chemali, in *Transactions of the SPWLA 46th Annual Logging Symposium*, New Orleans, USA, 2005, p. Paper HHH.
15. T. Kruspe, H. Thern, G. Kurz, M. Blanz, R. Akkurt, S. Ruwali, D. Seifert and A. F. Marsala, in *Transactions of the SPWLA 50th Annual Logging Symposium*, 2009, p. Paper EEE.
16. *U.S. 4,710,713A Pat.*, 1987.
17. M. N. Miller, Z. Paltiel, M. E. Gillen, J. Granot and J. C. Bouton, in *SPE Annual Technical Conference and Exhibition*, New Orleans, USA 1990 p. Paper SPE 20561.
18. M. G. Prammer, G. D. Goodman, S. K. Menger, M. Morys, S. Zannoni and J. H. Dudley, in *Transactions of the SPWLA 41th Annual Logging Symposium*, Dallas, USA 2000, p. Paper EEE.
19. R. L. Kleinberg, A. Sezginer, D. D. Griffin and M. Fukuhara, *J. Magn. Reson.*, 1992, **97**, 466–485.
20. L. DePavia, N. Heaton, D. Ayers, R. Freedman, R. Harris, B. Jorion, J. Kovats, B. Luong, N. Rajan, R. Taherian, K. Walter and D. Willis, in *SPE Annual Technical Conference and Exhibition*, Denver, USA, 2003 p. Paper SPE 84482.
21. S. Chen, E. Beard, M. Gillen, S. Fang and G. Zhang, in *Transactions of the SPWLA 44th Annual Logging Symposium*, Galveston, USA, 2003, p. Paper ZZ.
22. H. Y. Carr and E. M. Purcell, *Phys. Rev.*, 1954, **94**, 630–638.
23. S. Meiboom and D. Gill, *Rev. Sci. Instrum.*, 1958, **29**, 688–691.
24. R. Akkurt, H. J. Vinegar, P. N. Tutunjian and A. J. Guillory, in *Transactions of the SPWLA 36th Annual Logging Symposium*, Paris, France, 1995, p. Paper N.
25. M. D. Hürlimann, *J. Magn. Reson.*, 2001, **148**, 367–378.

26. M. D. Hürlimann and L. Venkataramanan, *J. Magn. Reson.*, 2002, **157**, 31–42.
27. M. D. Hürlimann, L. Venkataramanan, C. Flaum, P. Speier, C. Karmonik, R. Freedman and N. Heaton, in *Transactions of the SPWLA 43th Annual Logging Symposium*, Oiso, Japan, 2002, p. Paper FFF.
28. M. D. Hürlimann, L. Venkataramanan and C. Flaum, *J. Chem. Phys.*, 2002, **117**, 10223–10232.
29. W. E. Kenyon, P. I. Day, C. Straley and J. F. Willemsen, *Soc. Petrol. Eng. Form. Eval.*, 1988, **3**, 622–636, Paper SPE 15643-PA; Erratum: *Soc. Petrol. Eng. Form. Eval.*, 1989, **4**, 8.
30. Y.-Q. Song, L. Venkataramanan, M. D. Hürlimann, M. Flaum, P. Frulla and C. Straley, *J. Magn. Reson.*, 2002, **154**, 261–268.
31. N. Heaton, H. Bachman, C. C. Minh, E. Decoster, J. LaVigne, J. White and R. Carmona, *Petrophysics*, 2008, **49**, 172–186.
32. P. Speier, S. Crary, C. Flaum and R. L. Kleinberg, *Transactions of the SPWLA 40th Annual Logging Symposium*, 1999.
33. D. McKeon, C. Cao Minh, R. Freedman, R. Harris, D. Willis, D. Davies, G. Gubelin, R. Oldigs and M. D. Hürlimann, in *Transactions of the SPWLA 40th Annual Logging Symposium*, 1999, p. Paper CC.
34. C. E. Morriss, P. Deutch, R. Freedman, D. McKeon and R. L. Kleinberg, *Log Anal.*, 1996, **37**/**2**, 54–59.
35. C. Cao Minh, E. Bordon, M. D. Hürlimann, T. Smithson, J. Kovats, R. Badry, M. Claverie, Q. Haiyan, M. Davenport, H. Rose and G. Womack, *Proceedings of the SPE Annual Technical Conference and Exhibition*, 2005.
36. R. L. Kleinberg and H. J. Vinegar, *Log Anal.*, 1996, **37**/**6**, 20–32.
37. M. Appel, *Petrophysics*, 2004, **45**, 296–307.
38. R. Akkurt, D. L. Seifert, A. Al-Harbi, T. M. Al-Beaiji, T. Kruspe, H. Thern and A. Kroken, *Petrophysics*, 2009, **50**, 140–152.
39. C. E. Morriss, J. MacInnis, R. Freedman, J. Smaardyk, C. Straley, W. E. Kenyon, H. J. Vinegar and P. N. Tutunjian, in *Transactions of the SPWLA 34th SPWLA Annual Logging Symposium*, 1993, p. Paper GGG.
40. C. L. Epstein and J. Schotland, *SIAM Rev.*, 2008, **50**, 504–520.
41. A. A. Istratov and O. F. Vyvenko, *Rev. Sci. Instrum.*, 1999, **70**(2), 1233.
42. R. L. Parker and Y.-Q. Song, *J. Magn. Reson.*, 2005, **174**, 314–324.
43. M. Prange and Y.-Q. Song, *J. Magn. Reson.*, 2009, **196**, 54–60.
44. A. N. Tikhonov and V. Y. Arsenin, *Solutions of Ill-Posed Problems*, Wiley, New York, 1977.
45. Y.-Q. Song, L. Venkataramanan and L. Burcaw, *J. Chem. Phys.*, 2005, **122**, 104104.
46. J. Butler, J. Reeds and S. Dawson, *SIAM J. Numer. Anal.*, 1981, **18**, 381–397.
47. G. C. Borgia, R. J. S. Brown and P. Fantazzini, *J. Magn. Reson.*, 1998, **132**, 65–77.
48. R. Freedman and C. E. Morriss, in *Proceedings of the 70th SPE Annual Technical Conference and Exhibition*, ed. S. 30560, Dallas, Texas, USA, 1995, p. SPE 30560.

49. L. Venkataramanan, Y.-Q. Song and M. D. Hürlimann, *IEEE Trans. Signal Process.*, 2002, **50**, 1017–1026.
50. C. Cao Minh, N. Heaton, R. Ramamoorthy, E. Decoster, J. White, E. Junk, R. Eyvazzedeh, O. Al-Yousef, R. Fiorini and D. McLendon, in *SPE Annual Technical Conference and Exhibition*, Denver, USA, 2003, p. SPE 84478.
51. B. Sun, K. J. Dunn and S. Clinch, in *Transactions of the SPWLA 51st SPWLA Annual Logging Symposium*, 2010.
52. N. Heaton, C. Cao Minh, R. Freedman and C. Flaum, in *Transactions of the SPWLA 41st Annual Logging Symposium*, Dallas, TX USA, 2000, p. Paper V.
53. N. J. Heaton, R. Freedman, C. Karmonik, R. Taherian, K. Walter and L. DePavia, in *Proceedings of the 77th SPE Annual Technical Conference and Exhibition*, San Antonio, Texas, 2002, p. SPE 77400.
54. M. G. Prammer, E. D. Drack, J. C. Bouton and J. S. Gardner, in *The Log Analyst*, 1996, vol. November – December, pp. 61–69.
55. L. Venkataramanan, F. K. Gruber, J. LaVigne, T. Habashy, J. G. Iglesias, V. Anand, M. Rampurawala, V. Jain, N. Heaton, R. Akkurt, E. Rylander and R. Lewis, in *Transactions of the SPWLA 55th SPWLA Annual Logging Symposium*, Abu Dhabi, 2014.
56. H. N. Bachman, S. Crary, R. Heidler, J. LaVigne and R. Akkurt, in *Proceedings of the SPE Annual Technical Conference and Exhibition*, Anaheim, California, 2007, p. SPE 110803.
57. P. Hook, D. Fairhurst, E. Rylander, R. Badry, H. N. Bachman, S. Crary, K. Chatawanich and T. Taylor, in *Proceedings of the 86th SPE Annual Technical Conference and Exhibition*, Denver, Colorado, USA, 2011, p. SPE 146883.
58. R. J. Brown, *J. Magn. Reson.*, 1989, **82**, 539–561.
59. M. Prange and Y.-Q. Song, *J. Magn. Reson.*, 2010, **204**, 118–123.
60. N. J. Heaton, C. C. Minh, J. Kovats and U. Guru, in *SPE Annual Technical Conference and Exhibition*, Society of Petroleum Engineers, 2004, p. SPE 90564.
61. E. T. Jaynes, *Phys. Rev.*, 1957, **106**, 620.
62. R. Salazar-Tio and B. Sun, in *Transactions of the SPWLA 50th SPWLA Annual Logging Symposium*, 2009.
63. N. Heaton, J. LaVigne, H. N. Bachman, E. Decoster, R. Badry, J. Hemingway, C. Patino and J. Cuadros, in *Transactions of the SPWLA 53rd SPWLA Annual Logging Symposium*, Cartagena, 2012.
64. W. Slijkerman, W. Looyestijn, P. Hofstra and J. Hofman, *SPE Reservoir Eval. Eng.*, 2000, **3**, 492–497.
65. R. Freedman, S. Lo, M. Flaum, G. Hirasaki, A. Matteson and A. Sezginer, *SPE J.*, 2001, **753215**, 452–464.
66. S. Fang, S. Chen, R. Tauk, P. Fornage and D. Georgi, in *Presented at the 79th SPE Annual Technical Conference and Exhibition*, Houston, TX USA, 2004, p. SPE 90569.
67. A. R. Mutina and M. D. Hürlimann, *J. Phys. Chem. A*, 2008, **112**, 3291–3301.

68. E. A. Clerke, D. F. Allen, S. C. Crary, A. Srivastava, R. Ramamoorthy, D. Saldungaray, P. Heliot, J. Goswami and G. Bordakov, in *Transactions of the SPWLA 55rd SPWLA Annual Logging Symposium*, Cartagena, 2014, p. Paper D.
69. A. Sezginer, C. Cao Minh, G. Van Dort, M. Herron, N. Heaton and R. Freedman, in *Transactions of the SPWLA 40th SPWLA Annual Logging Symposium*, 1999, p. Paper NNN.
70. L. Venkataramanan, F. K. Gruber, T. M. Habashy and D. E. Freed, *J. Magn. Reson.*, 2010, **206**, 20–31.
71. F. K. Gruber, L. Venkataramanan, T. M. Habashy, P. M. Singer and D. E. Freed, *J. Magn. Reson.*, 2013, **228**, 95–103.
72. R. Freedman, *Petrophysics*, 2006, **47**, 93–111.
73. V. Anand and R. Freedman, *Petrophysics*, 2012, **53**, 256–271.
74. V. Anand, R. Freedman, S. Crary, C. Cao Minh and R. L. Terry, *SPE Reservoir Eval. Eng.*, 2011, **14**, 750–762.
75. R. Kausik, C. Cao Minh, L. Zielinski, B. Vissapragada, R. Akkurt, Y.-Q. Song, C. Liu, S. Jones and E. Blair, in *Proceedings of the SPE Annual Technical Conference and Exhibition*, Society of Petroleum Engineers, 2011.
76. K. E. Washburn and J. E. Birdwell, *J. Magn. Reson.*, 2013, **233**, 17–28.
77. R. F. Sigal and E. Odusina, *Petrophysics*, 2011, **52**, 32–49.
78. E. Rylander, P. M. Singer, T. Jiang, R. Lewis, M. R. and S. Sinclair, in *Presented at the 88th SPE Annual Technical Conference and Exhibition*, Woodlands, Texas, 2013, p. SPE 164554.
79. R. Kausik, K. Fellah, E. Rylander, P. M. Singer, R. E. Lewis and S. M. Sinclair, in *Presented at the International symposium of the Society of Core Analysts*, Avignon, France, 2014, pp. SCA2014–A2028.
80. C. Cao Minh, S. Crary, L. Zielinski, C. B. Liu, S. Jones and S. Jacobsen, in *Presented at the SPE Canadian Unconventional Resources Conference*, Alberta, Calgary, 2012, p. SPE 161578.
81. R. Freedman, C. Cao Minh, G. Gubelin, J. J. Freeman, T. McGinness and D. Rawlence, in *Transactions of the SPWLA 39th SPWLA Annual Logging Symposium*, 1998 p. Paper II.
82. N. Bloembergen, E. Purcell and R. Pound, *Phys. Rev.*, 1948, **73**, 679–712.
83. R. L. Kleinberg, *Methods in the Physics of Porous Media, Nuclear Magnetic Resonance*, Academic Press, 1999.
84. J. Korringa, D. O. Seevers and H. Torrey, *Phys. Rev.*, 1962, **127**, 1143–1150.
85. R. L. Kleinberg, W. E. Kenyon and P. P. Mitra, *J. Magn. Reson., Ser. A*, 1994, **108**, 206–214.
86. S. Godefroy, J.-P. Korb, M. Fleury and R. G. Bryant, *Phys. Rev. E*, 2001, **64**, 021605.
87. K. R. Brownstein and C. E. Tarr, *Phys. Rev.*, 1979, **19**, 2446–2453.
88. D. O. Seevers, in *Transactions of the SPWLA 7th Annual Logging Symposium*, 1966, vol. 138, p. 682.
89. J. D. Loren and J. D. Robinson, in *Transactions of the SPE 44th Annual Symposium*, 1970, vol. SPE 2529, pp. 269–278.

90. L. Venkataramanan, M. D. Hürlimann, J. A. Tarvin, K. Fellah, D. Acero-Allard and N. Seleznev, *Petrophysics*, 2014, **55**, 572–586.
91. M. D. Hürlimann, K. G. Helmer, L. Latour and C. H. Sotak, *J. Magn. Reson., Ser. A*, 1994, **111**, 169–178.
92. A. Souza, G. Carneiro, L. Zielinski, R. Polinski, L. Schwartz, M. D. Hürlimann, A. Boyd, E. H. Rios, B. Coutinho, W. A. Trevizan, V. Machado and R. Bagueira, in *Transactions of the SPWLA 54th Annual Logging Symposium*, 2013, p. Paper U.
93. L. L. Latour, R. L. Kleinberg and A. Sezginer, *J. Colloid Interface Sci.*, 1992, **150**, 535–548.
94. R. Kleinberg, S. Farooqui and M. Horsfield, *J. Colloid Interface Sci.*, 1993, **158**, 195–198.
95. C. Straley, C. E. Morriss, W. E. Kenyon and J. J. Howard, *Log Anal.*, 1995, **36**, 40–56.
96. A. Timur, *J. Pet. Technol.*, 1969, 775–786.
97. J. D. Loren, *J. Pet. Technol.*, 1972, **24**, 923–928.
98. C. R. Coates, D. Marschall, D. Mardon and J. Galford, *Log Anal.*, 1998, **39/1**, 51–63.
99. T. S. Ramakrishnan, L. M. Schwartz, E. J. Fordham, W. E. Kenyon and D. J. Wilkinson, *Log Anal.*, 1999, **40**, 260–270.
100. P. J. McDonald, J.-P. Korb, J. Mitchell and L. Monteilhet, *Phys. Rev. E*, 2005, **72**, 011409.
101. K. E. Washburn and P. T. Callaghan, *Phys. Rev. Lett.*, 2006, **97**, 175502.
102. R. J. S. Brown and I. Fatt, in *Fall Meeting of the Petroleum Branch of AIME*, Society of Petroleum Engineers, 1956, paper SPE-743-G.
103. W. J. Looyestijn, *Petrophysics*, 2008, **49**, 130–145.
104. C. E. Morriss, R. Freedman, C. Straley, M. Johnston, H. J. Vinegar and P. N. Tutunjian, *Log Anal.*, 1997, **38/2**, 44–59.
105. Y. Zhang, G. J. Hirasaki, W. V. House and R. Kobayashi, in *Transactions of the SPWLA 43th Annual Logging Symposium*, 2002, p. Paper HHH.
106. C. Straley, in *Transactions of the SPWLA 47th Annual Logging Symposium*, 2006, p. Paper AA.
107. G. A. LaTorraca, S. W. Stondard, P. R. Webber, R. M. Carlson and K. J. Dunn, in *Transactions of the SPWLA 40th Annual Logging Symposium*, 1999, p. Paper PPP.
108. J. Bryan, A. Kantzas and C. Bellehumeur, *SPE Reservoir Eval. Eng.*, 2005, **8**, 44–52.
109. L. M. Burcaw, R. L. Kleinberg, J. Bryan, A. Kantzas, Y. Cheng, A. Kharrat and R. Badry, in *Transactions of the SPWLA 49th SPWLA Annual Logging Symposium*, 2008.
110. D. E. Freed, *J. Chem. Phys.*, 2007, **126**, 174502.
111. D. E. Freed and M. D. Hürlimann, *C. R. Phys.*, 2010, **11**, 181–191.
112. L. Zielinski, I. Saha, D. E. Freed, M. D. Hürlimann and Y. Liu, *Langmuir*, 2010, **26**, 5014–5021.
113. L. Zielinski and M. D. Hürlimann, *Energy Fuels*, 2011, **25**, 5090–5099.
114. J. J. Chen, M. Hürlimann, J. Paulsen, D. Freed, S. Mandal and Y. Q. Song, *ChemPhysChem*, 2014, **15**, 2676–2681.

115. Y. Meridji, G. Hursan, M. Eid and R. Balliet, in *SPE Saudi Arabia Section Technical Symposium and Exhibition*, Society of Petroleum Engineers, 2013.
116. M. D. Hürlimann, *J. Magn. Reson.*, 1998, **131**, 232–240.
117. M. Hürlimann, A. Matteson, J. Massey, D. Allen, E. Fordham, F. Antonsen and H. Rueslatten, *Petrophysics*, 2004, **45**, 414–421.
118. D. E. Freed, L. Burcaw and Y.-Q. Song, *Phys. Rev. Lett.*, 2005, **94**, 067602.
119. M. H. Donaldson, D. E. Freed, Y. Q. Song and L. Zielinski, in *Transactions of the SPWLA 54th SPWLA Annual Logging Symposium*, 2013.
120. P. N. Sen, *Concepts Magn. Reson., Part A*, 2004, **23**, 1–21.
121. P. P. Mitra, P. N. Sen, L. M. Schwartz and P. LeDoussal, *Phys. Rev. Lett.*, 1992, **68**, 3555–3558.
122. M. D. Hürlimann, M. Flaum, L. Venkataramanan, C. Flaum, R. Freedman and G. J. Hirasaki, *Magn. Reson. Imaging*, 2003, **21**, 305–310.
123. R. Freedman, V. Anand, D. Catina, B. Grant, P. Tabrizi, R. Torres, K. Ganesan, C. Borman, C. Krueckl and D. Ryan, *Petrophysics*, 2013, **54**, 439–456.
124. A. Boyd, H. E. and W. G. Gubelin, in *Presented at the SPE Annual Technical Conference and Exhibition*, Austin, Texas, 1995, p. SPE 30561.
125. N. J. Heaton, V. Jain, B. Boling, D. Oliver, J.-M. Degrange, P. Ferraris, D. Hupp, H. Prabawa, M. T. Ribeiro, E. Vervest and I. Stockden, in *Presented at the 88th SPE Annual Technical Conference and Exhibition*, San Antonio, Texas, 2012, p. SPE 160022.
126. V. Jain, C. Cao Minh, N. Heaton, P. Ferraris, L. Ortenzi and M. T. Ribeiro, in *In Transactions SPWLA 54th Annual Logging Symposium, Paper TT*, 2013.
127. U. Guru, N. Heaton, H. N. Bachman, J. LaVigne, W. A. Hamed and R. Nelson, *Petrophysics*, 2008, **49**, 342–350.
128. E. Decoster and R. Carmona, in *Transactions of the SPWLA 49th Annual Logging Symposium*, 2008, p. Paper VVV.
129. *Single-Sided NMR*, ed. F. Casanova, J. Perlo and B. Blümich, Springer, 2011.

CHAPTER 3

Bench-top NMR—Food: Solid Fat Content Determination and Emulsion Droplet Sizing

M. A. VODA*[a] AND J. VAN DUYNHOVEN[b]

[a]Unilever Research and Development, Vlaardingen, The Netherlands; [b]Laboratory of Biophysics and Wageningen NMR Centre, Wageningen University, Wageningen, The Netherlands
*E-mail: adrian.voda@unilever.com

3.1 Introduction

Due to its versatility in assessment of compositions and structural features, nuclear magnetic resonance (NMR) has become a well-established measurement tool in food science and technology.[1] A unique advantage of NMR is its ability to quantify in a precise and accurate manner, both at a compositional and structural level. The non-invasive nature makes NMR ideally suited to investigating foods. For many of these measurements there is no compelling need to use expensive and complex high-field NMR equipment, and the first implementation of low-field instrumentation in food studies dates to 4 decades ago. In the early 1970s the food industry had a strong need to replace cumbersome dilatometric methods for determination of solid content in fats. A straightforward relaxometric measurement on a permanent magnet with a simple NMR spectrometer provided a cheap and robust alternative measurement of solid fat content (SFC) in fat blends.[2] In the following

New Developments in NMR No. 5
Mobile NMR and MRI: Developments and Applications
Edited by Michael L. Johns, Einar O. Fridjonsson, Sarah J. Vogt, and Agnes Haber

Published by the Royal Society of Chemistry, www.rsc.org

decades, relaxometry methods were also developed for fat content in chocolate and oil content in seeds.[3] In the late 1970s the industry recognized that NMR diffusometric assessment of droplet size distributions in food emulsions at high-field[4] were suitable for implementation in benchtop mode.[5] It was in the early 1980s when the first commercial benchtop NMR instrument for water droplet size distributions in oil-continuous food emulsions, such as margarines, appeared on the market.[5–7] Soon afterwards oil droplet sizing was also implemented on commercial benchtop NMR instrumentation.[8,9]

Benchtop NMR instruments have now reached a high level of sophistication in automation and standardisation. These instruments have also become very user-friendly, and can hence be operated by non-NMR experts with minimal training.[9] The wide application range of benchtop NMR instruments in foods is summarized in Figure 3.1. Although initially developed for routine quality and process control, the potential of benchtop NMR instruments for research and development purposes was rapidly recognized. In food research applications the unbiased nature of benchtop NMR allows for detection of a wide range of phase-compositional and microstructural features by diffusometric and relaxometric experiments. This has been exploited for rapidly generating and testing hypotheses regarding relationships between food structures and consumer-relevant properties, such as texture and shelf-life.[10] Here rapid and non-invasive assessment of structural and phase-compositional features in high-throughput mode is a particular advantage. In food development, progress often depends on rapid testing of pre-defined phase-compositional and structural features in product prototypes. Benchtop NMR equipment can deliver such information in a cost-effective and reliable manner. These are also advantages during scaling up to industrial manufacturing processes. Once in the industrial foods supply chain, the need for rapid and accurate quality and process control instruments can also be met by benchtop NMR instrumentation.[11,12] In the last few years we have also witnessed technological developments, such as unilateral

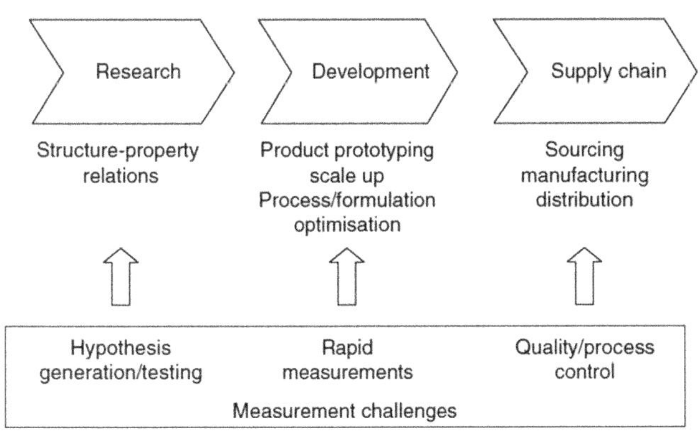

Figure 3.1 Application areas of benchtop NMR in food science and technology. Reproduced from ref. 9 and 16 with permission from Elsevier.

and portable NMR for non-invasive assessment of material properties, which bring on- and in-line process and quality control within reach.[13] One of the latest trends in the development of low-field benchtop NMR capability (with significant relevance for food) is towards achieving spectral resolution in the frequency domain by making use of magnets with homogeneous fields.[14,15]

3.2 Design of Commercial Benchtop NMR Equipment

The core of commonly used commercial benchtop NMR instrumentation in the foods industry is a permanent magnet consisting of a yoke holding two poles at a distance of a few cm. Typically, B_0 field strengths of 0.12–1.4 T can be obtained. With proton Larmor frequencies in the range of 5–60 MHz, sensitivity is sufficient for observing abundant species in foods, such as water and oil. The permanent magnets currently in use exhibit a relatively poor homogeneity, typically resulting in a T_2* of a few milliseconds for most liquids. Shimmed and novel designs of low-field permanent magnets with improved homogeneity are now being introduced, but are not common yet. This is also true for field lock units, for minimisation of field drift one now relies on maintaining the magnet at a constant temperature. The typical air gaps of magnets in benchtop NMR instrumentation is in the 25–100 mm range, and allows for sample (tube) diameters between 10 and 100 mm.

In the earliest commercial benchtop NMR instruments, electronic performance was strongly compromised by cost. Contemporary instruments have more advanced designs, allowing sophisticated multi-pulse experiments, NMR signal can be recorded with adequate digital resolution in a linear manner in phase-sensitive mode. Requirements for probehead design can be rather specific for different benchtop NMR applications. For many quantitative content measurements (oil, moisture) one needs to record the absolute value of the magnetization. This requires good rf homogeneity over a large sample volume, and for these so-called absolute measurements one can often accept a compromise on longer dead times and pulse widths. For relative measurements, only magnetization decays are considered and no absolute quantification is pursued. For such applications, short pulse widths and dead times are required. While favouring a low Q-factor, a compromise with respect to rf homogeneity is accepted. Many traditional benchtop NMR applications, such as SFC measurement, are still performed by probeheads that are simply maintained at magnet temperature. For such applications, rapid measurements are also required to prevent any significant changes in sample temperature.

For diffusometric applications, the current standard in commercial benchtop NMR instrumentation is the use of gradient amplifiers that are driven by capacitor discharges. Although relatively inexpensive, performance in terms of reproducibility is compromised, in particular in combination with non-shielded gradient coils. Consequently, time-consuming gradient balancing procedures need to be used. Benchtop NMR equipment with linear gradient amplifiers and shielded gradient coils has been introduced on the market, but increased performance was accompanied with an increased cost.

3.3 Industrial Routine Benchtop Assessment of SFC

3.3.1 Acquisition and Processing

The measurement and calculation of SFC relies on the difference in lifetimes of the NMR signals of the solid fat and liquid oil. In the case of the direct method, two sampling points of the NMR signal are recorded: at short (11 μs) and after a comparatively longer (70 μs) time after a single pulse excitation. As intuitively shown in Figure 3.2, the signal measured at short time reflects the magnetization of both solid and liquid phases, while the signal at long time provides the magnetization of the liquid phase only. Due to the dead time of the receiver, the magnetization at the short time after excitation has already suffered a decrease on the account of the fast relaxation of the solid phase. In order to get the absolute value of the total magnetization a so-called *F* factor has been introduced and the SFC can be calculated as:

$$\text{SFC(Direct)} = \frac{(E_{11} - E_{70})F}{E_{70} + (E_{11} - E_{70})F + D} \quad (3.1)$$

where *D* is the digital offset and E_{11} (E(11)) and E_{70} (E(70)) are the signal intensities as indicated in Figure 3.2. A third calibration factor, *K*, needs to be introduced to account for the effect of signal decay due to magnetic field inhomogeneity. A set of three calibration standard samples has been

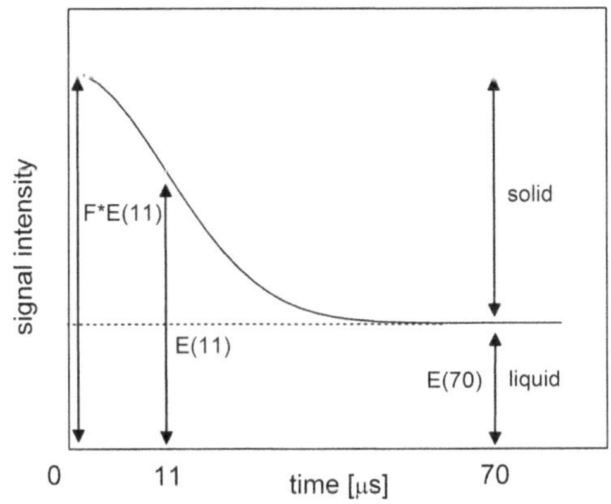

Figure 3.2 Schematic depiction of direct SFC method. A free induction decay is shown with the critical sampling times (intensities E(11), E(70) sampled at 11 μs and 70 μs, respectively). Reproduced from ref. 17 with permission from AOCS.

introduced to determine F, K and D. The calibration procedure is described in Section 3.3.2.

In the case of the indirect method the magnetization of the liquid fraction is sampled at two temperatures, a low one (E_T) where both solid and liquid phases exist, and one where the solids are completely melted (E_{60} at 60 °C, for example). After correcting for the difference in the NMR signal at different temperatures, the difference between the two magnetization values represents the SFC:

$$\mathrm{SFC(Indirect,T)} = 100 - \left(\frac{R_{60}}{E_{60}} \times \frac{E_T}{R_T}\right) \times 100 \tag{3.2}$$

where $R_{60,\mathrm{T}}$ represents the signals of a reference oil that remains liquid at both temperatures and is used for the temperature correction of the NMR signal.

3.3.2 Tempering Protocols

In product formulation and manufacturing, fat blends are designed and selected on the basis of their melting behavior. For this purpose, measuring the SFC value at different temperatures provides a rapid and routine solution. The SFC value however depends strongly on thermal history of the sample, hence tempering protocols are required to bring the blend in a defined phase-composition. In all tempering protocols the thermal history is removed by first melting the sample, typically at 80 °C, followed by short storage at 60 °C and a rapid quench at 0 °C. At this temperature, the fat typically first crystallizes in a metastable α polymorph. The rate and extent of the subsequent conversion to the more stable β and β′ polymorphs depends strongly on the blend composition.[18] The so-called non-stabilizing blends convert so rapidly that they can be measured after 1 hour of storage at 0 °C. For the so-called stabilizing or tempering fat blends, prolonged (>1 day) storage at 26 °C is required. Once a tempering scheme has been selected one will have to choose whether measurements are to be performed according to the parallel or serial method. In the parallel method, specimens from the same sample are prepared for each temperature where SFC is required. In the case of the serial method, only one sample is used and spans the whole temperature range in an incremental manner while SFC measurements take place after each tempering interval. A disadvantage of serial measurements is lower speed and interdependence of SFC values at different temperatures, hence parallel schemes are more commonly used.

3.3.3 Method Standardization

The NMR solid fat content measurement has not only become a critical method for research, development and manufacturing but also for verification of raw materials and product specifications during finalization of commercial contracts. The significant financial interests involved in material and product specifications by SFC have been recognized by bodies such as the American Oil Chemist Society (AOCS Cd 16b-93 and Cd 16-81), ISO (ISO 9292) and IUPAC (IUPAC 2.150), who all drafted official methods for the assessment of SFC by NMR. These methods showed some small but distinct differences, in particular with respect to tempering of samples before the actual NMR measurement.[18] Recently, however, we have witnessed harmonization of the different methods. The harmonization also pertained to specification of sample blocks used for sample storage during tempering.[18] A critical aspect of the direct SFC method is the use of reference standards to calculate the correction factor (F-factor) for machine-dependent dead times.[2] For daily machine calibration the direct SFC method currently relies on a set of plastic-in-oil (PIO) standards. These PIO standards are prepared in a proprietary manner by instrument manufacturers.[17] The PIO standards have been referenced in the late 1970s to a selection of commercial fats and blends. The current PIO standards are all references to master PIO sets that are all under the custody of instrument manufacturers. The AOCS has urged for a transparent reference system consisting of secondary PIO standards for daily use that are derived from a primary standard set that is fully traceable and independent of instrument manufacturers. Such a traceable primary standard system has meanwhile been designed by the EC Institute of Reference Materials and Measurement (IRMM) in a joint effort between instrument manufacturers and the food industry. The primary standard system consists of precisely weighed mixtures of liquid triolein (OOO) and solid tristearate (SSS). This primary standard system has been thoroughly tested in an inter-laboratory comparison involving different manufacturers. It was shown that the SSS/OOO mixtures can indeed be used as traceable primary standard system.[19] F-factors can be obtained that are independent of the SFC value of the SSS/OOO mixture and instrument used.

3.3.4 Application Scope and Performance of Routine SFC Methods

The two benchtop NMR methods that are currently in use for SFC determination mainly differ in throughput and method precision and accuracy. The direct method is most often employed, as it is easy to use and has a high throughput. The accuracy (or 'trueness') of the method is compromised by the use of a single F-factor for all fat blends. This can introduce small but significant errors, since the F-factor corrects for an average decay time and, in reality, these can differ for different fat crystal polymorphs. This is shown in Figure 3.3A, which indicates differences of up to 2% between measured direct SFC and the true value for samples that differ widely in fat composition. The individual differences can be explained by variation in the NMR transversal decay rate for different crystal polymorphs that can occur in different blends. Note that the average difference at room temperature is negligible. This is

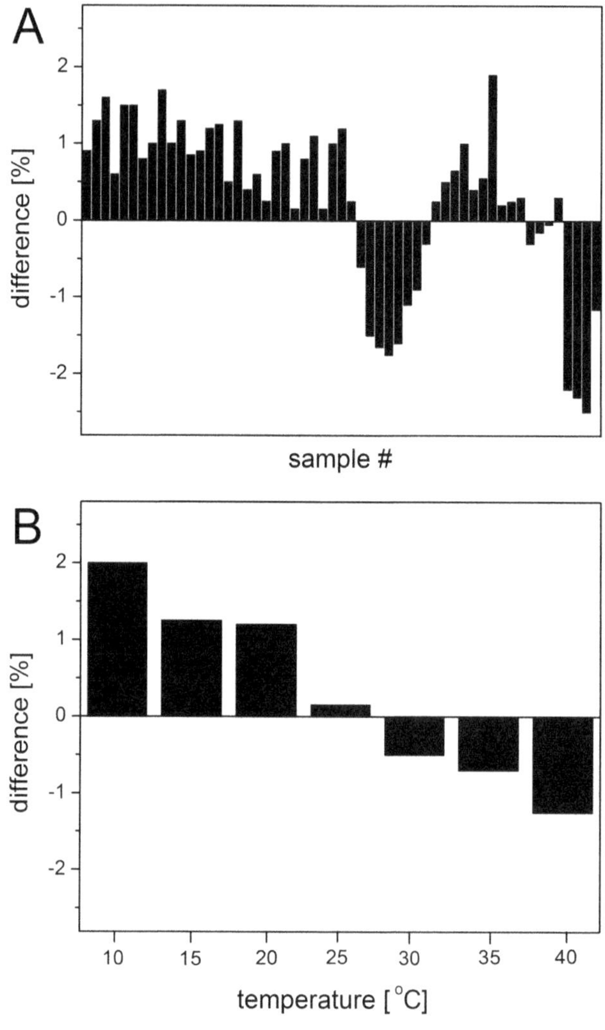

Figure 3.3 Method accuracy for direct SFC method. (A) Absolute error in SFC for 56 commercial fat blends. (B) Effect of temperature on averaged (over 56 samples) absolute error of direct SFC method. Reproduced from ref. 17 with permission from AOCS.

expected, since the PIO standards have been calibrated in such a manner to yield correct results for an 'average fat blend'. At different temperatures, however, small but significant errors occur (Figure 3.3B). These can also be explained by deviations in relaxation behavior of fat crystal polymorphs from average behavior at room temperature. The compromised accuracy of the direct SFC method is, however, compensated by a very good precision in terms of repeatability and reproducibility (Table 3.1). The repeatability *r* represents method reliability when measurements are carried out within one lab and on the same day, and with the same analyst and instrument.

Table 3.1 Performance of direct and indirect methods for assessment of SFC.[17]

	Direct	Indirect
Repeatability (r)	1.1	2.0
Reproducibility (R)	3.3	4.0

From an extensive inter-laboratory testing exercise a value of $r = 1.1\%$ has been reported,[17] and this has been deemed more than adequate for industrial and commercial purposes. The reproducibility R includes variation due to different laboratories and differs from r by a factor of 3, which is considered acceptable.

The indirect method is conceptually simple as it only requires measurements of the liquid phase at measurement temperature and at 60 °C where one can assume that all the solid phase has become molten. The method can be considered accurate, since no effects of crystal polymorph or machine calibration can occur. However, precision can be compromised due to sample filling and dilation of the sample in the measurement tube. This is reflected in the higher r and R values in Table 3.1. For reasons of low throughput and precision, the indirect method is less commonly used than the direct method.

3.4 Industrial Routine Droplet Size Measurement by Pulsed Field Gradient NMR

3.4.1 Acquisition

The spatial position of nuclear spins can be labelled through their Larmor frequency by means of a magnetic field gradient. This provides the underlying principle for measuring molecular self-diffusion by NMR. The NMR diffusion experiment can be performed by an application of a pair of pulsed field gradients into either a spin-echo (SE) or a stimulated echo (StE) rf pulse sequence. A first gradient pulse of magnitude g and duration δ is applied after the first 90° rf pulse and introduces a dephasing of the nuclear spins precession. A second gradient pulse is applied after the 180° (in the case of SE), or after the third 90° rf pulse (for StE), which partially refocuses the spins' phases. The phase difference is due to the diffusing spins that could not recover their initial phase and this is reflected in the intensity of the NMR echo. The signal loss in the case of free diffusion is proportional to the average root mean square displacements occurring between the two gradient pulses.[20] Differences between SE and StE rely on the fact that nuclear spins during the diffusion time are subject to either transverse relaxation (T_2) or longitudinal relaxation (T_1). This implicitly limits the duration of the diffusion time (Δ) and the applicability of SE, thus in practice StE is preferred (Figure 3.4).

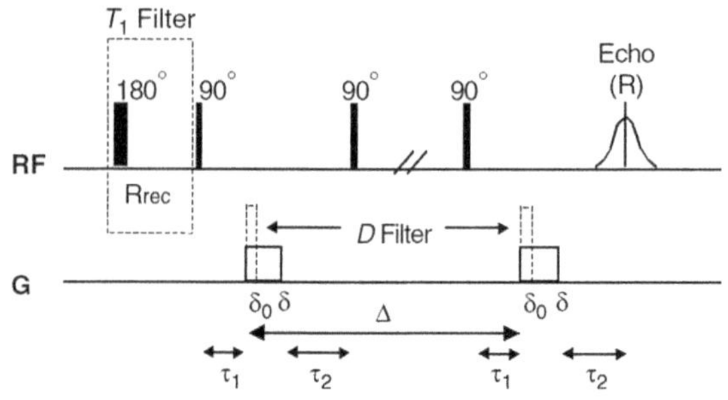

Figure 3.4 Schematic representation of StE sequence used to measure droplet size distribution. Time-domain filters to separate the water and oil signals are also indicated. Here δ is the gradient pulse duration, Δ is the diffusion time and G is the gradient strength. Reproduced from ref. 20 and 21 with permission from John Wiley & Sons.

3.4.1.1 *Application of Pulsed Field Gradients*

The application scope of pulsed field gradient (PFG) experiments is strongly determined by the performance of gradient amplifiers and gradient coils. Linear gradient amplifiers and shielded gradient coils allow for straightforward implementation of accurate and precise diffusometric experiments. However, it has been difficult to achieve a good compromise between performance and cost. Hence, the current standard in benchtop NMR is the use of inexpensive capacitor-driven gradient amplifiers and non-shielded gradient coils. Due to the non-linearity of these amplifiers, the diffusion encoding is performed by variation of the duration of the applied pulsed field gradient. Poor reproducibility of the field gradient pulses needed to be dealt with by a cumbersome and time-consuming balancing procedure for every PFG duration.

Echoes recorded by a PFG experiment on benchtop NMR instruments are broadened due to magnetic field inhomogeneities that affect refocusing of spins. In order to narrow the echo and to provide a more reproducible fit, a steady gradient can be applied during the PFG experiment.[8,22,23] To ensure consistency among different spectrometers, the steady gradient is adjusted to a value where the half-life of a water FID is 0.5 ms.[21]

The balancing of gradients for each PFG duration is a time-consuming task that is repeated for each sample analysis. An alternative automated procedure for gradient calibration was developed to avoid this repetitive balancing.[22] It uses variable gradient strength to encode diffusion. In this setup, a gradient calibration procedure consisting of three automated steps is realized, which is schematically shown in Figure 3.5. The first step is the setting of the steady gradient. The second step consists of gradients calibration by performing a whole experiment on samples with known diffusion coefficients. Gradient ranges also differ for W/O and O/W applications in order to obtain optimum diffusion decays. Depending on the performance

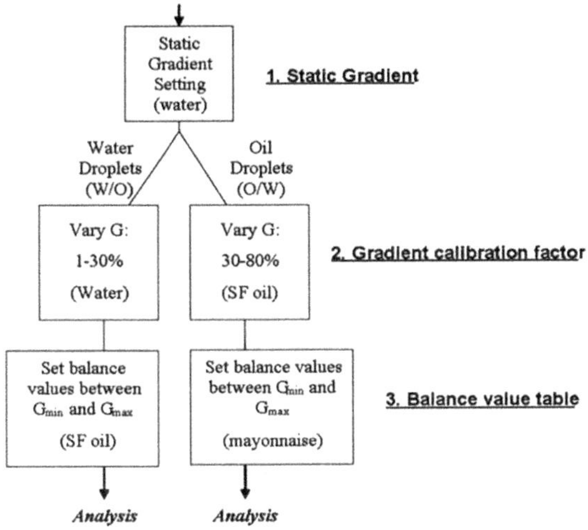

Figure 3.5 Schematic outline of the gradient calibration routine for droplet size distribution (DSD) in food emulsions by PFG-NMR. Reproduced from ref. 22 with permission from Elsevier.

of the gradient system to deliver consistent values at the upper end, one may set a working limit of 80% of the available power in order to avoid artifacts. In the third calibration step, the balance values are determined for the gradient ranges used for W/O and O/W experiments. A table with balance values is automatically created and is used for measurement of various samples without the need for repeating this task for each sample. It is common that balance values depend on the gradient strength and on the magnet used.[22] The dependence on the gradient strength can be explained by eddy currents and heating of the coils.

3.4.1.2 Selection of Water and Oil Phases: Time-Domain Filters

For the determination of droplet size distributions in W/O and O/W emulsions two different acquisition approaches are used. Due to the lack of spectral resolution to resolve the water and the oil frequencies, time-domain filters have been implemented in order to suppress the signal of continuous phase of the emulsion. After the filter, different acquisition parameters are used for the two types of emulsions. The suppression of the oil signal in W/O emulsion systems is realized by a T_1-relaxation filter, since the life time of water signal is much longer. The filter consists of a π pulse and a delay corresponding to the zero-value of the z-magnetization of oil. In the case of O/W emulsions, water in practice typically exhibits a dispersion of relaxation times and therefore a single zero-magnetization time cannot be defined. Instead, a diffusion filter is best suited to separate water and oil signals, taking into account that the molecular diffusivities differ by 2 orders of magnitude. The gradient duration is set to suppress the water signal but ensuring there is still enough oil signal left for the required PFG measurement. Although for standard applications in food industry the time-domain filters generally perform well, in specific cases practical complications are encountered. The relaxation filter assumes a unique and non-dispersed T_1 value for oil, which is not always the case. In the case of oil filtering by diffusion, there is also a residual water signal. Furthermore, both relaxation and diffusion filters reduce the echo intensity of interest, thus compromising S/N. These issues can be circumvented by switching to the frequency domain, but the required spectral resolution has however so far only been achieved on a custom made magnet equipped with rf micro-coil and PFG capability.[24]

3.4.1.3 Data Processing

Diffusion of molecules in a confined environment can be treated in terms of different mathematical approaches.[25–29] In its current routine implementation, the droplet sizing method makes use of the Gaussian Phase Distribution approach (GPD),[30] where the assumption is that nuclear spins in a magnetic field gradient accumulate a Gaussian distributed phase. Murday and Cotts[26] derived the expression for the PFG attenuation of the NMR echo decay of liquid molecules confined in a spherical droplet:

$$\frac{S}{S_0} = R(\Delta,\delta,g,a) = \exp\left[-2\gamma^2 g^2 \sum_{m=1}^{\infty} \frac{1}{\alpha_m^2\left(\alpha_m^2 a^2 - 2\right)} \times \left(\frac{2\delta}{\alpha_m^2 D} - \frac{2 + e^{-\alpha_m^2 D(\Delta-\delta)} - 2e^{-\alpha_m^2 D\Delta} - 2e^{-\alpha_m^2 D\delta} + e^{-\alpha_m^2 D(\Delta+\delta)}}{\left(\alpha_m^2 D\right)^2}\right)\right] \tag{3.3}$$

where a is radius of the droplets and α_m is the mth positive root of the Bessel function equation $J_{3/2}(\alpha a)/\alpha a = J_{5/2}(\alpha a)$. S and S_0 represent NMR signal intensities with and without the gradient pulse of amplitude g and duration δ, D is the bulk diffusion coefficient of the dispersed phase, and Δ is the diffusion time in the StE sequence.

Until recently, the Murday–Cotts equation based on the GPD approximation was preferred as it is the most accurate to describe the restricted diffusion in a spherical geometry.[31] However, the GPD approximation does not account for diffraction effects in small droplets.[32,33] An improved method that relies on the block gradient pulse approximation (BGP) was recently developed for describing restricted diffusion within spherical geometries.[29,34] Its performance and feasibility for routine implementation were recently assessed[24] and are discussed further in Section 3.5.5.

Food emulsions are typically polydispersed and PFG decays cannot simply be described by expressions for a uniform droplet size. A distribution of droplet sizes can be accounted for[4] by:

$$R_{\text{obs}}(\Delta,g,\delta,a) = \frac{\int_0^\infty a^3 P(a) R(\Delta,\delta,g,a)\,\mathrm{d}a}{\int_0^\infty a^3 P(a)\,\mathrm{d}a}, \tag{3.4}$$

where $R(\Delta,\delta,g,a)$ is given by eqn (3.3). For many food applications, one can assume a log-normal distribution of droplet sizes:

$$P(a) = \frac{1}{\sqrt{2a\sigma(2\pi)}} \frac{1}{\exp\left[\frac{(\ln 2a - \ln D_{3,3})^2}{2\sigma^2}\right]}, \tag{3.5}$$

in which case the distribution is characterized by the volume weighted geometric mean diameter $D_{3,3}$ and the standard deviation (width) σ. With only two parameters required to match the equation, a significant simplification of the numerical inversion required to extract $P(a)$ from the experimental data is realized.

3.4.1.4 Application Scope and Performance of Routine Droplet Sizing Analysis

Food emulsions typically consist of droplets with sizes ranging from 0.1 up to 100 μm.[35,36] Different classes of techniques can be deployed to study an emulsion's microstructure and to retrieve droplet size distributions: light scattering, microscopy combined with image analysis, electric sensing, and ultrasound spectrometry. These methods rely on different physical principles to sense droplet size. Furthermore, the sensitivity of each method is dependent on experimental conditions and sample preparation, and so far, no unambiguous measurement approach exists. Light scattering is the most widely used method in sizing emulsion droplets.[35] It can probe a wide range of sizes (see Table 3.2) but an obvious requirement here is the optical transparency of the sample. Diluting the system may not always be feasible without perturbing the size of the droplets. The same issues are encountered by microscopy techniques. Microscopy combined with image analysis is a more direct approach since it relies on visualisation of droplets for quantification of their size distributions.[37,38] However, a large number of images need to be acquired for multiple sample specimens to assure representative sampling of the droplet size distribution. Concentrated emulsions can be analysed by ultrasound spectrometry. The main impediments here are the errors introduced by the presence of gas bubbles in the emulsion and the required analysis of experimental data, which requires a large number of thermo-physical properties of the emulsion components. Electrical sensing is limited by the required sample perturbing preparations and emulsion disruption during the experiment.[35,39]

In contrast to the above methods, NMR is non-perturbing and non-invasive. Any system can be investigated, irrespective of refraction index and transparency. Furthermore, samples with a wide range of viscosity can be measured. NMR is fast (minutes to tens of minutes), highly reproducible, and can be deployed for routine use without profound NMR expertise. The main complicating factors of the NMR droplet sizing method can be circumvented in a relatively straightforward manner. Droplet self-diffusion and inter-droplet

Table 3.2 Method overview for droplet sizes.[35,36] Reproduced from ref. 36 with permission from Elsevier.

Technique	Physical principle	Theoretical range (μm)	Experimental complications
PFG NMR	Restricted diffusion	0.2–100	Droplet self-diffusion Inter-droplet diffusion Internal structure in droplets Multiple emulsions
EPS	Conductivity	0.4–1200	Droplet clustering Non-droplet particles Concentrated emulsions Multiple emulsions
LS	Diffraction of light	0.1–1000	Droplet clustering Non-droplet particles Concentrated emulsions Multiple emulsions
LM/IA	Optical contrast	0.5–macro-scale	Optical contrast Small sampling volume Multiple emulsions
US	Ultrasonic attenuation/velocity	0.01–1000	Droplet clustering Non-droplet particles Concentrated emulsions Multiple emulsions

diffusion can respectively be suppressed by manipulation of the viscosity of the continuous phase and setting an adequate measurement temperature. Internal structures in droplet sizes can also compromise assessment of droplet size distribution (DSD), in particular crystal networks due to (partial) crystallization of fat in oil droplets. Setting the measurement temperature value above the melting temperature of the crystal network can make the system consistent with the liquid droplet assumption of the NMR droplet sizing method. As was argued previously, the systematic errors introduced by measuring on systems with partially crystallized oil droplets are still relatively small.[21] Countermeasures against experimental complications are arguably less straightforward in the case of other droplet sizing methods.[35] Limitations of the PFG-NMR method exist in the length scale that can be probed, in particular for oil droplets, where maximum achievable gradient strength is one of the limiting factors. The practical limit of the routine method for oil droplet diameter depends on the amount of dispersed phase (sensitivity) and lifetime of the signal, as longer diffusion times are needed. The operational range calculated for a commercial benchtop NMR instrument operating at 20 MHz is presented in Figure 3.6. Based on the longitudinal relaxation time of oil (200 ms), the plot reveals the calculated size that can be probed as a function of dispersed oil phase concentration and diffusion time, Δ. The signal in Figure 3.6 is normalized to the pure oil signal (100%). It was estimated that reliable DSDs can be obtained from intensities above a threshold of 3% (denoted by the interrupted line). For

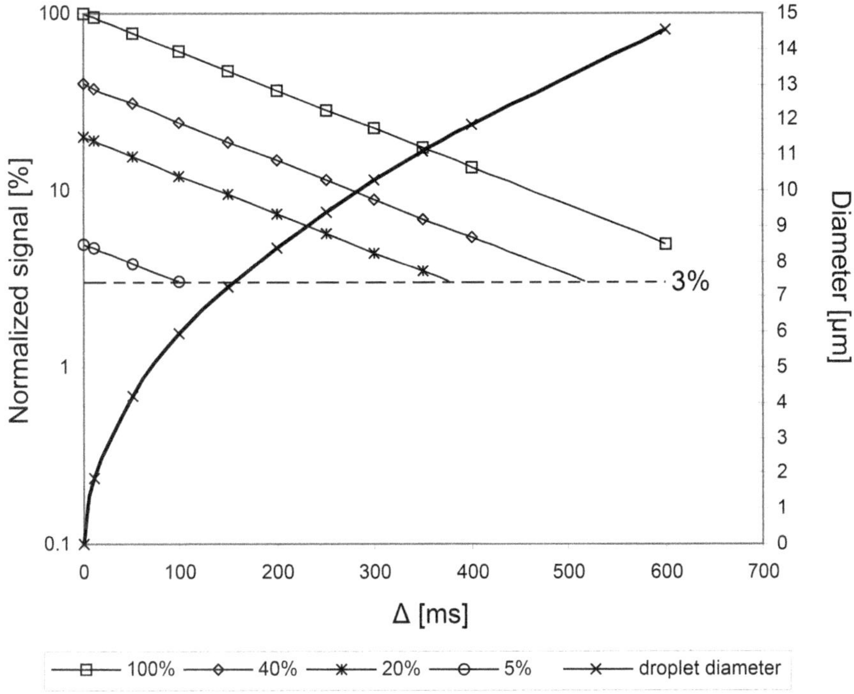

Figure 3.6 Simulated operational window for assessment of oil (O/W) DSDs by PFG NMR at 20 MHz for a range of dispersed phase concentrations. The calculations were based on the typical sensitivity of a Bruker minispec MQ20. Courtesy of Gert-Jan Goudappel, Unilever Discover Vlaardingen.

example, at a given Δ value of 200 ms one can measure a maximum of about 8 µm droplet size but not in emulsions containing 5% dispersed oil phase. It can be observed that the maximum attainable size in a 40% dispersion is around 10 µm. Fortunately, the size of droplets in O/W food emulsions currently on the market do not exceed this value. Larger droplet sizes can be probed by research instruments operating at higher static magnetic field strengths equipped with stronger pulsed field gradients.

The performance of the routine PFG-NMR method to determine droplet size distributions in food emulsions has been extensively reviewed.[36] Accuracy ("trueness") was tested by comparing the results of PFG-NMR with those obtained by several of the alternative methods, and precision by testing the repeatability of DSDs under practical circumstances. Generally, good agreement is obtained for both water and oil mean droplet sizes, and an illustration for a range of O/W emulsions is shown in Figure 3.7. With respect to precision of the determination of the mean droplet size, NMR outperforms other methods although the distribution width (σ) is typically assessed in a less reproducible manner.

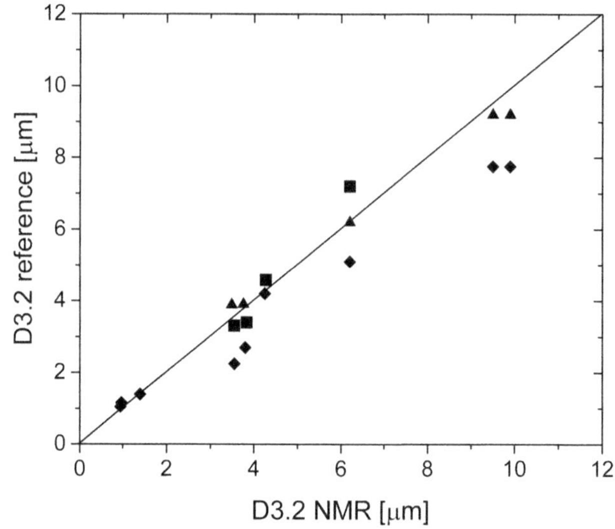

Figure 3.7 Comparison of mean droplet size diameter obtained by NMR and (■) LM/IA, (▲) EPS and (♦) LS for a range of O/W food emulsions. Reproduced from ref. 21 with permission from Elsevier.

3.5 Non-Routine Methodologies

3.5.1 Full Lineshape Analysis of Transversal Decay Curves

In their current implementation for routine commercial applications, the SFC methods neglect the richness in phase-compositional information that is contained in transversal relaxation decays. An example is shown in Figure 3.8, which shows distinct differences between the decays of common fat crystal polymorphs.[40] The existence of such polymorphs is neglected in the SFC method, although it is known that this introduces small systematic errors.[17] However, in foods, the interconversion of crystal polymorphs is a common phenomenon and lies at the basis of many textural and shelf-life properties of lipid-based products. Several studies have attempted to describe phase-compositional complexity using transversal NMR relaxometry on benchtop NMR instruments. In a first attempt, it was shown that crystal polymorphism in lipid blends can be described in a quantitative manner.[40] By concatenating a FID with a CPMG sequence, a transversal decay could be obtained that spanned a wide range of time scales (10 μs–500 ms). This allowed for fitting of the relative contributions of crystal polymorphs, semi-solid and liquid phases.[41]

3.5.2 Rapid Single-Shot Relaxometric Measurements

Multiple modulation-multiple echo (MMME) is an elegant ultrafast single-shot technique to measure relaxation times and diffusivities without the need for phase cycling.[42] MMME is rather an NMR acquisition scheme

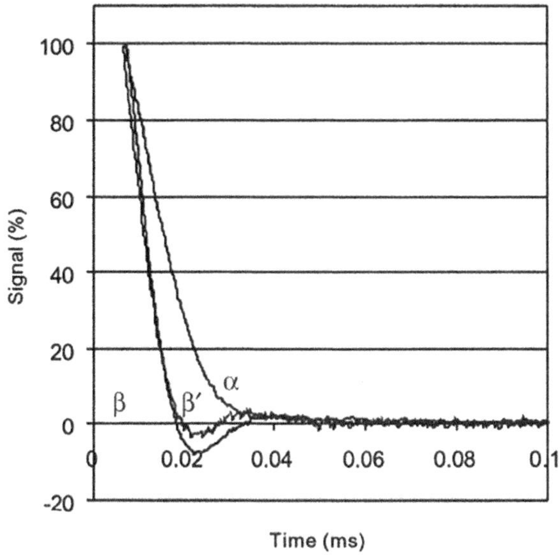

Figure 3.8 Example of transversal relaxation decays of α, β and β′ fat crystal polymorphs. Reproduced from ref. 40 with permission from Elsevier.

portfolio than a single pulse sequence, as it is able to provide rapid measurements of diffusion, flow and imaging. The sequence consists of a set of rf pulses with unequal time spacings that generate a number of spin-echoes. Due to the fact that information about relaxation and diffusion is available from a one-shot measurement, it is expected that this approach will soon enter the field of food science and technology.

The continuous wave free precession (CWFP) technique has been demonstrated in feasibility studies on seeds and meat,[43] but other foods applications can be envisaged. CWFP emerged from the steady-state free precession (SSFP) technique where the time between pulses is shorter than T_2*. At this limit, the overlap of the FID and echo signals in SSFP results in no decay during the inter-pulse delays.[44] In the CWFP regime the continuous amplitude signal is preserved between pulses and the signal does not depend on T_1. Therefore, fast pulsing in CWFP provides valuable sensitivity enhancement.[43]

3.5.3 On-Line Measurements

The manufacturing efficiency of many fat-structured food products would benefit significantly when SFC could be monitored on-line in a real-time manner. SFC measurements using closed benchtop magnets without open access have been deemed as not feasible, hence use of one-sided magnets such as the NMR-MOUSE has been proposed.[45] A good correlation was observed between the SFC values obtained by the NMR-MOUSE and a conventional benchtop NMR instrument in the case of static crystallization conditions and off-line measurements. Furthermore, the NMR-MOUSE was used to monitor the variation of SFC on-line during crystallization of a fat sample.[45] Measurements in static conditions were found to be reliable, whereas dynamic conditions such as stirring during crystallization, resulted in inaccurate SFC estimations. Although corrections for temperature variations were applied when performing such analysis, significant differences were still observed compared to the standard SFC-NMR method.

3.5.4 Rheo-SFC NMR

Fat crystallization phenomena during industrial manufacturing of foods mostly take place under dynamic processing conditions. This not only involves variation of temperature, but often also flow/shear. It is well known that flow and shear can have a profound effect on fat crystallization and final product properties, such as texture or shelf-life stability.[46,47] *In situ* shear cells have been designed that allow for direct assessment of SFC under shear.[48] Both direct and indirect SFC measurements have been applied, initially at magnet temperature,[48] but recently implementation of a variable temperature Rheo-NMR shear cell on a benchtop 20 MHz NMR instrument has also been described.[49] This investigation was further pursued to gain new insights on the hierarchical organization of nano-scale fat crystal networks.[47]

3.5.5 Model-Free Approaches for Droplet Size Distributions

Currently implemented approaches to obtain quantitative droplet size distributions make explicit assumptions on the distribution function. Typically, a unimodal or lognormal distribution is assumed. In order to avoid such assumptions model-free approaches have been proposed. In the early work, Monte-Carlo simulations were used to obtain DSD histograms, but this approach was briefly described and not further pursued in the literature.[7] It has been demonstrated that $P(R)$ in eqn (3.5) can be approximated with high accuracy without prior assumption on its analytical form.[50] Two methods were found to be effective:[51] the generating function method and a numerical routine based on a solution of the Fredholm integral to solve the distribution function. In the first approach, the unknown distribution is approximated by a generating function series. The coefficients of the generating function are calculated by a least square procedure. The method is able to describe the distribution function in detail based on experimental PFG data. The method gives good results when the generating function does not depart greatly from the true distribution.[52] The second approach describes the echo decay as a Fredholm integral equation and it solves it numerically in the case of a known diffusion coefficient of the dispersed phase. The methods were successfully applied to uni- and bi-modal droplet size distributions. Furthermore, for complex emulsions where the self-diffusion coefficient of the dispersed phase is unknown,[51] a procedure has been reported that is able to provide both diffusion coefficient and distribution function directly from experimental data. An alternative method of inverting eqn (3.5) based on regularization techniques was introduced by Johns *et al.*[53] This method was recently employed in the characterization of double emulsions (although the study employed non-food systems).[54] A recent study[24] has shown that data inversion *via* regularization in combination with the new BGP approximation[27,34] resulted in better accuracy compared to previously developed short gradient pulse (SGP) and Gaussian phase distribution (GPD) methods. The work was carried out on a custom-made system operating at 1.1 T (permanent magnet) and using an rf micro-coil that allowed the spectral separation of oil and water signals.[24] This is the first chemical shift resolved DSD reported on a low-field benchtop NMR. Concatenated with the spectral resolution performance, the authors made use of the BGP approach to improve the numerical data inversion. Simulations of random walks in spherical confinement were conducted in order to compare the accuracy of the GPD, SGP and BGP approximations. The BGP assumption provided the best performance, with an error in estimating the droplet diameter of a maximum of 0.4%, while SGP and GPD deviated by 2–12% and 2–4%, respectively, depending on the parameter space explored (Δ/δ, $D_0\delta/a^2$). DSDs obtained by analysis of frequency domain data, by diffusion filtering, and by optical microscopy were compared and are shown in Figure 3.9.[24] Clearly, the chemical shift resolved data reveals a better agreement with the microscopy result. A clear distinctive advantage of the spectroscopic resolution was highlighted in the case of a water-in-crude oil emulsion, which impeded DSD assessment in the time-domain.

3.5.6 Double Emulsions

Multiple emulsions are systems where the droplets of the dispersed phase themselves contain smaller dispersed droplets. From the perspective of food technology, the most interesting are W/O/W emulsions since these systems offer the possibility of compartmentalizing otherwise incompatible water-soluble ingredients. Important structural and functional parameters of multiple emulsions like DSD, DPR (disperse phase ratios), the encapsulation efficiency and controlled release of encapsulated ingredients can be investigated by means of NMR. Initially, PFG-NMR was sporadically used to determine DSD in food double emulsions.[55–57] The difference in relaxation and diffusion behavior could be exploited to separate the signals from dispersed and continuous water phase.[55] Modelling approaches were employed[57] to analyse W/O/W (water–soybean oil–water) emulsions that underwent inter-compartmental exchange and thus the volume fraction of the inner water compartment and the DSD of the inner water phase could be obtained. A more detailed investigation of W/O/W emulsions with a wide DSD range was presented in,[58] where the Murday–Cotts formalism, a log-normal distribution, and the exchange model of Pfeuffer and Hindmarsh[57,59] were combined. The exchange process between the quasi-continuous and the inner dispersed phases in the W/O/W system was studied along with its

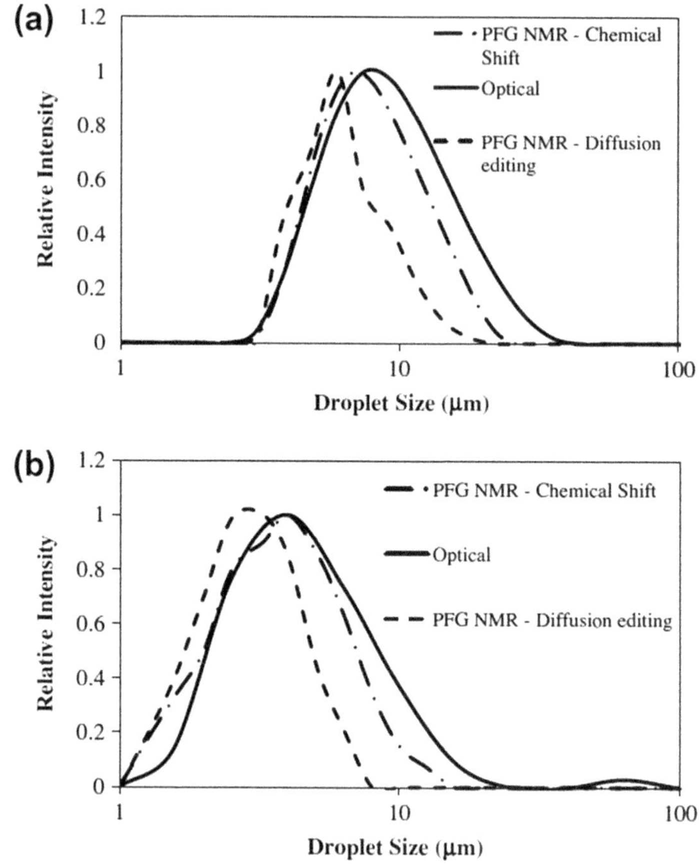

Figure 3.9 Droplet size distributions obtained by PFG NMR on the 1.1 T system with the BGP regularization method, compared with optical counting measurements for (a) emulsion A (decane-in-water) and (b) emulsion B (toluene-in-water). Reproduced from ref. 24 with permission from Elsevier.

contribution to the signal attenuation as a function of gradient strength.[57,60] Release kinetics, DSD, and molecular exchange in W/O/W emulsions were recently analysed in the context of an acquisition scheme that provides a higher S/N by multiple-echo sampling of a CPMG train added to the PFG-StE sequence,[61] carried out at medium resolution at 20 MHz using a Bruker custom made system.[62] The DSD was not retrieved by exploiting spectral resolution, yet still by time-domain filters, due to the fact that field inhomogeneities were dominating the natural line width and unambiguous spectral separation of oil and water was not possible. Spectrally resolved longitudinal relaxation experiments[61] were applied to investigate the DPR and molecular exchange within double emulsions, by using relaxation agents in the outer water phase.

3.6 Perspectives and Conclusions

3.6.1 Conclusions

The routine applications for SFC and DSD, fully automated and implemented on affordable benchtop NMR equipment, play an important role in the quality and process control for spreads and dressings in the food supply chain. Many of the first shortcomings of the implementation on low-field benchtop instruments have been circumvented by innovations in pulse programming, data-processing and analysis. These routine methods are attractive for applications in both food laboratories and in manufacturing environments. The NMR SFC method is widespread in fat technology laboratories and factories. Its applicability to in-/on-line measurements is however limited and it strongly depends on process parameters related to factory environments. In the case of routine DSD by NMR, the selection of optimal acquisition parameters and assessment of droplets with arbitrary shapes remain open issues. Recent data analysis approaches like model-free regularization and BGP approximation yielded improved accuracy, but are not yet exploited in routine applications.

3.6.2 Perspectives

3.6.2.1 Hardware Developments

Since they were released on the market, hardware improvements in commercial NMR benchtop instruments mainly concerned electronics to gain sensitivity. The magnet itself and its field homogeneity saw only minor improvements and resolution always remained far below that available *via* conventional NMR spectroscopy. A significant number of time-domain applications were developed to serve various needs of different industries, and benchtop NMR was presented as a downsized version of a high-field instrument. While industry was enjoying the first fruit of NMR miniaturization, a few seeds sprouted in academic research to develop mobile NMR instrumentation. This quest started more than three decades ago and a comprehensive review on hardware, methodology, and applications of mobile NMR is available.[13] Besides the commercial benchtops with closed magnet design, single sided and mobile surface explorers like the NMR-MOUSE[63] offer flexibility to operate on large intact samples, to investigate food products from the shelf through packaging, and to monitor in- and on-line processing. The applicability of the NMR-MOUSE for SFC determination through packaging was demonstrated[64] for products like margarine, mayonnaise and coffee cream, and for on-line detection of SFC in model fat blends.[45] The permanent field gradient of the NMR-MOUSE makes it suitable to measure diffusion and therefore it could also be applied to study food emulsions. Diffusion weighted CPMG decays recorded with the NMR-MOUSE made possible the quantification of oil/water ratios.[65,66] Thus far the concept of single sided magnets has not been pursued in the food industry as an on/in-line quality control tool. Practical limitations in factory environments, such as hygiene constraints, limit the robustness and application scope for on-line and/or mobile instrumentation. In many cases, it is sufficient to deploy benchtop NMR for end-quality control in off-line mode and there is no strong drive for innovation in on-line and/or mobile instrumentation. An interesting example in advances of mobile NMR is the earth-field's approach. A commercial instrument (Terranova-MRI made by Magritek) was employed to obtain DSD in W/O emulsions.[67] A particular feature of a mobile earth-field NMR device is the fact that it uses ultra-low frequency electromagnetic waves that can penetrate metal media like pipes and containers and therefore may be used in hostile factory environments.

3.6.2.2 From Time-Domain to Frequency-Domain

It was in 2000 when NMR at inhomogeneous magnetic fields witnessed an important move towards spectroscopy by the introduction of the so-called field matching approach. The key insight here is that a mixed echo can retrieve chemical shift information if certain conditions are met to match B_1 and B_0 field profiles.[68,69] The continuing developments in single sided NMR twisted the field matching method towards a field shimming approach that blossomed with the achievement of a proton spectral resolution of about 0.25 ppm in an *ex-situ* manner by a portable open magnet.[14] Along with these innovations in single-sided/*ex-situ*, closed magnets were also considered for spectroscopy. Examples of Halbach design were demonstrated during recent years and furthermore, improved shimming and frequency locking were implemented.[24,62,70–72] Driven by the outcome of the pioneering work, some instruments manufacturers have already developed benchtop NMR equipment with medium- and high-spectral resolution, available on the market at competitive prices (Teachspin, Magritek, Oxford Instruments, Bruker, Nanalysis). When equipped with pulsed field gradients that allow for droplet size measurement, this could open new avenues for the assessment of DSD in food emulsions.

Acknowledgements

Gert-Jan Goudappel (Unilever R&D) is acknowledged for providing Figure 3.6 in this chapter.

References

1. J. P. M. van Duynhoven, in *Encylopedia of Spectroscopy and Spectrometry*, Elsevier, Oxford, 2010, vol. 1, pp. 663–670.
2. K. P. A. M. Van Putte and J. Van den Enden, *J. Phys. E: Sci. Instrum.*, 1973, **6**, 910.
3. P. N. Tiwari and W. Burk, *J. Am. Oil Chem. Soc.*, 1980, **57**, 119–121.

4. K. J. Packer and C. Rees, *J. Colloid Interface Sci.*, 1972, **40**, 206–218.
5. J. C. Van den Enden, D. Waddington, H. Van Aalst, C. G. Van Kralingen and K. J. Packer, *J. Colloid Interface Sci.*, 1990, **140**, 105–113.
6. I. Fourel, J. P. Guillement and D. Le Botlan, *J. Colloid Interface Sci.*, 1994, **164**, 48–53.
7. B. Balinov, O. Söderman and T. Wärnheim, *J. Am. Oil Chem. Soc.*, 1994, **71**, 513–518.
8. G. J. W. Goudappel, J. P. M. Van Duynhoven and M. M. W. Mooren, *J. Colloid Interface Sci.*, 2001, **239**, 535–542.
9. J. Van Duynhoven, A. Voda, M. Witek and H. Van As, *Annu. Rep. NMR Spectrosc.*, 2010, **69**, 145–197.
10. A. M. Haiduc, J. P. M. van Duynhoven, P. Heussen, A. A. Reszka and C. Reiffers-Magnani, *Food Res. Int.*, 2007, **40**, 425–434.
11. H. Todt, W. Burk, G. Guthausen, A. Guthausen, A. Kamlowski and D. Schmalbein, *Eur. J. Lipid Sci. Technol.*, 2001, **103**, 835–840.
12. S. M. Jepsen, H. T. Pedersen and S. B. Engelsen, *J. Sci. Food Agric.*, 1999, **79**, 1793–1802.
13. B. Blümich, J. Perlo and F. Casanova, *Prog. Nucl. Magn. Reson. Spectrosc.*, 2008, **52**, 197–269.
14. J. Perlo, F. Casanova and B. Blümich, *Science*, 2007, **315**, 1110–1112.
15. F. Dalitz, M. Cudaj, M. Maiwald and G. Guthausen, *Prog. Nucl. Magn. Reson. Spectrosc.*, 2012, **60**, 52–70.
16. S. K. Küster, E. Danieli, B. Blümich and F. Casanova, *Phys. Chem. Chem. Phys.*, 2011, **13**, 13172–13176.
17. G. J. W. Goudappel, M. C. M. Gribnau, V. K. S. Shukla and J. P. M. van Duynhoven, *AOCS Inform*, 1999, **10**, 479.
18. M. C. M. Gribnau, *Trends Food Sci. Technol.*, 1992, **3**, 186–190.
19. A. Bernreuther, H. Schimmel and J. P. M. van Duynhoven, in *Magnetic Resonance in Food Science: Food for Thought*, ed. J. van Duynhoven, H. As, P. S. Belton and G. A. Webb, Royal Society of Chemistry, 2013, pp. 150–155.
20. J. E. Tanner and E. O. Stejskal, *J. Chem. Phys.*, 1968, **49**, 1768–1777.
21. J. P. M. Van Duynhoven, G. J. W. Goudappel, G. Van Dalen, P. C. Van Bruggen, J. C. G. Blonk and A. P. A. M. Eijkelenboom, *Magn. Reson. Chem.*, 2002, **40**, S51–S59.
22. J. P. M. van Duynhoven, B. Maillet, J. Schell, M. Tronquet, G. J. W. Goudappel, E. Trezza, A. Bulbarello and D. van Dusschoten, *Eur. J. Lipid Sci. Technol.*, 2007, **109**, 1095–1103.
23. W. S. Price, *Concepts Magn. Reson.*, 1998, **10**, 197–237.
24. I. A. Lingwood, T. C. Chandrasekera, J. Kolz, E. O. Fridjonsson and M. L. Johns, *J. Magn. Reson.*, 2012, **214**, 281–288.
25. B. Balinov, B. Jonsson, P. Linse and O. Soderman, *J. Magn. Reson., Ser. A*, 1993, **104**, 17–25.
26. J. S. Murday and R. M. Cotts, *J. Chem. Phys.*, 1968, **48**, 4938–4945.
27. D. S. Grebenkov, *Concepts Magn. Reson., Part A*, 2009, **34**, 264–296.
28. D. S. Grebenkov, *Concepts Magn. Reson., Part A*, 2008, **32**, 277–301.
29. D. S. Grebenkov, *Rev. Mod. Phys.*, 2007, **79**, 1077.

30. D. C. Douglass and D. W. McCall, *J. Phys. Chem.*, 1958, **62**, 1102–1107.
31. M. L. Johns and K. G. Hollingsworth, *Prog. Nucl. Magn. Reson. Spectrosc.*, 2007, **50**, 51–70.
32. A. Guillermo and M. Bardet, *Anal. Chem.*, 2007, **79**, 6718–6726.
33. W. S. Price and O. Soederman, *Isr. J. Chem.*, 2003, **43**, 25–32.
34. A. V. Barzykin, *J. Magn. Reson.*, 1999, **139**, 342–353.
35. D. J. McClements, *Crit. Rev. Food Sci. Nutr.*, 2007, **47**, 611–649.
36. M. A. Voda and J. Van Duynhoven, *Trends Food Sci. Technol.*, 2009, **20**, 533–543.
37. S. Schuster, R. Bernewitz, G. Guthausen, J. Zapp, A. M. Greiner, K. Köhler and H. P. Schuchmann, *Chem. Eng. Sci.*, 2012, **81**, 84–90.
38. G. Van Dalen, *J. Microsc.*, 2002, **208**, 116–133.
39. D. J. McClements and J. N. Coupland, *Colloids Surf., A*, 1996, **117**, 161–170.
40. J. van Duynhoven, I. Dubourg, G. J. Goudappel and E. Roijers, *J. Am. Oil Chem. Soc.*, 2002, **79**, 383–388.
41. E. Trezza, A. M. Haiduc, G. J. W. Goudappel and J. P. M. Van Duynhoven, *Magn. Reson. Chem.*, 2006, **44**, 1023–1030.
42. E. E. Sigmund, H. Cho and Y. Q. Song, *Concepts Magn. Reson., Part A*, 2007, **30**, 358–377.
43. L. A. Colnago, M. Engelsberg, A. A. Souza and L. L. Barbosa, *Anal. Chem.*, 2007, **79**, 1271–1274.
44. R. Bagueira de Vasconcelos Azeredo, L. A. Colnago and M. Engelsberg, *Anal. Chem.*, 2000, **72**, 2401–2405.
45. S. Martini, M. L. Herrera and A. Marangoni, *J. Am. Oil Chem. Soc.*, 2005, **82**, 313–317.
46. A. G. Marangoni, N. Acevedo, F. Maleky, F. Peyronel, G. Mazzanti, B. Quinn and D. Pink, *Soft Matter*, 2012, **8**, 1275–1300.
47. G. Mazzanti, M. Li, A. G. Marangoni and S. H. J. Idziak, *Cryst. Growth Des.*, 2011, **11**, 4544–4550.
48. G. Mazzanti, E. M. Mudge and E. Y. Anom, *J. Am. Oil Chem. Soc.*, 2008, **85**, 405–412.
49. E. M. Mudge and G. Mazzanti, *Cryst. Growth Des.*, 2009, **9**, 3111–3118.
50. L. Ambrosone, A. Ceglie, G. Colafemmina and G. Palazzo, *J. Chem. Phys.*, 1997, **107**, 10756–10763.
51. L. Ambrosone, A. Ceglie, G. Colafemmina and G. Palazzo, *Langmuir*, 1999, **15**, 6775–6780.
52. L. Ambrosone, G. Colafemmina, M. Giustini, G. Palazzo and A. Ceglie, in *Trends in Colloid and Interface Science XIII*, Springer, 1999, pp. 86–88.
53. K. G. Hollingsworth and M. L. Johns, *J. Colloid Interface Sci.*, 2003, **258**, 383–389.
54. E. Hughes, A. A. Maan, S. Acquistapace, A. Burbidge, M. L. Johns, D. Z. Gunes, P. Clausen, A. Syrbe, J. Hugo and K. Schroen, *J. Colloid Interface Sci.*, 2013, **389**, 147–156.
55. I. Lönnqvist, B. Håkansson, B. Balinov and O. Söderman, *J. Colloid Interface Sci.*, 1997, **192**, 66–73.
56. R. Mezzenga, B. M. Folmer and E. Hughes, *Langmuir*, 2004, **20**, 3574–3582.

57. J. P. Hindmarsh, J. Su, J. Flanagan and H. Singh, *Langmuir*, 2005, **21**, 9076–9084.
58. F. Wolf, L. Hecht, H. P. Schuchmann, E. H. Hardy and G. Guthausen, *Eur. J. Lipid Sci. Technol.*, 2009, **111**, 730–742.
59. J. Pfeuffer, U. Flögel, W. Dreher and D. Leibfritz, *NMR Biomed.*, 1998, **11**, 19–31.
60. X. Guan, K. Hailu, G. Guthausen, F. Wolf, R. Bernewitz and H. P. Schuchmann, *Eur. J. Lipid Sci. Technol.*, 2010, **112**, 828–837.
61. R. Bernewitz, F. Dalitz, K. Köhler, H. P. Schuchmann and G. Guthausen, *Microporous Mesoporous Mater.*, 2013, **178**, 69–73.
62. M. Cudaj, G. Guthausen, A. Kamlowski, D. Maier, T. Hofe and M. Wilhelm, *Nachr. Chem.*, 2010, **58**, 1155–1157.
63. G. Eidmann, R. Savelsberg, P. Blümler and B. Blümich, *J. Magn. Reson., Ser. A*, 1996, **122**, 104–109.
64. A. Guthausen, G. Guthausen, A. Kamlowski, H. Todt, W. Burk and D. Schmalbein, *J. Am. Oil Chem. Soc.*, 2004, **81**, 727–731.
65. H. T. Pedersen, S. Ablett, D. R. Martin, M. J. D. Mallett and S. B. Engelsen, *J. Magn. Reson.*, 2003, **165**, 49–58.
66. D. R. Martin, S. Ablett, H. T. Pedersen and M. J. D. Mallett, in *Magnetic Resonance in Food Science: Latest Developments*, ed. P. S. Belton, A. Gil, G. A. Webb and D. Rutledge, Royal Society of Chemistry, 2003, vol. 286, pp. 54–61.
67. E. O. Fridjonsson, L. S. Flux and M. L. Johns, *J. Magn. Reson.*, 2012, **221**, 97–102.
68. C. A. Meriles, D. Sakellariou, H. Heise, A. J. Moulé and A. Pines, *Science*, 2001, **293**, 82–85.
69. I. Ardelean, R. Kimmich and A. Klemm, *J. Magn. Reson.*, 2000, **146**, 43–48.
70. A. McDowell and E. Fukushima, *Appl. Magn. Reson.*, 2008, **35**, 185–195.
71. E. Danieli, J. Perlo, B. Blümich and F. Casanova, *Angew. Chem., Int. Ed.*, 2010, **49**, 4133–4135.
72. E. Danicli, J. Mauler, J. Perlo, B. Blümich and F. Casanova, *J. Magn. Reson.*, 2009, **198**, 80–87.

CHAPTER 4

Hardware Developments: Single-Sided Magnets

PETER BLÜMLER*[a] AND FEDERICO CASANOVA*[b]

[a]University of Mainz, Institute of Physics, 55099 Mainz, Germany;
[b]Magritek, Aachen, Germany
*E-mail: bluemler@uni-mainz.de, federico@magritek.com

4.1 Introduction

Single-sided NMR was introduced shortly after the oil industry[1,2] proved that NMR signals from arbitrarily large samples can be measured by placing them in the stray field of open magnets. This concept requires that the magnet and radiofrequency (rf) probe project magnetic and radiofrequency fields to volumes external to the sensor, where the object is located. This was the approach followed to build the NMR-MOUSE.[3] In its initial design[4] it utilized a magnet in the shape of a "U" or a horseshoe to generate a magnetic field of about half a tesla with a gradient of about 20 T m^{-1} in a volume located right above the magnets. Single-sided NMR lifted the constraints on the sample size by placing the equipment on one side of a surface, which shall be called the "equipment side". The NMR experiment is then conducted on the other, the "sample side".

Various magnet designs have been reported for single-sided NMR within the past decades. They can be classified by geometry (*e.g.* U-shape, bar or barrel magnets), the direction of the flux relative to the surface or by the shape of the magnetic field (gradient *vs.* sweet spot). In the presence of a

New Developments in NMR No. 5
Mobile NMR and MRI: Developments and Applications
Edited by Michael L. Johns, Einar O. Fridjonsson, Sarah J. Vogt, and Agnes Haber

Published by the Royal Society of Chemistry, www.rsc.org

static gradient, the excited volume approaches a slice parallel to the sensor surface. This is convenient to obtain depth localization into the object. On the other hand, magnets generating a sweet spot, or a point in space where at least the first-order derivatives of the magnetic field magnitude are zero, sacrifice depth resolution to maximize sensitivity. By cancelling the static gradient, the size of the excited volume can be considerably increased. Even though volume, V, maximization is achieved at the expense of magnetic field strength, B_0, the product VB_0^2 that defines the sensitivity of the magnet, is typically the largest for these type of magnets. Similar general rules are valid for the rf excitation of the spins in such a volume and the limitations on the achievable signal-to-noise ratio (SNR).

In view of many excellent reviews[5–7] on this subject the authors decided to give a more general and "hands-on" discussion of the hardware involved in single-sided NMR. It is a more a summary of simple and straight-forward thoughts which might guide the conceptual engineering of new hardware developments rather than a detailed analysis. For the same reason we only mention a few applications (more can be found in the rest of this book and the reviews mentioned above).

4.2 Single-Sided Magnet Geometries

4.2.1 Basic Designs

Figure 4.1 shows some exemplarily basic magnet designs with the resulting flux lines on the sample side. It illustrates the designs and how rather simple modifications alter the flux strength and decay on the sample side. Comparable to the flux density of an infinitely small magnetic dipole, the decay of flux in the direction of its magnetization is not different to that perpendicular (it only changes sign; the magnitude is identical in any direction). Similarly, the decay of flux on the sample side doesn't significantly change when the same cubic magnet is oriented with its magnetization direction parallel (Figure 4.1b) or perpendicular (Figure 4.1a) to the surface. This decay of flux with increasing distance to the surface is somewhat unavoidable. Since we restricted the magnetic material to the other side of this surface, the magnetic flux must decay in the absence of other flux sources. It is not straight-forward to prove this *via* Gauss' law:

$$\nabla \cdot \mathbf{B} = 0 \tag{4.1}$$

since we are dealing with a surface that is not enclosing the flux source. However, it can be considered infinitely large compared to it, so every flux line leaving the surface must come back without crossing another, hence they must dilute with distance. This renders the problem of all magnetic fields involved in single-sided NMR. In order to give a bit of an over-simplified rule of thumb: the flux decays more or less exponentially with distance from the surface into the sample with a decay constant less than the size of magnets placed on the other side. The shape of this decay, however, is under control

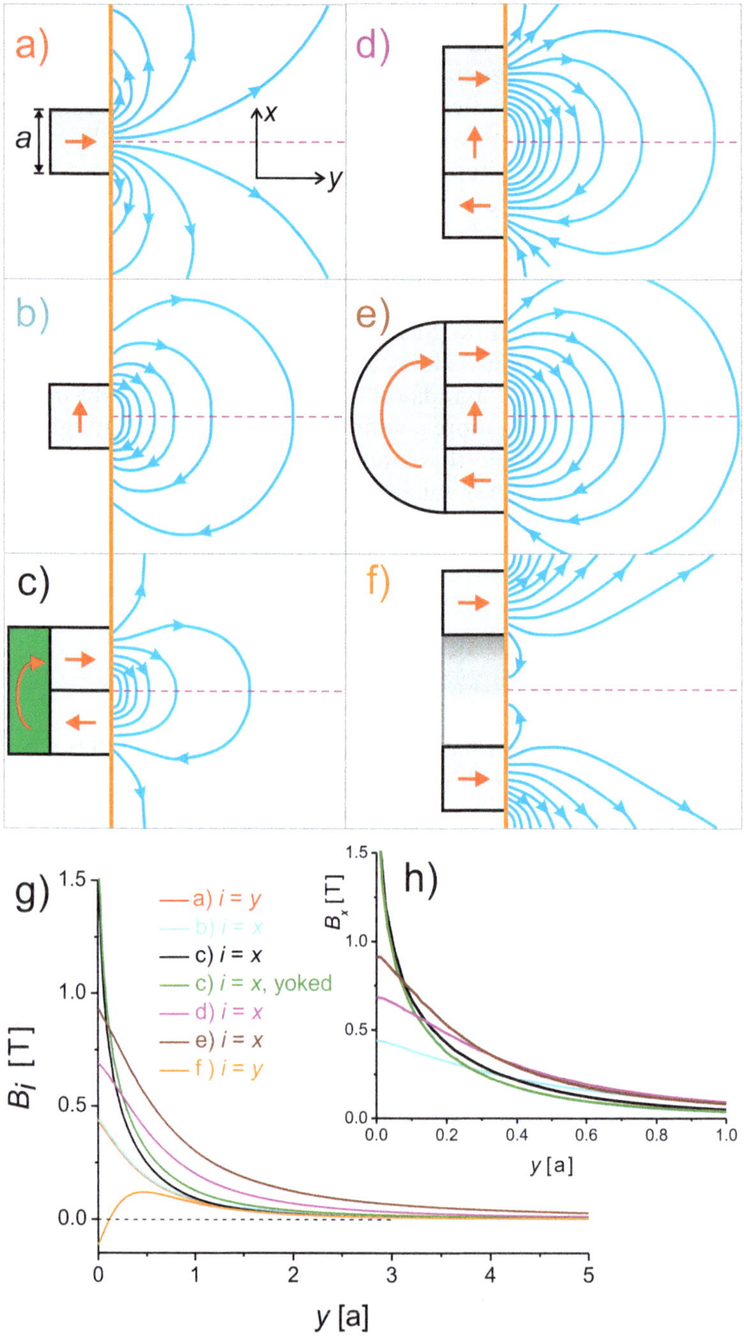

Figure 4.1 Comparison of the flux generated on the sample side by various single-sided magnet designs. The inserted figures (a)–(f) show the principal designs of the most elementary magnets. The thick orange line divides the "equipment" on the left from the "sample" side on the right.

of the arrangement of magnets, but at best local maxima and saddle points can be generated.[8] Because the SNR of NMR signals depends on the polarizing magnetic flux, the problem is further dramatized and this is still not considering the necessary rf field for excitation. Hence, the investigation of sample material far away from the surface suffers from poor SNR or requires very large magnets.

However, these general considerations only hold at far distances and the flux in the direct vicinity of the magnets can be shaped in various ways, as will be discussed in the next sections. The different designs are categorized by the direction of the main flux component relative to the surface.[5]

A single square magnet placed directly below the surface with its magnetization parallel to it (see Figure 4.1b) produces roughly the same field decay above that surface as a magnet with its magnetization perpendicular to the surface (Figure 4.1a, compare also Figure 4.1g). The flux is increased if the magnets are combined, as in Figure 4.1c, where there is one magnet pushing the flux out of the surface right next to another pulling it back. This simple superposition of flux almost triples the field strength of the magnet assembly and can be even further intensified by using an iron yoke (green block in Figure 4.1c), which concentrates and redirects the flux on the rear of the magnet. This magnet is then conceptually identical to a "U" or horseshoe magnet. If the decay rate of the magnet in Figure 4.1c is compared to that of Figure 4.1b, one also notices that it increases dramatically in the center of the magnet. Depending on the application in mind this can be an advantage or disadvantage, as will be discussed later.

If the latter two magnets are combined by inserting the magnet from Figure 4.1b in-between the antiparallel magnets of Figure 4.1c, one gets a section of a linear Halbach-array (Figure 4.1d). While this reduces the flux at the surface center, it certainly pushes the flux further out into the sample resembling roughly the decay rate of Figure 4.1b. This assembly produces theoretically no flux on its rear side (when extended to a Halbach array);

The blue lines and arrows are the flux lines on the sample side. The gray blocks represent magnet material polarized in the direction of the red arrows inside. (a) Simple magnetic block of size a with its magnetization perpendicular to the surface. (b) The same as (a) but with the magnetization parallel to the surface. (c) A double block of antiparallel magnetization but perpendicular to the surface. The green block represents an iron yoke to redirect and intensify the flux to the sample side. (d) Part of a linear Halbach array and (e) the same structure but with a redirecting magnet on its back. (f) A cross-section through a hollow magnetized cylinder (aka "barrel magnet"). (g) Strongest flux component (B_i with $i = x, y$; see coordinate system in (a)) central to the magnet and along x on the sample side of each magnet design, as indicated by the dashed purple lines. The units of y are roughly given in sizes a of a single magnet block (see (a)). (h) The same as (g) but only for magnets with the main flux component parallel to the surface and for normalized magnet dimensions. See ref. 8 and text for more details.

therefore, a yoke is of no use here. However, the flux can be increased by adding an "active yoke" on this rear side, as shown in Figure 4.1e. This roughly doubles the flux of Figure 4.1b without increasing the decay rate, resulting in the strongest flux at a deeper penetration depth.

Such a line of argumentation is convincing and often applied in this field of research, when various magnet designs are compared in terms of performance. However, a candid comparison must normalize all these magnets to a unit volume (or mass), as done in Figure 4.1h. The normalization leaves the flux density at the surface untouched but levels the decays relative to the unit size of the magnetic arrangements. The same arguments as before are valid but the differences become less dramatic.

Typically, the application of such single-sided magnets can be distinguished in investigations of an extended volume or a thin slice in the sample. For the first, a large volume of modest homogeneity ("sweet spot") is needed that can be excited by the bandwidth of the applied rf pulse. Therefore, low decay rates are preferred for this application. A typical example of this is the so-called "barrel magnet" (see Figure 4.1f), which is discussed in more detail in Section 4.2.4.1. Here the flux direction changes sign along the central line to create a relatively flat or homogeneous maximum before the unavoidable decay at farther distances.

Contrarily, the excitation of thin slices needs strong field gradients or a strong decay rate, which preferably should be close to linear in the region of interest. Such considerations are important to determine the optimal design due to the flux decay in the direction normal to the surface (y-direction). The flux dependence in the other two dimensions is however generally favored to be as homogeneous as possible to have a good spatial definition of the volume the signal originates from. This is crucial for sample profiling where the signals from thin slices inside the sample are detected. Hence, any curvature in the flux parallel to the surface will blur the resolution.

The easiest way to straighten the flux is to separate its sources. This is illustrated on the magnet from Figure 4.1c without a yoke. We start with a design made from two magnets, and it is assumed that the magnets can be simplified by dipoles positioned symmetrically around the gap (see sketch in Figure 4.2a and 4.2b).

The magnetic flux generated by a dipole of magnetization **m** at distance **r** is then given by:

$$\mathbf{B}(\mathbf{r}) = \frac{\mu_0}{4\pi}\left[\frac{3\mathbf{r}(\mathbf{m}\cdot\mathbf{r})}{r^5} - \frac{\mathbf{m}}{r^3}\right] \tag{4.2}$$

To gain a flat or homogeneous field in between two dipoles (at locations $\mathbf{x}_{1,2} = [\pm x_0, 0]$ and magnetization vectors given in red, see Figure 4.2a) at the position of interest $\mathbf{r}_c \equiv [0,1]$, the sum eqn (4.2) for both dipoles has to be calculated:

$$\mathbf{B}_c = \sum \mathbf{B}(\mathbf{r}_c \pm \mathbf{x}_{1,2}) = \frac{\mu_0}{4\pi}\left[\frac{6mx_0}{(x_0^2 + 1)^{5/2}}, 0\right] = [B_x^c, 0]. \tag{4.3}$$

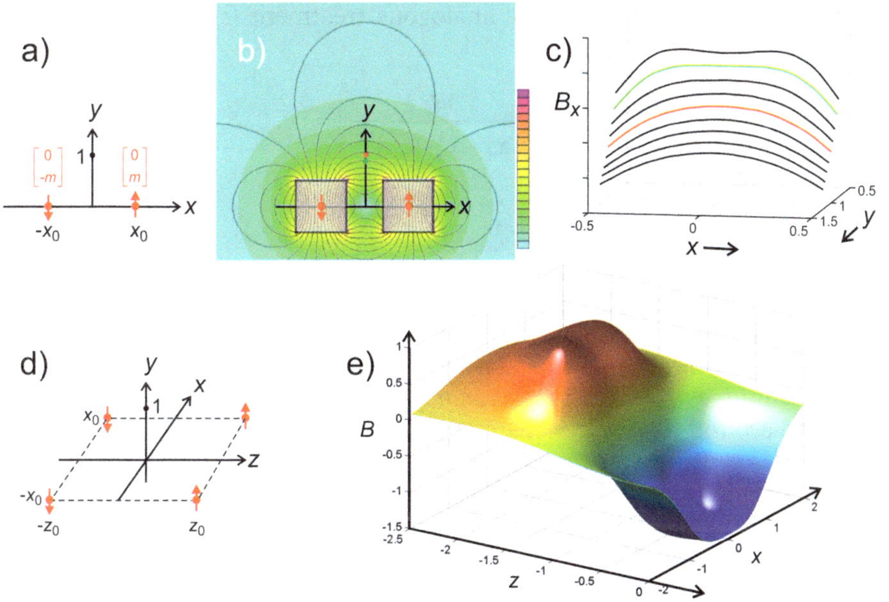

Figure 4.2 Field homogenization by magnet separation: (a) Simple dipolar two-dimensional model, where two magnetic dipoles (red arrows) are placed at positions $\pm x_0$. The flux should be homogeneous at position [0,1] marked by a dot on the y-axis. (b) Finite Element Method (FEM) simulation of two strong magnets of side length 1 positioned at $x_0 = \pm\sqrt{3}/2$. Shown is the absolute flux density in colors as given in the color-bar at the right side and the flux lines as thin black lines. (c) Display of the flux across the gap at different heights $y = 0.8$–1.2. Red line: $y = 1$, green line: $y = 0.85$. The same positions are also indicated in (b) by colored dots on the y-axis. (d) The same as (a) but in three dimensions. (e) Calculation of B_z generated at height $y = 1$ above four dipoles positioned at $x_0 = \pm 1/\sqrt{3}$ and $z_0 = \pm 1$. Only half of the symmetric flux surface is shown along z. The flux axis is in units of $\mu_0 m/4\pi$.

As expected from the geometry there is only an x-component of the flux at the centre (B_x^c). In order to homogenize the field the gradients of this component along the x direction must be minimized, hence the root of the second partial derivative of B_x^c along x has to be found:

$$\frac{\partial^2 B_x^c}{\partial x^2} = 0 \quad \Rightarrow \quad x_0 = \frac{\sqrt{3}}{2}. \tag{4.4}$$

Of course, this is only a first approximation for infinitely small magnets. In order to compare this with realistic finite magnets, a FEM simulation of magnets with a square cross-section of unit side length was performed (see Figure 4.2b) and B_x profiles extracted at certain distances above the center of the magnets (not the surface) to be comparable with the dipolar model. Figure 4.2c shows that even under these circumstances the value of the best position is only *ca.* 15% away from eqn (4.4). If the problem is extended by another dimension, completely analogous treatment (*cf.* Figure 4.2d and 4.2e) yields:

$$\mathbf{B}_c = \frac{\mu_0}{4\pi}\left[0,\,0,\,\frac{12my_0z_0}{\left(x_0^2+y_0^2+z_0^2\right)^{5/2}}\right] = \left[0,\,0,\,B_z^c\right] \tag{4.5}$$

$$\frac{\partial B_z^c}{\partial x^2} = \frac{\partial B_z^c}{\partial x^2} = 0 \;\Rightarrow\; x_0 = \frac{1}{\sqrt{3}} \wedge z_0 = 1. \tag{4.6}$$

These simple examples illustrate how conveniently a magnetic problem can be initially approximated by using dipoles. Typically, it allows analytical optimization with excellent starting points for further refinements using FEM techniques. Alternatively, the magnets can be represented by coils wrapped around the surfaces parallel to its polarization direction and using the Biot–Savart law to calculate the field distribution, which—at least for simple geometries—is favorable because it may yield analytical solutions (see for instance eqn (4.7)).

4.2.2 Realized Devices

4.2.2.1 *External Field Parallel to Surface*

Most magnets designed for single-sided NMR use fields parallel to the surface. This is probably because this configuration offers more flexibility in the arrangement of the magnets. Flux parallel to the surface can be generated either by the magnet's polarization parallel to the surface (Figure 4.1b) or by two magnets with anti-perpendicular polarization (Figure 4.1c) or their combination (Figure 4.1d or 4.1e). The discussion above on how to place or space magnets already illustrates the ability to control the shape of the magnetic field in this principal configuration. Further refinements using additional correction magnets or pole pieces give probably the biggest set of parameters to match the application's requirements.

Instead of discussing each design with its pro and cons in detail, we decided to summarize the properties of conceptually different devices in Table 4.1. We hope this allows an easy comparison of size, weight, achieved flux and gradients or homogeneity.

4.2.2.2 *External Field Perpendicular to Surface*

"Focusing" and shaping flux perpendicular to the surface of single-sided magnet arrangements is more difficult because the flux cannot easily be directed due to the exclusion of magnetic material beyond the surface. As antiparallel magnetization will mainly create flux parallel to the surface (see Section 4.2.2.1) the main design feature is the use of one and none (voids) magnetization. Although this will obviously not increase the flux, the sensitivity of such designs is typically still very high due to the decrease of the

Table 4.1 This table summarizes various designs of single-sided magnets with a resulting field parallel to the surface. The first column relates the magnet geometry conceptually to the ones shown in Figure 4.1, while the second column gives a short reference followed by size and mass. The fifth column refers to the magnetic flux generated at a certain height and the last one gives values for either the gradient, *G*, or homogeneity, *H*. Note the different units for *G* and *H*. Please refer to ref. 5 and 7.

Figure 4.1	Short description and references	Size [cm^3]	m [kg]	B_{surf} [T] @ height [cm]	G [T m^{-1}], H [ppm mm^{-1}]
(b)	Simple bar magnet[9]	14.5 × 10 × 8	1.8	0.28 @ 0	$G = 10$
(b)	Inserted shim magnets[10–13]	11.5 × 10 × 6	5	0.11 @ 1	$G = 0.3–2.5$ [12]
(b)	Shim magnet on top[14]	5 × 5 × 5	1.8	0.075 @ 0	$H = 3600$
(c)	Original NMR-MOUSE[4,15–18]	5 × 5 × 4	2.5	0.5	$G = 10$
(c)	Profile-MOUSE[19] (see Figure 4.3a)	10 × 10 × 7	7	0.4	$G = 20$
(c)	Homogenized MOUSE[20] for spectroscopy[21] (see Figures 4.4 and 4.6)	28 × 12 × 28	≈35	>0.25	$G = 2.5$ $H = 0.05$
(c)	2 pairs + pole pieces[22–24]	15 × 15 × 7	≈10	*ca.* 0.1	$G = 0.13$ [25]
(d)	flat cylindrical Halbach[22,24]	10 × 10 × 5	2	0.26	$G = 7–14$

associated gradients. Therefore, this flux direction is mainly used to create magnets with "sweet spots".

The majority of the magnet designs in this section (see Table 4.2) can be sufficiently analyzed by combining cylindrical magnets (*e.g.* the barrel magnet is a subtraction of two cylinders of different diameters, see Figure 4.5). The flux-component normal to the surface, B_y, of such a cylinder polarized in the same direction and of radius, r, and length, l (see Figure 4.1a) is calculated analogously to a solenoid coil of identical geometry using the Biot–Savart law:

$$B_y(y) = \frac{B_R}{2}\left[\frac{l+y}{\sqrt{r^2+(l+y)^2}} - \frac{y}{\sqrt{r^2+y^2}}\right] \tag{4.7}$$

with B_R as the remanence of the permanent magnet material. From this equation it can be seen that the flux is inversely proportional to the radius and increases almost linearly with l for $l < r$ to saturate at about $l \gtrsim r$; see also ref. 26.

4.2.3 Magnets Generating a Uniform Gradient

The early design of the NMR-MOUSE offered a typical depth resolution of about a millimeter, while the maximum penetration depth was less than five millimetres.[4] Under these conditions, the analysis hardly implied depth

Table 4.2 Like Table 4.1 this is a summary of single-sided magnet designs with the field perpendicular to the surface. The nomenclature is the same as in Table 4.1. The reference to Figure 4.1a′ refers to two such simple magnets facing each other and creating zero flux in their center. Figure 4.1f′ takes into account that the magnet there is similar to a cross-section through a barrel magnet. Note that this might describe the used geometry but not the flux as shown in Figure 4.1. Please refer to ref. 5 and 7.

Figure 4.1	Short description and references	Size [cm^3]	m [kg]	B_{surf} [T] @ height [cm]	G [T m^{-1}], H [ppm mm^{-1}]
(a)	Simple bar	5 × 5 × 5	2	0.45 @ 0	G = 20
	magnet[27,28] with shimming[29]	20 × 20 × 12	35	0.3 @ 0	H = 100
(a′)	Anti-pair[30,31]	10.6 × 13		0.1	
(f)	Barrel magnet[29,32,33]	5 × 5 × 9	6	0.25 @ 2.4	G = 5
	(see Figure 4.5) MOLE[34]	16 ×16 × 5	6	0.08 @ 1.5	H = 1250
(f′)	"Sliced barrel"[35,36]	10 × 10 × 15		0.05 @ 2	
	with pole piece[37]	15 × 15 × 8	14	0.1	G = 0.31

profiling but rather only crude depth selection. To improve the uniformity of the static gradient, several magnet arrays[19,37–39] have been optimized during the last decade (see Table 4.1). By means of numerical simulations, the size, position, shape, and polarization direction of the multiple magnet blocks forming the arrays were varied to maximize the depth range and lateral extension where a uniform gradient is generated. This last feature is very important when the sensor is used for depth profiling and slice-selective imaging.[20]

4.2.3.1 *Depth Profiling*

In the presence of an ideal uniform static gradient (G_0), a depth profile of an object can be measured using G_0 as a read gradient. However, the strong gradients generated, in practice, by magnets restricted to a single side reduce the excited depth range to a slice hardly thicker than 1 mm, and the efficiency of the rf coil quickly decays with depth, thereby complicating the uniform excitation of thick slices. Thus, two approaches have been adopted to scan a large depth range. In magnets generating a strong gradient, the rf circuit can be tuned to excite the different Larmor frequencies representing the slices in the desired depth range in successive experiments. Even though it is simple and fast, the main limitation of this approach is the systematic contrast variation across the sample obtained in the profiles. This problem stems from the fact that the contrast parameters, such as relaxation times or self-diffusion coefficients, depend on the B_0 and B_1 values, which are likely to change for each new frequency and circuit tuning. A different way of obtaining depth profiles is by changing the relative position of the sample with respect to the sensitive slice[16,19] while keeping the excitation frequency constant as in STRay Field Imaging (STRAFI)[40] (see Figure 4.3). It is a truly

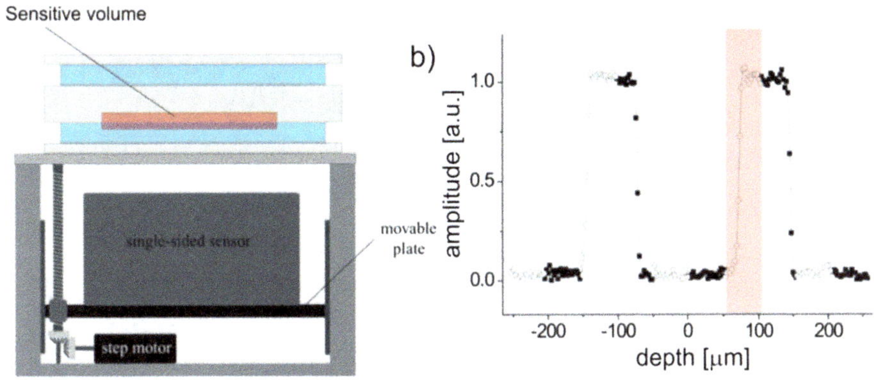

Figure 4.3 (a) Lift used to reposition the sensor with respect to the sample. (b) Profile of a sample made of two 70 μm thick latex layers separated by a 150 μm thick glass piece. The complete profile was obtained as the catenation of 50 μm profiles obtained as the FT of the echo signal measured every 50 μm by moving the sensor with a nominal resolution of about 4 μm.[19]

distortion-free procedure, where the whole profile is measured under exactly the same magnetic and radiofrequency field distributions. Furthermore, it also has important implications on the sensor design. Both the magnet and the rf coil can be optimized to work only at a fixed depth. Therefore, better performance and a simpler geometry are expected. An example of this is the *profile*-NMR-MOUSE,[19] which achieves micrometer resolution over a large depth range while maintaining simplicity and high sensitivity.

4.2.3.2 Lateral Imaging Using Pulsed Gradients

Spatial resolution along the lateral directions of the sensitive slice is achieved by implementing pure phase-encoding imaging methods working in the presence of the slice-selecting static gradient. To generate the pulsed gradients required by these methods, flat gradient coils have been designed and mounted on different single-sided sensors.[20,41,42] Prado *et al.* implemented a single-point spin-echo imaging method[41] that combines a Hahn-echo sequence, applied to refocus the phase spreading due to the static background gradient, with gradient pulses applied during the free evolution periods to encode the spatial information in the phase of the echo. To demonstrate the technique they used a U-shaped magnet with two solenoidal coils to generate a constant pulsed gradient along the gap of the magnet (Figure 4.4a). The technique was extended in 2002 to measure 2D images by using a bar magnet generating a symmetric sensitive volume and two pairs of rectangular coils set at the side of the magnet to produce pulsed gradients along the two lateral directions (Figure 4.4b). The relatively uniform gradient created by this magnet along the depth direction was used to demonstrate

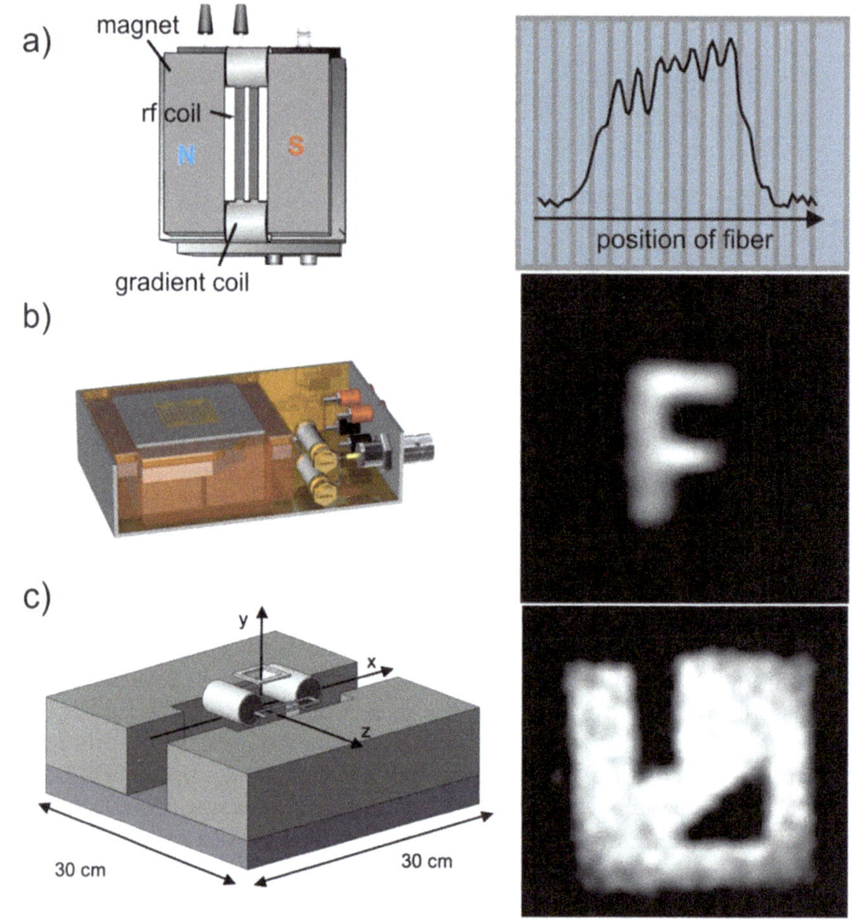

Figure 4.4 (a) NMR-MOUSE for 1D lateral imaging[41] with gradient coils in the magnet gap and an elongated rf coil. The graph on the right shows a spatial profile of a phantom made from several rubber bands (1 mm thick and separated by 0.3 mm) wound around a PTFE plug. (b) Small unilateral MRI scanner with a bar magnet (left) and image (right) of the rubber letter F on top of the rf coil.[20] (c) Gradient coil arrangement mounted in the gap of the unilateral MRI tomography (left) and slice-selective NMR image of a rubber phantom (right).

that flat slices could be selected at different depths into the object by changing the excitation frequency.[42]

An important advance in the applicability of imaging techniques with single-sided sensors was achieved when multi-echo imaging became available. In this way, long trains of echoes created by applying a Carr–Purcell–Meiboom–Gill (CPMG) sequence could be co-added in order to increase the sensitivity during detection. By doing so, the experimental time could be reduced by a factor proportional to the number of echoes acquired in the train. As the

CPMG[43,44] sequence preserves only the component of the magnetization set by the 90° pulse along the refocusing train, two experiments need to be combined to acquire both quadrature components. The method was demonstrated on an improved U-shaped magnet generating a uniform static gradient along depth (Figure 4.4c). The sequence can be set to enhance the contrast in the image using the Hahn-echo sequence as a diffusion or T_2^{Hahn} filter, or directly by adding different parts of the CPMG decay to introduce T_2^{CPMG} weighting. 3D spatial resolution is achieved by combining the 2D phase encoding method with slice selection. After calibrating the frequency dependence with depth, 1 mm thick slices were selected over a depth range covering 10 mm.

4.2.4 Magnets Generating a Volume of Homogeneous Fields

4.2.4.1 Sweet-Spot Magnets

For a given magnet geometry, the static gradient (normal to the surface) can be cancelled over a restricted volume by combining two units of the same type with opposite polarization and different sizes. Taking the case of the bar magnet geometry, two bar magnets with different diameters and opposite polarization can be concentrically arranged to generate a point where the total gradient is zero at a certain distance from the sensor's surface. This is the basic idea behind the barrel magnet,[45,46] which can be understood as the combination of two cylinders with opposite polarization defining a hollow cylinder (see Figure 4.5a). While the magnetic fields of the big and small magnet are the same at the surface, the gradient of the smaller block is much larger and decays faster with depth. In this way, the two gradients match at a certain depth to generate a zero crossing of the total gradient (position of the sweet spot). The depth where the sweet spot is generated depends on the dimensions of the inner and outer magnets, and can be analytically obtained by combining eqn (4.7) with different signs and radii. The optimization of this geometry focuses on maximizing the sensitivity of the magnet given by B_0^2V*, where $V*$ is the volume excited for a defined bandwidth.[7] A discrete version of the barrel magnet is the NMR-MOLE (MObile Lateral Explorer).[34]

The U-shaped magnet geometry can also be modified to generate a sweet spot. To do this, a pair of small magnets arranged in an anti-parallel configuration has to be placed in the gap of the magnet with their polarization opposing the one of the main magnet. As for the barrel magnet, the size, gaps and position of all blocks need to be optimized to maximize the sensitivity. This geometry has been also optimized using the scalar potential method[25] (see Table 4.2).

4.2.4.2 Shimming the Magnetic Field to High Homogeneity

Single-sided sensors provide unlimited access to study samples of arbitrary size by working in the stray field of open magnets. For many years it was accepted that the magnetic field generated in a volume external to the

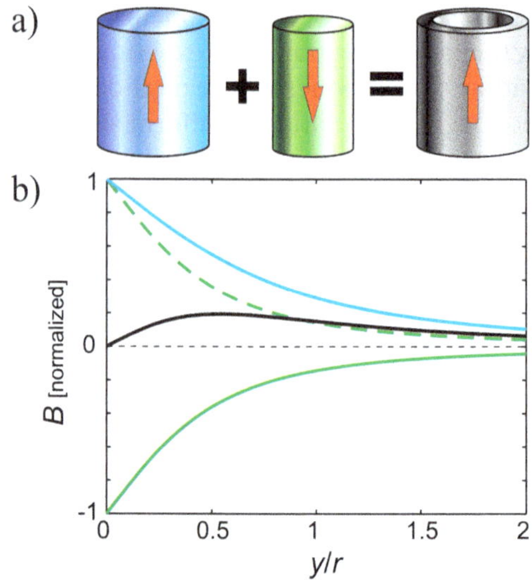

Figure 4.5 (a) A barrel (grey on the right) magnet can be thought as a combination of a larger cylinder with up magnetization (blue) and a smaller cylinder with down magnetization (green). (b) Using eqn (4.7), the resulting fields (for infinitely long cylinders) are calculated *versus* the central axis normalized by the radius of the larger cylinder. Blue line: flux of the larger cylinder. Green solid line: flux of the smaller cylinder. Green dashed line: negative flux of the smaller cylinder. Black line: resulting field of the barrel magnet with a smooth region at about $y/r = 0.5$.

magnet array must be inherently inhomogeneous. Although some magnet geometries were proposed to reduce the gradient of the magnetic field along the depth direction,[46] they were just considered as a way to increase the size of the excited volume to improve the SNR. Moreover, the standard approach to improve the homogeneity of closed magnet arrays based on the use of resistive coils[47] had to be discarded due to the excessive currents required to correct the inhomogeneities of stray fields. By accepting inhomogeneous magnetic fields as unavoidable, only relaxation times and diffusion coefficients could be used for sample characterization. The most important step towards generating a magnetic field with sufficient homogeneity to measure high resolution ^{1}H spectra with an open magnet was the realization that small magnets properly placed in the main magnet array can be used to tune the spatial dependence of the magnetic field B_0. This concept was used to shim the intrinsic inhomogeneities of a U-shaped magnet. Instead of adjusting the current in a coil to generate the required harmonic field correction, the positions of the shim magnets need to be adjusted mechanically. In this way, it was possible to homogenize the magnetic field of a single-sided sensor to the sub-ppm ($<10^{-6}$) regime, and the first proton spectra could be acquired in the stray field of a single-sided magnet.[21]

As shown in Section 4.2.2, several magnet geometries can be used to build a single-sided sensor. The spatial dependence of the magnetic field generated by any of these geometries can be modified by incorporating into the main array a set of small movable permanent magnets forming a "shim unit". Such a "shim unit" must be specially designed to correct the spatial dependence of the field generated by the particular main geometry and for this reason it will not necessarily work for different magnet geometries. The shim unit is numerically optimized to reproduce the inhomogeneity of the main magnet in such a way that once it is mounted in the magnet array the inhomogeneity of the shim unit cancels the one of the main unit. To correct for deviations observed in the space distribution of the magnetic field when moving from the simulation to the real magnet, the position of the magnet blocks forming the shim unit needs to be adjustable. Small displacements of the magnet blocks around their optimum positions allow then to experimentally reach the performance calculated numerically.

This shimming strategy has been demonstrated choosing a U-shaped magnet as the "main" unit.[20,38] Setting the coordinate system as shown in Figure 4.6a and assuming that the region of interest is located at a depth y_0, the magnetic field can be considered to be oriented along the z axis with its dominant gradient component along the depth direction. Due to the symmetry of the magnet along z, the spatial dependence of the field along this direction has only even terms in its Taylor expansion. As a function of position, the field has a minimum at the center of the magnet ($z = 0$) and increases towards the permanent magnets blocks. The field along x is an even function, but in this case the field decreases when moving away of the center. Taking into account terms up to second order, the field described above can formally be expressed by expanding the magnitude of the magnetic field $B_0(\mathbf{r})$ around $\mathbf{r}_0 = [0, y_0, 0]$ as:

$$B_0(\mathbf{r}) = B_{00} + G(y_0)(y - y_0) + \alpha_z(y_0)z^2 + \alpha_x(y_0)x^2,$$

$$\text{where } B_{00} = B_0(\mathbf{r}_0),\ G_y(y_0) = \left.\frac{\partial B_0(\mathbf{r}_0)}{\partial y}\right|_{\mathbf{r}_0} \text{ and } \alpha_k(y_0) = \left.\frac{\partial^2 B_0(\mathbf{r}_0)}{\partial k^2}\right|_{\mathbf{r}_0} \text{ with } k = x, z. \tag{4.8}$$

As the field generated by the shim unit must reproduce the spatial dependence of the main field having the smallest possible average field strength, one possibility is to build the shim unit as a miniaturized version of the geometry of the main magnet (in this case, a U-shaped geometry but with different aspect ratio and smaller size, *cf.* Figure 4.6a). By setting the polarization of the shim unit opposite to that of the main magnet, the inhomogeneities of the main field can be corrected while the total field strength is maintained at an acceptable magnitude.

Eqn (4.8) contains three coefficients, G_y, α_z, and α_x, that characterize the field inhomogeneities associated with the main unit. A similar equation holds for the shim unit. Assisted by numerical simulation tools the size and position of the blocks forming the shim unit can be varied until the

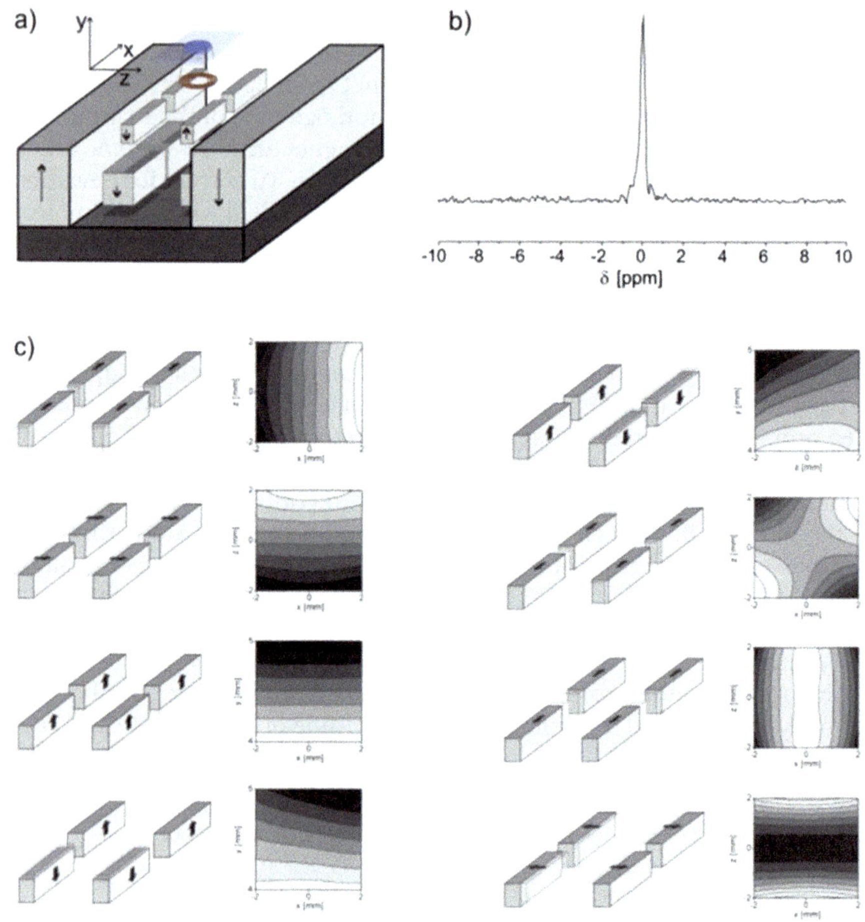

Figure 4.6 (a) U-shaped magnet shimmed using two pair of magnet blocks also mounted in U-shaped configuration but without the iron yoke. The arrows show the direction of polarization of the magnets. The magnet built in ref. 21 works at a depth $y_0 = 5$ mm, where a field of 0.25 T is generated. (b) Spectrum of a large water sample measured on a shimmed magnet. (c) The figure shows two sets of four magnets. The lower shim unit was included to cancel the strong gradient along the depth direction. The positions of these magnet blocks were fixed so that, in practice, this set could be considered to be part of the main unit with reduced gradient along the depth direction. The upper shim unit was placed directly underneath the sensor surface. These four blocks can be moved along the three Cartesian directions in order to generate the required shim components. The effect of moving the shim magnets in the directions indicated by the arrows is shown by means of 2D maps of the total magnetic field strength.

coefficients describing its magnetic field reproduces the main field in the largest volume possible. This step will lead to a shim unit that compensates for the field inhomogeneity of an ideal U-shaped magnet. However, imperfections in the magnet blocks used to build either the main magnet or the shim unit are expected to lead to variations in the values of G_y, α_z, and α_x. Besides needing control variables to adjust the terms y, x^2, and z^2, such imperfections unavoidably introduce inhomogeneities that are not considered in eqn (4.8). For instance, the symmetry arguments used to assume an even dependence of the field along the z-axis and x-axis will not hold and, for example, constant gradients along all three directions are expected to be present. In general, any imperfection in the magnet pieces introduces an asymmetry in the field that can be characterized (up to second order) by linear components x and z, and by cross terms xy, xz, and yz. The crucial step towards matching the gradient of both units at one depth in a real magnet requires a variable to control the field dependence of at least one of the units. Such control can be acquired by moving the blocks of the shim unit, as shown in Figure 4.6c.

As high homogeneity can only be achieved in a volume with limited size, a selective volume excitation method is needed. This was achieved by combining the natural lateral selection of the rf coil with a 90° soft rf pulse applied in the presence of a pulsed gradient along the depth. The sensor was also equipped with three single-sided shim coils that can produce linear corrections along the Cartesian directions. These coils were utilized in the final stage of the shimming procedure. Once the shim units were mounted and placed in their optimum positions (calculated), the final shim configuration was found iteratively by scanning the magnetic field from the resonance frequency of a 1 mm^3 water sample. Figure 4.7b shows the spectrum of a water/oil sample much larger than the sensitive volume (an arbitrarily large sample) placed on top of the sensor, whereby the line-width is 2.2 Hz, corresponding to a spectral resolution of about 0.25 ppm. This resolution allows the use of *ex situ* NMR to determine the molecular composition of liquids by analyzing the ^{1}H NMR spectra.

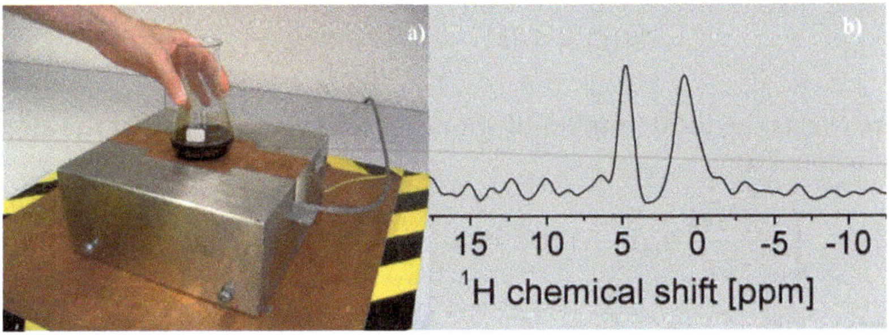

Figure 4.7 (a) Picture of the shimmed single-sided magnet measuring the spectrum of a large oil sample in a flask. (b) Spectrum of a mixture of oil and water showing that the resolution achieved at this stage is sufficient to separate this two main components.[21]

4.3 Signal-to-Noise Ratio for Single-Sided NMR

4.3.1 The NMR Signal in a Single-Sided Setup

Following the concept in,[48] the spatial dependence of both magnetic fields involved in an NMR experiment have to be known to calculate its signal; *i.e.* the static magnetic field, $\mathbf{B}_0$ (as discussed in the previous section), and the radio-frequency field, $\mathbf{B}_1$.

To excite the nuclear spins at a coordinate $\mathbf{r}$, only that component of the local $\mathbf{B}_1$-vector will be effective which is perpendicular to the local $\mathbf{B}_0$-vector, hence the effective excitation field is:

$$B_1^{\text{eff}}(\mathbf{r}) = \frac{|\mathbf{B}_0(\mathbf{r}) \times \mathbf{B}_1(\mathbf{r})|}{B_0(\mathbf{r})}. \tag{4.9}$$

This effective field not only defines the local excitation of the nuclear spins but due to the principle of reciprocity[49] also defines its contribution to the detected NMR signal. This is because this effective field excites a local magnetization $M(\mathbf{r})$ that oscillates with the local Larmor frequency $\omega_0(\mathbf{r})$ and induces a signal contribution $S(\mathbf{r})$ in the same coil given by:

$$S(\mathbf{r}) = -\frac{\partial}{\partial t}\left[\frac{B_1^{\text{eff}}(\mathbf{r})}{I_{\text{coil}}} M(\mathbf{r}) \exp\left(i\omega_0(\mathbf{r})t\right)\right], \tag{4.10}$$

where the rf field is given per unit current, I_{coil}. The total NMR signal, $S(t)$, yields from integration over the entire sample volume. Furthermore, we are only interested in the amplitude of the signal, hence the oscillating time-dependent term in eqn (4.10) can be omitted:

$$S(t) \propto \int_{\text{sample}} \frac{B_1^{\text{eff}}(\mathbf{r})}{I_{\text{coil}}} M(\mathbf{r})\, \omega_0(\mathbf{r})\, \mathrm{d}^3 r. \tag{4.11}$$

The magnetization, $M(\mathbf{r})$, in thermal equilibrium is given by:[50]

$$M(\mathbf{r}) = N(\mathbf{r}) \frac{\gamma^2 \hbar^2}{3 k_B T} I(I+1)\, B_0(\mathbf{r}), \tag{4.12}$$

where $N(\mathbf{r})$ is the local number of spins per unit volume, so that the signal is proportional to:

$$S(t) \propto \int_{\text{sample}} \frac{B_1^{\text{eff}}(\mathbf{r})}{I_{\text{coil}}} B_0^2(\mathbf{r})\, N(\mathbf{r})\, \mathrm{d}^3 r \propto \int_{\text{sample}} |\mathbf{B}_0(\mathbf{r}) \times \mathbf{B}_1(\mathbf{r})|\, B_0(\mathbf{r})\, \mathrm{d}^3 r \tag{4.13}$$

For the last proportionality, eqn (4.9) was substituted and a uniform spin-density assumed. If so, eqn (4.13) can be further simplified as:

$$S \propto B_0^2\, B_1\, V \approx B_0^2\, B_1\, \Delta y = B_0^2\, B_1\, \frac{\Delta\omega_{rf}}{\gamma\, G_0} \tag{4.14}$$

where the first simplification uses B_1 as the proportion of the rf field that is perpendicular to B_0 and V is the excited volume. If it is further assumed that there are no other field inhomogeneities other than that along depth (y), the volume can be replaced by the thickness of the excited slice, Δy. The latter is given by the ratio of the rf excitation bandwidth, $\Delta\omega_{\text{rf}}$, and the strength of the local B_0 gradient, $G_0 = \partial B_0/\partial y$. Eqn (4.14) is quite important to compare the sensitivity of different single-sided setups, keeping in mind that this has to be evaluated at a certain distance above the surface where the rf excitation frequency matches γB_0.

While eqn (4.14) estimates the NMR signal amplitude, the noise is usually estimated by the amplitude of the thermal fluctuations per unit frequency bandwidth, Δf, in an NMR experiment:[50]

$$\sigma = \sqrt{4k_B T_{\text{coil}}\,\Delta f\,R_{\text{coil}}}\,, \tag{4.15}$$

where T_{coil} is the temperature and R_{coil} is the resistance of the coil. If we assume that the detection bandwidth, Δf, is set proportionally to the excitation bandwidth, $\Delta\omega_{\text{rf}}$, the SNR is then given by:

$$\text{SNR} = \frac{S(t)}{\sigma} \propto \frac{B_0^2\,B_1}{G_0}\sqrt{\Delta\omega_{rf}}\,. \tag{4.16}$$

Typically, the bandwidth, $\Delta\omega_{\text{rf}}$, is inversely proportional to the duration of the rf excitation, *i.e.* pulse length, as long as it is narrower than the bandwidth of the rf circuit. The latter is quite often the limiting factor at the lower frequencies used here because it has to be compromised with the quality factor, Q, of the coil which is synonymous with B_1. In case the bandwidth is determined by the duration of the rf pulse, which together with the B_1 strength is also defining the tip angle, α:

$$\alpha = \gamma B_1 t_p \approx \left.\frac{\pi}{2}\right|_{\Delta y} \quad \text{or} \quad \Delta\omega_{\text{rf}} \propto \frac{1}{t_p} \propto B_1, \tag{4.17}$$

where it is assumed that this tip angle is roughly constant and optimized over the slice thickness, Δy.

If the excitation bandwidth is however limited by the bandwidth of the circuit:

$$\Delta\omega_{\text{rf}} = \omega_0/Q \propto B_0. \tag{4.18}$$

Both eqn (4.17) or eqn (4.18) can then give the equation for the sensitivity (eqn (4.16)) a different weight on the associated magnetic fields; however, in practice pulse length and bandwidth are mainly determined by the application. To optimize the SNR *via* the bandwidth of the used rf circuit again is a compromise between rf sensitivity (high Q) and bandwidth or excited volume (low Q). Naturally, this depends on many parameters and cannot easily be reduced to a rule of thumb. For a deeper analysis please refer to ref. 7.

Finally, eqn (4.16) is further complicated when applying experiments with multiple pulses.[48] Either way, eqn (4.16) states that the signal will be determined by the product of the magnetic fields, which decay rapidly with depth. Hence, choosing a slice further away from the surface will be penalized with significantly lower signals from it.

It has also been shown that the unavoidable field inhomogeneities in B_0 can be reduced significantly when B_1 can be matched to have roughly the same spatial dependencies over the sampled region.[51] However, the coherence loss in such gradients needs proper rephasing by a suitable pulse sequence, for which the linewidth of the spectrum must be much smaller than the excitation bandwidth. Hence, the magnetic field must be already modestly homogeneous, like in a shimmed "sweet spot" magnet. When this is provided, the application of this technique can reduce the linewidth by a factor of ~forty.[52]

4.3.2 Rf Coils for External Noise Cancellation

So far, only the thermal noise contributions to the NMR-signal were considered (see eqn (4.15)), which is the case for a coil perfectly shielded from the environment. Unlike normal NMR coils, which encase the sample, the rf coils used in single-sided NMR setup cannot be shielded *per se* or only if the sample allows it to be shielded, which makes them very susceptible to environmental noise. This problem can be reduced by the use of gradiometer coils, which cancel far field noise.[53–55] The most simple and conceptually easiest is a so-called "figure-eight" coil (see Figure 4.8a), which consists of two counter-wound coils,[15,56] loops or a semi-toroid.[57,58] Noise from a distance will induce essentially the same current in both loops but with opposite phase and hence cancel out.[15] When both loops are fed with the same current the resulting magnetic field will point parallel to the plane of the coil from one loop to the other and the excited spins will accordingly induce signals of opposite curl in the loops, which then will add up coherently in the detector. To fulfill this function, the gradiometer coils need at least two loops of the same area wound in opposite directions or more generally:

$$\sum_i n_i^+ A_i^+ = \sum_j n_j^- A_j^- \tag{4.19}$$

saying that the induced signal from all clockwise windings (with n_i^+ turns of area A_i^+) must be equal to that of all anti-clockwise coils (with index −).[59] Figure 4.8 shows different layouts of this principal idea.

Two or more such "figure-eight" coils can be combined to form arrays[28] (see *e.g.* Figure 4.8c), or one loop can be split up into smaller loops with the same total area. This is done in Figure 4.8d with the clockwise loop, which is split into four loops of half radius. If these smaller outer loops are bent around the inner coil while keeping their area constant, one might arrive at a layout as in Figure 4.8e. Here the current on the inner side of the red loops is parallel to the current direction in the most inner blue loop and the current directions cancel in radial connections to the outer side. Hence these radial parts can be omitted and one finally ends up with a concentric gradiometer

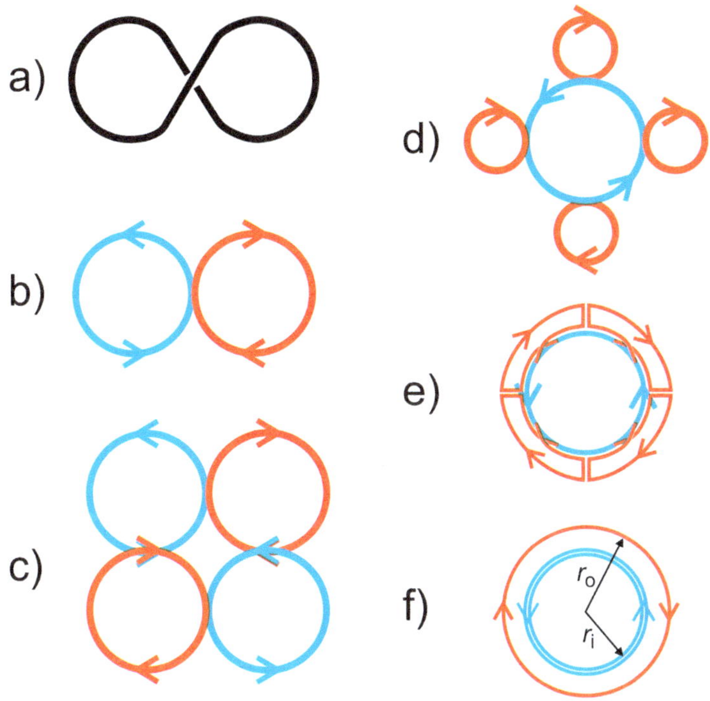

Figure 4.8 Different gradiometer coils: (a) concept of a "figure-eight" coil. (b) The same as (a) but displayed as a blue loop for counter-clockwise and a red loop for clockwise sense of current flow or winding. (c) Combination of two "figure-eight" coils.[26,28] (d) similar to (b) but with the red loop split into four smaller loops each with a fourth of the original area or half the radius. (e) Deformation of (d) to get to (f) the concentric gradiometer coil.[59] See text for details.

coil (Figure 4.8f), where the inner loop has twice the number of windings or twice the current of the counter-wound outer loop and the ratio of their radii is $r_o/r_i = \sqrt{2}$ (*cf.* eqn (4.19)).[59]

While the coils in Figure 4.8a mainly produce B_1 fields parallel to the surface and are therefore very suitable to be used with magnets that have the B_0 field perpendicular to the surface (see Section 4.2.2.2), the concentric gradiometer coil (Figure 4.8f) is much more sensitive when B_1 fields perpendicular to the surface are needed, *e.g.* with magnets with B_0 parallel to the surface (although in principle the "figure-eight" configuration, Figure 4.8a, can be used as well, when set with its short axis parallel to B_0).

4.4 Conclusions

Unlike in conventional NMR, where the sample is always prepared to fit the instrumentation, single-sided NMR is most successful where the hardware can be optimized to fit the application. This argument drives the development of single-sided hardware resulting in so many different designs which all have their pro and cons in view of fulfilling different tasks. This variety makes a detailed review of each prototype cumbersome; therefore, we have chosen the style of this manuscript as a "nuts and bolts" tutorial that summarizes concepts, methods and some other unpublished thoughts, which may be used as a guideline in the initial phase of new designs.

Acknowledgements

Both authors want to thank foremost their mentor and friend Bernhard Blümich, who pushed the field of single-sided NMR by numerous important ideas and never allowed the fire to weaken, which kept us all under steam. PB wants to thanks Mike Mallett for his good ideas and all the stimulating scientific and less scientific discussions. His actual supervisor, W. Heil, has to be acknowledged for agreeing to a research stay in Cambridge during which most of this manuscript was written.

References

1. R. J. S. Brown, R. Chandler, J. A. Jackson, R. L. Kleinberg, M. N. Miller, Z. Paltiel and M. G. Prammer, *Concepts Magn. Reson.*, 2001, **13**, 335–413.
2. D. E. Woessner, *Concepts Magn. Reson.*, 2001, **13**, 77–102.
3. B. Blümich, *Sci. Am.*, 2008, **299**, 68–73.
4. G. Eidmann, R. Savelsberg, P. Blümler and B. Blümich, *J. Magn. Reson., Ser. A*, 1996, **122**, 104–109.
5. B. Blümich, J. Perlo and F. Casanova, *Prog. Nucl. Magn. Reson. Spectrosc.*, 2008, **52**, 197–269.
6. B. Blümich, F. Casanova, M. Dabrowski, E. Danieli, L. Evertz, A. Haber, M. Van Landeghem, S. Haber-Pohlmeier, A. Olaru, J. Perlo and O. Sucre, *New J. Phys.*, 2011, **13**, 015003.
7. *Single-Sided NMR*, ed. F. Casanova, J. Perlo and B. Blümich, Springer, 2011.
8. Y. M. Pulyer, *IEEE Trans. Magn.*, 2002, **38**, 1553–1563.
9. W.-H. Chang, J.-H. Chen and L.-P. Hwang, *Magn. Reson. Imaging*, 2010, **28**, 129–138.
10. J. L. Paulsen, L. S. Bouchard, D. Graziani, B. Blümich and A. Pines, *Proc. Natl. Acad. Sci. U. S. A.*, 2008, **105**, 20601–20604.
11. A. E. Marble, I. V. Mastikhin, B. G. Colpitts and B. J. Balcom, *J. Magn. Reson.*, 2007, **186**, 100–104.
12. J. C. Garcia-Naranjo, I. V. Mastikhin, B. G. Colpitts and B. J. Balcom, *J. Magn. Reson.*, 2010, **207**, 337–344.
13. H. Kato, K. Kishi, N. Takahashi, J. I. Asaumi, Y. Honda, Y. Yanagi and M. Aoki, *Concepts Magn. Reson., Part B*, 2008, **33B**, U201–U202.
14. W.-H. Chang, C.-Y. Chung, J.-H. Chen, D. W. Hwang, C.-H. Hsu, C. Yao and L.-P. Hwang, *Magn. Reson. Imaging*, 2011, **29**, 869–876.

15. S. Anferova, V. Anferov, M. Adams, P. Blümler, N. Routley, K. Hailu, K. Kupferschläger, M. J. D. Mallett, G. Schroeder, S. Sharma and B. Blümich, *Concepts Magn. Reson.*, 2002, **15**, 15–25.
16. P. J. McDonald, P. S. Aptaker, J. Mitchell and M. Mulheron, *J. Magn. Reson.*, 2007, **185**, 1–11.
17. M. C. A. Brown, D. A. Verganelakis, M. J. D. Mallett, J. Mitchell and P. Blümler, *J. Magn. Reson.*, 2004, **169**, 308–312.
18. M. Ciarrocchi, V. Di Miccoli, M. Alecci, A. Sotgiu and A. Galante, *Meas. Sci. Technol.*, 2009, **20**, 017002.
19. J. Perlo, F. Casanova and B. Blümich, *J. Magn. Reson.*, 2005, **176**, 64–70.
20. J. Perlo, F. Casanova and B. Blümich, *J. Magn. Reson.*, 2004, **166**, 228–235.
21. J. Perlo, F. Casanova and B. Blümich, *Science*, 2007, **315**, 1110–1112.
22. E. Potenziani II, J. P. Clarke and H. A. Leupold, *J. Appl. Phys.*, 1987, **61**, 3466–3467.
23. Y. Zhang and D. Xie, *J. Electromagn. Waves Appl.*, 2006, **20**, 747–760.
24. W.-H. Chang, J.-H. Chen and L.-P. Hwang, *Magn. Reson. Imaging*, 2006, **24**, 1095–1102.
25. A. E. Marble, I. V. Mastikhin, B. G. Colpitts and B. J. Balcom, *J. Magn. Reson.*, 2005, **174**, 78–87.
26. J. Perlo, in *Single-Sided NMR*, ed. F. Casanova, J. Perlo and B. Blümich, Springer, Heidelberg, 2011, pp. 87–110.
27. B. Blümich, V. Anferov, S. Anferova, M. Klein and R. Fechete, *Macromol. Mater. Eng.*, 2003, **288**, 312–317.
28. B. Blümich, V. Anferov, S. Anferova, M. Klein, R. Fechete, M. Adams and F. Casanova, *Concepts Magn. Reson.*, 2002, **15**, 255–261.
29. C. Hugon, G. Aubert and D. Sakellariou, *J. Magn. Reson.*, 2010, **214**, 124–134.
30. M. Locatelli, *Rev. Inst. Fr. Pet.*, 1998, **53**, 477–481.
31. J. A. Jackson, L. J. Burnett and J. F. Harmon, *J. Magn. Reson.*, 1980, **41**, 411–421.
32. S. Rahmatallah, Y. Li, H. C. Seton, I. S. Mackenzie, J. S. Gregory and R. M. Aspden, *J. Magn. Reson.*, 2005, **173**, 23–28.
33. A. R. Rath, S. B. W. Roeder and E. Fukushima, *Rev. Sci. Instrum.*, 1985, **56**, 402–410.
34. B. Manz, A. Coy, R. Dykstra, C. D. Eccles, M. W. Hunter, B. J. Parkinson and P. T. Callaghan, *J. Magn. Reson.*, 2006, **183**, 25–31.
35. J. C. Goswami, A. Sezginer and B. Luong, *IEEE Trans. Antennas Propag.*, 2000, **48**, 1393–1402.
36. R. L. Kleinberg, A. Sezginer, D. D. Griffin and M. Fukuhara, *J. Magn. Reson.*, 1992, **97**, 466–485.
37. A. E. Marble, I. V. Mastikhin, B. G. Colpitts and B. J. Balcom, *J. Magn. Reson.*, 2006, **183**, 228–234.
38. H. Popella and G. Henneberger, *COMPEL-Int. J. Comput. Math. Electr. Electron. Eng.*, 2001, **20**, 269–278.
39. P. J. Prado, *Magn. Reson. Imaging*, 2003, **21**, 397–400.
40. P. J. McDonald and B. Newling, *Rep. Prog. Phys.*, 1998, **61**, 1441–1493.

41. P. J. Prado, B. Blümich and U. Schmitz, *J. Magn. Reson.*, 2000, **144**, 200–206.
42. F. Casanova and B. Blümich, *J. Magn. Reson.*, 2003, **163**, 38–45.
43. S. Meiboom and D. Gill, *Rev. Sci. Instrum.*, 1958, **29**, 688–691.
44. M. D. Hurlimann and D. D. Griffin, *J. Magn. Reson.*, 2000, **143**, 120–135.
45. E. Fukushima and J. A. Jackson, US Patent, 6,489,872, 1999.
46. E. Fukushima and J. A. Jackson, US Patent, 6,828,892 B1, 2004.
47. M. J. E. Golay, *Rev. Sci. Instrum.*, 1958, **29**, 313–315.
48. F. Casanova and J. Perlo, in *Single-Sided NMR*, ed. F. Casanova, J. Perlo and B. Blümich, Springer, Heidelberg, 2011.
49. D. I. Hoult and R. E. Richards, *J. Magn. Reson.*, 1976, **24**, 71–85.
50. P. T. Callaghan, *Principles of Nuclear Magnetic Resonance Microscopy*, Oxford Science Pub., 1994.
51. C. A. Meriles, D. Sakellariou, H. Heise, A. J. Moule and A. Pines, *Science*, 2001, **293**, 82–85.
52. J. Perlo, V. Demas, F. Casanova, C. A. Meriles, J. Reimer, A. Pines and B. Blümich, *Science*, 2005, **308**, 1279.
53. B. H. Suits, A. N. Garroway and J. B. Miller, *J. Magn. Reson.*, 1998, **131**, 154–158.
54. B. H. Suits, A. N. Garroway and J. B. Miller, *J. Magn. Reson.*, 1998, **135**, 373–379.
55. B. H. Suits and A. N. Garroway, *J. Appl. Phys.*, 2003, **94**, 4170–4178.
56. D. V. Trushkin, O. A. Shushakov and A. V. Legchenko, *Geophys. Prospect.*, 1994, **42**, 855–862.
57. E. Fukushima, S. B. W. Roeder, R. A. Assink and A. A. V. Gibson, US Patent, 4,590,427, May 20, 1986.
58. R. A. Assink, E. Fukushima, A. A. V. Gibson, A. R. Rath and S. B. W. Roeder, *J. Magn. Reson.*, 1986, **66**, 176–181.
59. E. Fukushima and P. Blümler, USPTO Utility Patent Application, 14/558,575, Dec 4, 2014.

CHAPTER 5

Hardware Developments: Halbach Magnet Arrays

PETER BLÜMLER*[a] AND FEDERICO CASANOVA*[b]

[a]University of Mainz, Institute of Physics, 55099 Mainz, Germany;
[b]Magritek GmbH, Pauwelsstr. 19, 52074 Aachen, Germany
*E-mail: bluemler@uni-mainz.de, federico@magritek.com

5.1 Concept and Theory

The term Halbach array (also known as "*magic rings*") refers to a certain arrangement of permanent magnets that increases the magnetic flux on one side while reducing or even cancelling it on the opposite side. Such two-dimensional magnetisation patterns were first proposed theoretically in 1973 by John C. Mallinson.[1] These patterns were referred to at the time as "magnetic curiosities", but later and independently realised by Klaus Halbach[2,3] who developed this idea further and constructed yokeless arrangements of permanent multipole magnets, nowadays known as Halbach arrays. As shown in Figure 5.1 they can have linear, circular or spherical geometry and generate dipolar or multipolar magnetic fields inside or outside the structure.

The conceptual and somewhat counter-intuitive idea of magnets that generate flux only on one side is nicely explained elsewhere.[4] Linear Halbach arrays (Figure 5.1a) are generated by a magnetisation pattern following:

$$\mathbf{M}(x) = \begin{pmatrix} M_x(x) \\ M_y(x) \end{pmatrix} = M_0 \begin{pmatrix} \sin kx \\ \cos kx \end{pmatrix}. \tag{5.1}$$

New Developments in NMR No. 5
Mobile NMR and MRI: Developments and Applications
Edited by Michael L. Johns, Einar O. Fridjonsson, Sarah J. Vogt, and Agnes Haber

Published by the Royal Society of Chemistry, www.rsc.org

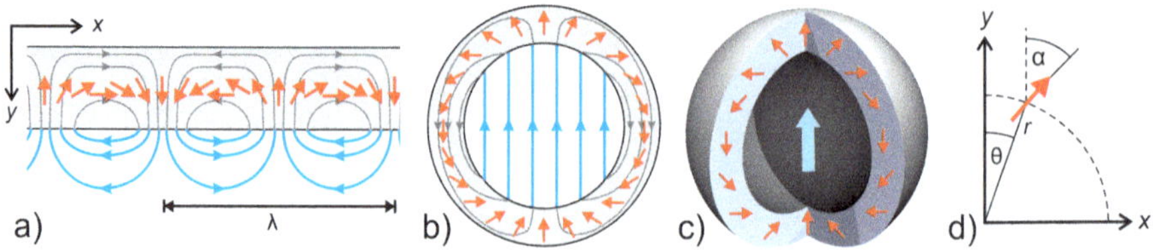

Figure 5.1 Schematic illustration of (a) linear, (b) cylindrical and (c) spherical Halbach arrays. The magnetisation direction changes continuously as indicated by the red arrows. In (a) and (b) the external flux lines are blue and internal flux lines are gray. They were omitted in (c). (d) Sketch of the coordinate system and the angles used for the construction of (b) and (c). The magnetisation (red arrow) is located on a circle of radius r and at an angle θ while pointing at an angle α.

This results in no flux on one side and enhanced flux on the other side, which oscillates with a wavelength $\lambda = 2\pi/k$. These arrays are best known as "refrigerator" and pin board magnets. A circular or cylindrical array (Figure 5.1b) is formed when one wavelength of a linear array is bent to a circle. Then the magnetisation obeys (see Figure 5.1d):

$$\mathbf{M}(R,\theta) = M_0 \begin{pmatrix} \sin\alpha \\ \cos\alpha \end{pmatrix} \quad \text{with } \alpha = (1+m)\theta \text{ and } m \in \mathbb{Z} \tag{5.2}$$

where m is the number of pole-pairs or modes, *e.g.* $m = \pm 1$ for a dipolar, $m = \pm 2$ for a quadrupolar field and so forth.[5] The sign of m directs the flux either inwards ($m > 0$) or outwards ($m < 0$). Halbach spheres are then made by azimuthally rotating this circular pattern, as shown in Figure 5.1c.

In order to generate the main polarizing and detection field for magnetic resonance experiments only dipolar fields are advantageous, which typically should be as strong and homogeneous as possible, hence $m = +1$ in eqn (5.2). Therefore, this publication will restrict the discussion to cylindrical and spherical Halbach arrays that generate homogeneous dipolar fields inside. It might be interesting to note that multipolar fields have many applications in engineering as magnetic bearings and motors[6,7] or in nuclear physics as in accelerators,[8] free electron lasers and neutron traps,[9] to name a few examples, but they will not be considered here.

The ideal dipolar Halbach magnet:

- Is defined as a magnet with its magnetisation changing continuously at twice the angle as the angle enclosed to a central axis (eqn (5.2) with $m = 1$), which will then be the direction of the resulting field.
- Generates the strongest homogeneous field possible per used magnet mass[10,11] ideally with perfect homogeneity, which makes them excellent magnets for mobile applications.[12] Using the latest very hard permanent magnetic materials magnetic flux densities exceeding 5 T can principally be obtained.[13]

- Has a circular and cylindrical arrangement (not spherical) and no external stray fields, allowing placement next to equipment sensitive to magnetic fields,[14] or nesting without strong torques.
- Is designed so that it does not require the use of yokes. Therefore, no saturation effects have to be taken into account during the magnet design and the magnetic field of the final magnet is in very good approximation the sum of the fields of its pieces,[15] which greatly simplifies the design and optimization.
- Has a generated magnetic field transverse to the cylinder axis, which allows the use of simple and efficient solenoidal rf coils for excitation and also allows NMR experiments where the direction of the magnetic field is rotated relative to the sample.[16,17]

Because magnetic resonance experiments have not been carried out with spherical Halbach arrangements we discuss their pros and cons at the very end of this review, which leaves us with Halbach systems of cylindrical or similar shapes for most of this review.

5.2 Practical Realization

5.2.1 Discretisation and Truncation of the Ideal Halbach

The ideal cylindrical Halbach inner dipole is sketched in Figure 5.2a. It is an infinitely long tube with a continuously changing magnetisation angle inside its wall. Obviously, such a continuously polarised material is hard to realise; therefore, various discrete versions of this principle were suggested to approximate it, some of which are displayed in Figure 5.2.

The most straightforward discretisation of the ideal Halbach is division of the ring into segments of identical shape but different magnetisation direction, which if individually made add up their production errors[20] (*cf.* Figure 5.2b and c). Therefore, another realization method is cutting a homogeneously magnetised hollow cylinder into identical segments and rearranging them (Figure 5.2c and d). However, machining magnetised material might be even more demanding than producing high quality magnets for assembly of the other schemes. A design from crescent shaped segments that can be made from only two sabre-shaped pieces has been suggested to tackle this problem.[21]

The design as wedges (Figure 5.2f) was originally proposed for MRI or other magnets that could benefit from polygonal cavities.[15] It is an optimal design in terms of used magnet material and maximal field achieved by multi-layered structures.[22] Moreover, it compensates for non-uniformities in the large chunks of magnetic material.[23] They are in many aspects comparable to segmented ideal Halbach cylinders, but much easier to assemble.[24] It was also suggested that they can easily be modified to introduce an off-centre sweet spot in the magnetic field, for example to improve MRI of body parts displaced from the geometric centre of the magnet.[25]

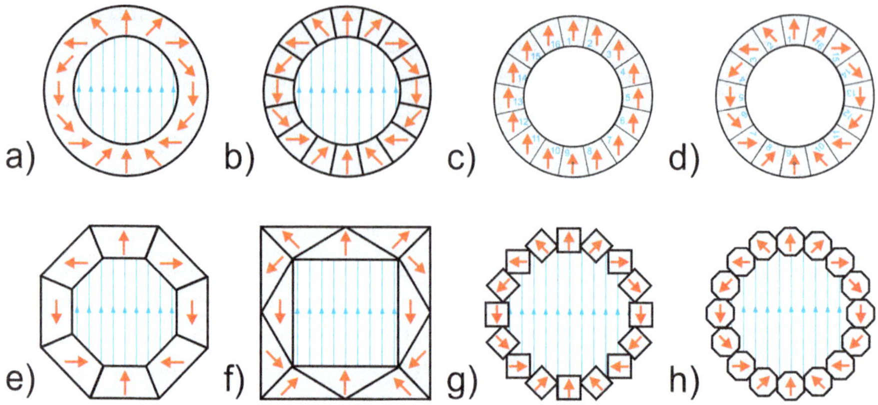

Figure 5.2 Different realizations of annular Halbach dipole: (a) ideal magnet, (b) discretised version of (a), (c) and (d) show how (b) could be made by cutting and rearranging pieces (indicated by a blue number) from a homogeneously magnetised cylinder (c) by swapping segments symmetric to the field axis (d);[15] (e) octagonal magnet from trapezoidal pieces,[7] (f) wedge design,[18] (g) NMR-Mandhala (Magnet Arrangements for Novel Discrete Halbach Layout) with 16 elements of quadratic cross-section,[5] (h) NMR-Mandhala with 16 elements of polygonal (octagonal) cross-section.[19] (Flux lines are omitted in (c) and (d)).

Another construction method uses magnets of identical shape and magnetisation direction,[26] *i.e.* NMR-Mandhalas (Magnet Arrangements for Novel Discrete Halbach Layout).[5] Originally, these arrays were proposed to consist of a circular arrangement of identical magnets with a squared cross-section (Figure 5.2g), which not only greatly simplifies the construction, but also reduces production costs and the influence of production errors. A generalization of this concept to magnets of polygonal (Figure 5.2h) or circular cross-section[27,28] was later included.[19] A similar conceptual design further bends a Mandhala of four magnets in 3D.[29]

The fact that the flux of an ideal Halbach is entirely enclosed in the structure lets such a magnet appear "non-magnetic" from the outside. While an infinitely long Halbach cylinder has no stray fields, any truncation in length will of course reintroduce external fields. However, they are small enough to still place Halbach magnets safely in the direct vicinity of other magnets[14] with manageable mutual forces and field distortions, which also makes them ideal candidates for pre-polarisation[12] or magnetic storage.[5] The additional flux on the outside comes along with a loss of flux inside the truncated structure. While the appearance of external field gradients might be acceptable, they are a nuisance or problem inside. To overcome this, essentially four strategies are proposed. The first seeks for an optimal length of truncation with respect to field strength and homogeneity.[30] The second simply stacks Halbach rings of optimised height at optimal distances.[19] While for two (infinitesimal thin) rings an analytical solution for the optimal distance of $\sqrt{2/3}$ of their radius can be found, the parameters need to be numerically optimised for many rings of finite length.[20] The big disadvantage of this method is that leaving spaces between individual rings might homogenise the field along the central axis, however it generates strong inhomogeneities towards the bore. They can to some extent be smoothed using iron sheets,[31] but this can be avoided if the tube is given a shape with a wall thickness increasing towards its ends.[32] This third method generates overall homogeneous but weaker fields. Finally, the fourth method allows for compensating this gradient as well as such in the plane of the rings by moveable parts, to be explained in Section 5.3.1.

The effect of truncation can also be greatly alleviated at one side by combining or closing the tubular structure of a Halbach cylinder by a Halbach hemisphere resulting in a magnet with the shape of a test tube.[33]

5.2.2 Achievable Flux Density

The magnetic flux density in an ideal Halbach dipole (see Figure 5.2a) made from a magnetic material with remanence B_R is given by:

$$B = B_R \ln\frac{r_o}{r_i} \tag{5.3}$$

where r_o is the outer and r_i the inner radius of the cylinder.

Therefore, the resulting flux in such a magnet can easily exceed the remanence of the magnet pieces and reach the limit given by the coercivity field strength[13] to resist de- or re-magnetisation by the high flux it generates. Therefore, flux densities up to 5 T were achieved using FeNdB alloys of very high-energy products and large wall thicknesses.[13]

NMR-Mandhalas (see Figure 5.2g and h) generate a comparably smaller field because the space between the outer radius, r_o, and the inner radius, r_i, is not completely filled with magnetic material. This simple geometric fact accounts for a flux reduction of about 50% when magnets with a square cross-section (see Figure 5.2g) are used and only *ca.* 25% flux reduction for Mandhalas built from magnets with polygonal (see Figure 5.2h) or circular cross-section. Therefore, when NMR-Mandhalas and Halbachs are compared in terms of produced flux density per mass of magnetic material they are almost identical.

5.2.3 Achievable Homogeneity

The ideal Halbach (see Figure 5.2a) has ideal homogeneity, but making real magnets (*cf.* Figure 5.2b) from discrete parts introduces imperfections and hence field inhomogeneities. If the homogeneity of such a discrete "ideal" magnet (Figure 5.2b) is then compared with NMR-Mandhalas, it turns out that NMR-Mandhalas made from polygonal magnets (*cf.* Figure 5.2h) have *ca.* 20–30% worse homogeneities, while those made from magnets with square sections (*cf.* Figure 5.2g) are even an order of magnitude worse. These values depend on the number of magnets used in the discretisation and almost exponentially on the radius of the volume over which the homogeneity is determined. Although all geometric improvements are typically tested in computer "experiments" using finite element methods (FEM), such comparisons are not very realistic since real magnetic materials show significant variations in their magnetic properties due to production and polarization. In order to get a more realistic picture, Figure 5.3a compares the homogeneity of the almost ideal discrete Halbach (*cf.* Figure 5.2b) made from 64 segments with varying remanence. It clearly shows that even sub-percent variations in

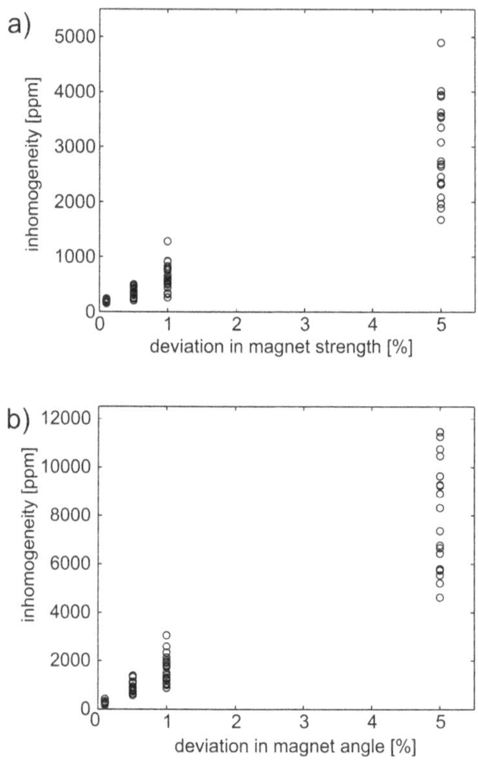

Figure 5.3 FEM simulation of an infinitely long Halbach cylinder made from 64 segments as shown in Figure 5.2b. (a) Influence of variations of the remanence ("strength") of the individual magnets on the overall homogeneity of the assembled discrete Halbach magnet. Shown is the resulting inhomogeneity in ppm of the average flux along a circle of half the inner radius. Each assembly is made of 64 segments with variations in the remanence normally distributed with a standard deviation σ shown on the abscissa in percent. 20 sets of such distributions were simulated and plotted individually to illustrate that the resulting field variations are not necessarily normally distributed. (b) The same scenario as in (a) but varying the magnetisation angle of the segments. The angular deviations were normally distributed with a standard deviation σ shown on the abscissa (scaled to a maximal deviation of $\pm 90°$). Again, 20 sets of such distributions were simulated and plotted as before.

the magnetic properties decrease the homogeneity by orders of magnitude. The used variations for Figure 5.3a are very realistic for rare-earth magnets. We measured standard variations in high quality permanent magnets, *e.g.* FeNdB (grade 48–53), ranging from 0.1% to 5% depending on supplier, price and batch.

The situation becomes even worse when it is considered that for constructing an almost perfect Halbach, segments with different magnetisation directions are needed (*cf.* Figure 5.2b). Variations in the magnetisation angle are therefore another source of inhomogeneities. Figure 5.3b shows a result similar to Figure 5.3a with about four times larger inhomogeneities. Again, these variations are absolutely realistic, *e.g.* an angular deviation of 1% or 0.9° corresponds to a misalignment of about 500 μm for a magnet of 30 mm size. Of course, the errors in strength and angular variation have to be combined in a realistic scenario.

Fortunately, there are ways to reduce such errors. For instance, the NMR-Mandhala design reduces the angular error, except for mechanical misalignments, because it is made from identical magnets that in principle should all have the same magnetisation direction or at least the same error. Furthermore, the use of identical magnets allows the magnets to be arranged in such a way that their inhomogeneities mutually cancel.[19] Another way to avoid larger angular errors is to magnetise a ring in a homogeneous magnetic field, then cut it into segments and reassemble, as suggested by Figure 5.2c and d. However, even if all these methods are carefully applied, it is hard to imagine building a Halbach system with homogeneities much better than 10 ppm over 10% of its inner diameter.

5.3 Advanced Designs

5.3.1 Shimmed Magnets

The crucial question that arises is how to build a magnet with homogeneities better than 10^{-5} from magnet material that has variations in its magnetisation on the order of 10^{-4}, or worse. A straightforward solution is to buy many building blocks, measure their magnetic properties very carefully, sort their values and use a subset with minimal variations to construct the magnet. However, the high symmetry of the Halbach arrangement allows the use of combinations of magnets in such a way that their deviation cancel in the resulting magnet.[19] Therefore, it might be advantageous to use the full range of variations but in carefully chosen pairs. Either way, during construction the mechanical impacts and imprecise positioning of the individual magnetic blocks will introduce errors that will remain unless their strength or position can still be varied in the finished magnet array.

For this reason, Halbach arrays with high homogeneities are realised in a two-stage process. The magnet geometry must first be optimized with the help of numerical simulation tools in order to achieve the highest geometrical homogeneity. Inhomogeneities due to the previously described discretisation and truncation of ideal Halbach cylinders must be minimised until the targeted homogeneity is achieved numerically. In a second step, the design must be revised to include variables to modify the magnetic field once the magnet has been assembled. This is necessary because the strength and angle of polarization as well as the geometry of the magnet blocks vary considerably from piece to piece due to production errors. On top of this, the limited mechanical precision to position the blocks in the final array also introduce important deviations from the results achieved in the numerical simulations.

Different strategies to improve ("shim") the field homogeneity either by carefully positioning additional smaller permanent magnets[34–36] or ferromagnetic pieces[31] inside the magnet array have been described recently. Once the main magnet structure has been built, the magnetic field is mapped in order to determine the spatial dependence of the required corrections. Afterwards, the homogeneity is improved either by inserting small magnetic blocks at defined positions or by moving them in an iterative process. The first attempt to shim a portable Halbach magnet was implemented by Moresi and Magin.[27] They built a Halbach array using SmCo cylindrical rods 15 cm long and 2 cm in diameter (Figure 5.4a). Once the magnets were mounted at the corresponding positions, the homogeneity of the system was improved by slightly adjusting the angular orientation of each rod. Using this approach, a homogeneity of 40 ppm was achieved in a volume of 3 mm diameter and 5 mm length together with a magnetic flux of 0.6 T. In 2007, Jachmann *et al.* proposed a shim method using so-called harmonic corrector rings.[34] They are made of concentric permanent magnet rods that can be considered magnetic dipoles whose overall and relative orientations can be used to adjust the field strength, orientation, and multipole order (Figure 5.4b). The authors demonstrated the approach by including such a corrector ring generating quadrupolar fields to correct the first derivative of the magnetic field. By applying this method, the NMR linewidth inside a cylindrical volume of 2 mm in diameter was reduced from 10 500 Hz (450 ppm) to 1200 Hz (56 ppm). Although this homogeneity was sufficient to resolve fluorinated compounds, additional corrector rings would be needed to resolve typical proton spectra.

A shim strategy that aims to correct the field inhomogeneities in a single shimming step was proposed by Hugon *et al.*[36] The method calculates from a measured field map the needed corrections for passive shimming, which has to be performed only once in a way very similar to shimming the cryo-cooled shims in superconducting magnets right after the magnet construction. The performance of this approach was demonstrated by shimming a magnet built based on another geometry made from two rings radially polarized in opposite directions (also known as "Aubert" design) and generating a magnetic field of 0.12 T. Each ring used 12 cubes arranged cylindrically and anti-symmetrically (Figure 5.4c). After carefully positioning the blocks, this magnet was expected to offer a homogeneous field up to order 4. However, due to the effect of position errors of the different

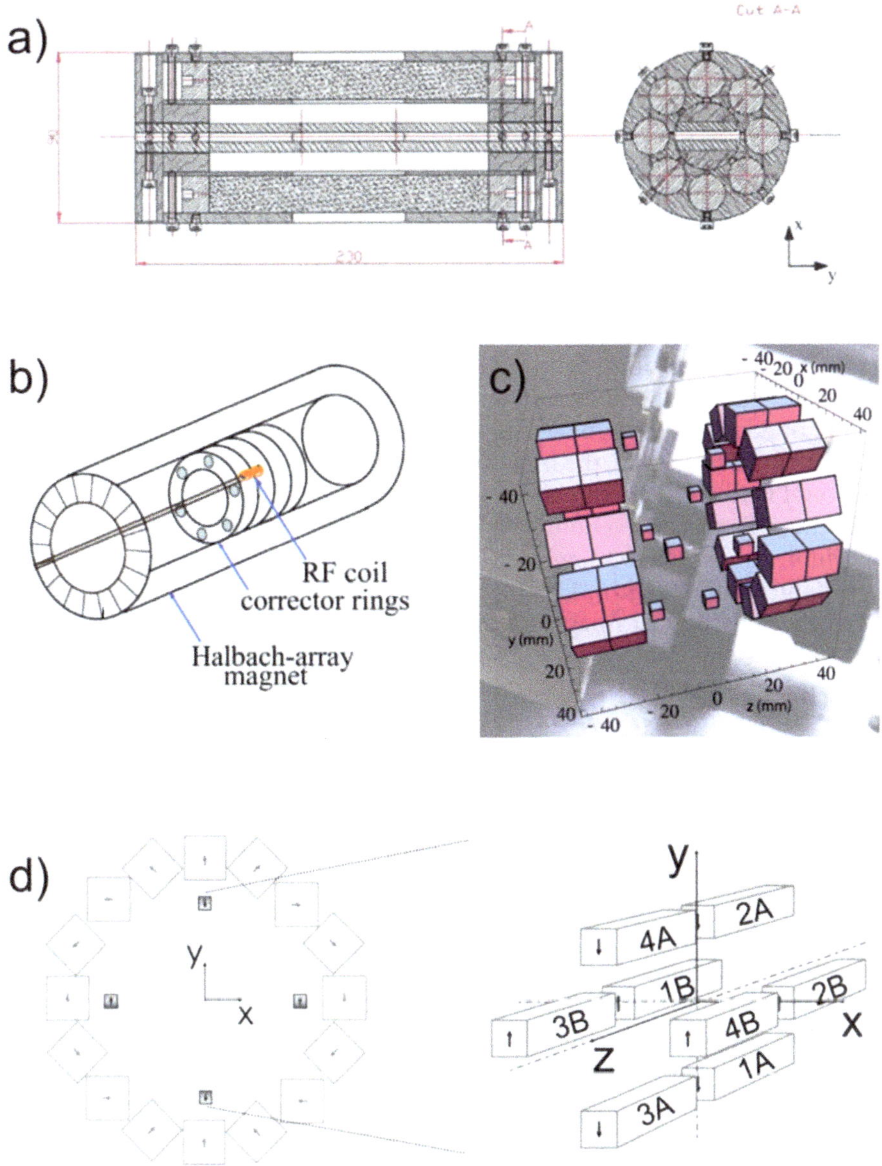

Figure 5.4 Different Halbach arrays including movable pieces or correcting blocks for shimming the total magnetic field in iterative ways. (a) Halbach ring built from cylindrical rods that can be rotated to correct the field inhomogeneities.[27] (b) Halbach magnet furnished with a corrector ring to reduce linear field components. Further rings can be added to correct for higher orders.[34] (c) Use of small blocks positioned at specific positions determined after mapping the magnetic field of the main magnet to shim a magnet built following the Aubert design.[36] (d) Shim unit designed to generate first and second order terms by moving eight small blocks positioned inside a Halbach magnet.[35]

magnetic parts, the measured homogeneity was worse than predicted. For example, an error of 0.05 mm on the gap between the rings was calculated to worsen the inhomogeneity by 15 ppm on the same diameter. Nevertheless, the NMR linewidth was improved from 40 ppm to 12 ppm for a sample of 3 mm^3.

A similar concept of mechanical shim units built from small movable permanent magnets was already demonstrated in 2007 by Perlo *et al.* on single-sided magnets.[37] Danieli *et al.* extended this idea to Halbach magnets[35] and they demonstrated that all the terms up to second order in the spherical harmonic expansion can be generated in a controlled way with a set of eight movable magnets placed inside a cylindrical Halbach array built from identical pieces (Figure 5.4d). The key of the method is to set the shim unit to reproduce the spatial dependence of the main field with the smallest average field strength possible. Then, by setting the polarization of the shim unit opposite to that of the main field, the inhomogeneities of the latter are corrected while the total field strength is maintained at an acceptable magnitude. By implementing this method, a homogeneity better than 10 ppm was achieved over a cylindrical volume of 4 cm in diameter and length (on a magnet with 30 cm diameter and length). This homogeneity allowed the authors to measure 3D images using fast imaging techniques without appreciable distortions. By reducing the size of the sample, sub-ppm resolution has been obtained in standard 10 mm NMR tubes, allowing the acquisition of ^1H chemical shift resolved spectra.

A disadvantage of previously described methods is that the extra magnets used for shimming reduce the effective bore of the main magnet. It must be noted that when the shim unit is placed outside the main magnet its efficiency is simply not sufficient to correct the typical inhomogeneities of the magnetic field. This issue was solved recently by constructing Halbach cylinders as a combination of magnetic blocks with different transverse geometries. In this way, the movable pieces used to correct the field inhomogeneities were incorporated in the main ring without sacrificing compactness or appreciably reducing the internal bore size. Figure 5.5a illustrates this magnet geometry. It consists of a stack of three cylindrical Halbach rings with different geometric proportions optimized to account for the field distortions along the cylinder axis due to the finite magnet length.[38] In order to correct for the inhomogeneities left after the assembly process, each ring is composed of fixed trapezoidal elements leaving parallel gaps between them. When rectangular magnets are filled in these gaps they can be radially moved to mechanically adjust the magnetic field in a highly efficient and accurate way (Figure 5.5b). By displacing the rectangular blocks in each ring with defined angular modulations and amplitudes, it is possible to independently generate the set of spherically harmonic corrections to the magnetic field up to an order $N/2$, where N is the number of rectangular blocks in the ring. As an example, Figure 5.5b shows the required modulation for the displacements of the rectangular magnets to generate tesseral corrections of the order and degree three (top) and four

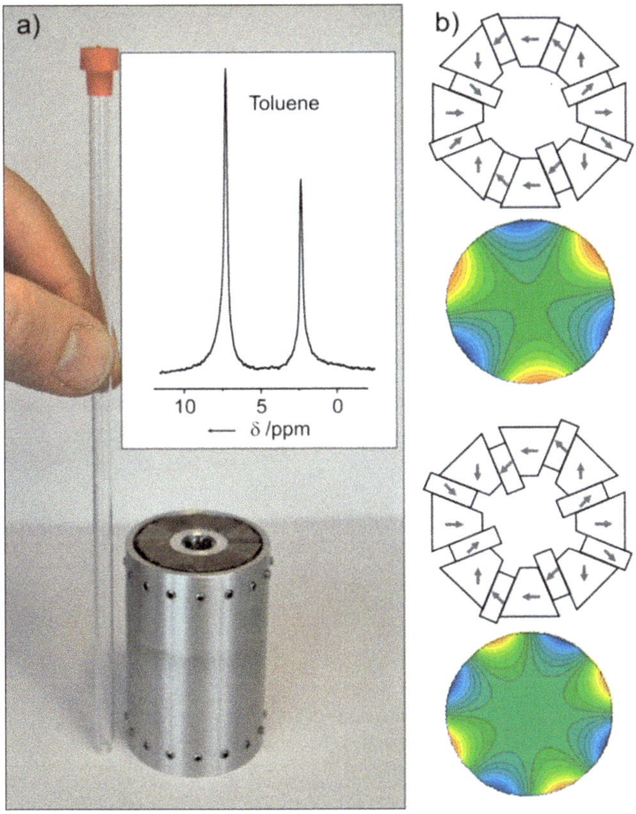

Figure 5.5 Halbach ring built from trapezoidal and rectangular blocks. While the trapezoidal magnets are fixed, the rectangular ones can be moved in- and out-wards to generate the different terms in the spherical harmonic expansion of the total magnetic field in a controlled way. (a) Picture of the magnet built to measure high-resolution spectra of samples in standard 5 mm tubes. (b) Modulation of the position of the rectangular magnets required to generate a third order (top) and a fourth order (bottom). Below each magnet arrangement is the resulting field map to illustrate the effect.[39]

(bottom) together with the corresponding two-dimensional field maps. The homogeneity attained with a first prototype of this magnet (generating a field of 0.7 Tesla) can be appreciated from the spectra of toluene acquired in a 5 mm NMR tube (inset Figure 5.5a). The resolution measured for a water sample was 4.5 Hz (0.15 ppm) at half height. This homogeneity was achieved using additional gradient coils for finer field adjustment of the first order terms only.[39]

Recently, a second magnet generation was built by scaling up the original prototype to generate a field of 1 Tesla. Besides increasing the number of shim magnets to allow for mechanical shimming up to a higher order, this magnet was furnished with shim coils generating up to third order terms.

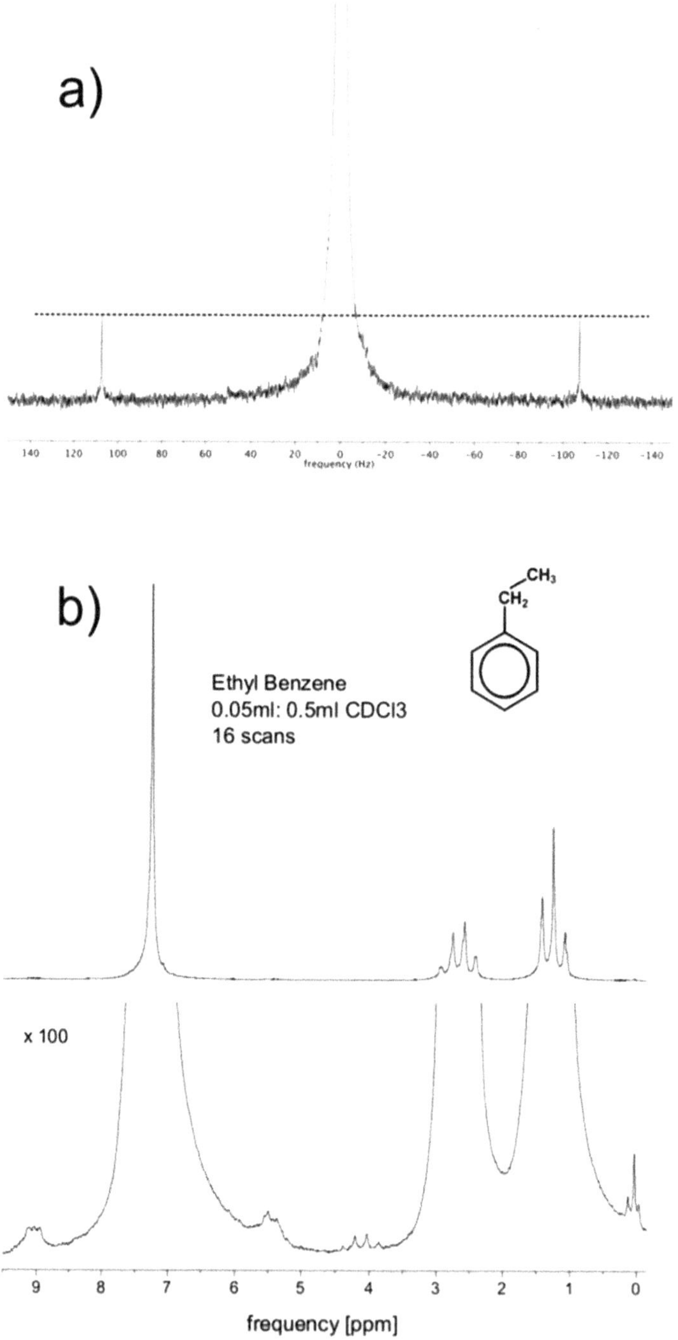

Figure 5.6 (a) Spectrum of 20% $CHCl_3$ dissolved in $CDCl_3$ measured in a single scan. The ^{13}C satellites can be used as a reference (horizontal line) to measure the resolution achieved at 0.055% height of the main line. (b) Spectrum of 10% ethylbenzene dissolved in $CDCl_3$. 16 scans were accumulated. A zoom of the spectrum shows the ^{13}C satellites clearly resolved.

The spectral resolution was improved to achieve 0.17 Hz at 50% and 17 Hz at 0.055% of the signal height. Figure 5.6a shows a zoom of the line measured in a single scan for 20% $CHCl_3$ dissolved in $CDCl_3$. The high sensitivity of the magnet even allows the observation of the ^{13}C satellites with a sensitivity that can be used as a reference to define the 0.055% amplitude. Figure 5.6b shows a spectrum of 10% ethylbenzene in $CDCl_3$. A zoom of this spectrum is shown at the bottom, where the ^{13}C satellites can be clearly resolved with a very high signal-to-noise ratio.

5.3.2 Nested Rings

As an infinitely long Halbach cylinder has no outer (stray) field, it does not appear as a magnet on the outside while the flux inside can be described as a single vector. Consequently, such circular arrays can be concentrically nested[40] one inside the other with the possibility of performing mechanical rotations with only small torque relative to each other by angular displacement θ.[41,42] As a result, the total magnetic field at its centre varies as the vector sum of this angle (see Figure 5.7). If two circular arrays are designed in such a way that their magnetic fluxes have the same magnitude B (in Figure 5.7: $B_1 = B_2 = B$ for $r_2 = \sqrt{r_1 r_3}$) it is in principle possible to vary the total magnetic flux from zero, when the vectors are antiparallel ($\theta = 180°$, Figure 5.7d), up to $2B$ for parallel flux vectors ($\theta = 0°$, Figure 5.7c). In reality, however, imperfections in construction, dimensional tolerances and homogeneity of the permanent magnets limit these ideal values.

The possibility to mechanically adjust or even sweep the magnetic field makes such nested arrays very attractive for magnetic resonance experiments at fixed Larmor frequencies. For example, Electron Paramagnetic Resonance (EPR)[43] or Dynamic Nuclear Polarization (DNP)[44,45] experiments where the microwave frequency typically cannot be varied easily. Additionally, such magnets might be interesting for relaxation measurements at variable magnetic fields up to about 1 T.[46] They can easily replace resistive electromagnets by a desktop size device that neither needs strong power-supplies nor any coolants (see Figure 5.8).

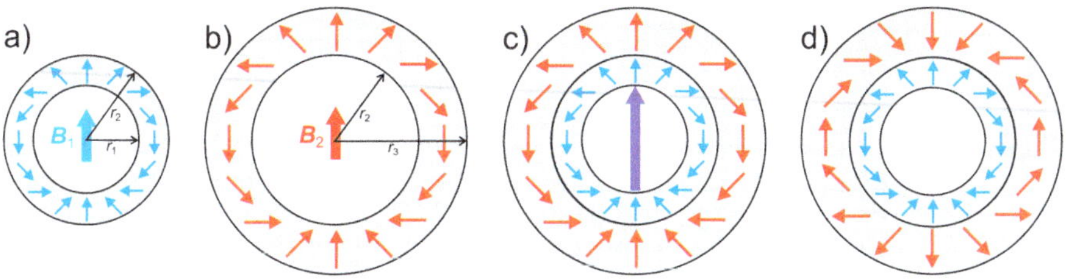

Figure 5.7 Concept of two nested Halbach dipoles. (a) Inner ring producing field B_1 and (b) outer ring producing field B_2. (c) Nested arrangement of the upper two rings with no angular displacement ($\theta = 0$) resulting in a field $B = B_1 + B_2$. (d) As (c) but with maximal angular displacement ($\theta = 180°$) resulting in a field $B = B_1 - B_2$, here shown as zero.

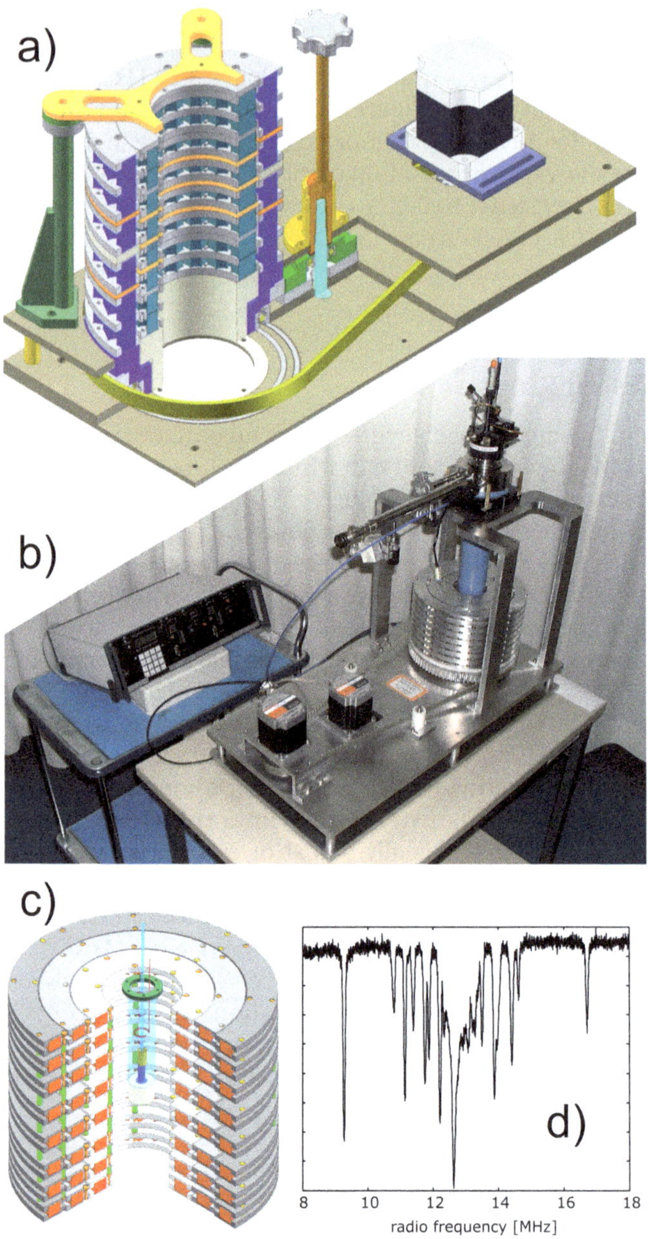

Figure 5.8 (a) Device made from two nested rings to vary a magnetic field continuously from about 0.1 to 0.45 T by mechanical rotation of the outer ring (outer ring: magnets in darker blue, inner ring: magnets in lighter blue). A stepper motor allows automatic field adjustments and sweeps by moving the outer ring *via* a belt. This particular device was designed for DNP.[45] (b) EPR desktop spectrometer with a magnet made from three concentric Mandhalas (as shown in (c)). The outermost cylinder was used for continuous field sweeps to acquire broad band EPR and Electron Nuclear DOuble Resonance (ENDOR) spectra like (d) Mims ENDOR spectrum taken at the maximum of the $m_I(^{63,65}Cu) = 3/2$ line in a glycine single crystal doped with 1% copper nitrate.[43]

5.3.3 Force-Free Opening Magnets

Whoever has assembled a larger structure from permanent magnets will never forget the tremendous and often dangerous attracting and repelling forces resulting from modern permanent magnet alloys. For short distances, they can be approximated by Maxwell's pulling formula:

$$F \approx \frac{AB_R^2}{2\mu_0} \approx 398 \frac{kN}{T^2 m^2} AB_R^2. \tag{5.4}$$

This means that, for instance, two small permanent magnets made of FeNdB, with a remanence of $B_R = 1.3$ T and side length of 1 cm (area $A = 10^{-4}$ m^2 and μ_0 is the vacuum permeability) exert a force, F, of more than 60 N on each other. When the side length is increased to 3 cm, the force even exceeds 600 N.

In spite of the strong attractive and repulsive forces between the magnetic pieces that comprise the array, it might be advantageous for some applications to open this magnetic structure. Halbach cylinders could be equipped with a hinge to clamp them around an object. To open magnet arrays with manageable forces would be especially useful for Halbach spheres to easily access their inner volume.

Considering the field strengths and sizes that are required for NMR the construction of such an openable magnet would thus, at first glance, appear to be an impossible proposition due to the forces that need to be overcome. However, the forces between magnets in discrete Halbachs show an angular dependence, which make it possible to choose an orientation for which the magnets show no or minimal forces along their connecting line.

5.3.3.1 Halbach Cylinders: NMR-CUFF

It was shown (Windt *et al.*, 2011) that an infinitesimally thin and infinitely long ideal Halbach cylinder can be split at the magic angle normal to the inner flux direction, that is 35.3° off its poles, (see Figure 5.9a) without any attractive or repulsive forces.[31] However, forces in the third dimension are maximal at this position, causing a very strong shear, which can readily be compensated by using strong hinges.

Figures 5.9b and c show a prototype (NMR-CUFF acronym for "CUt open Force Free"[31]) of such a magnet, which from its principle layout is a NMR-Mandhala with four magnets of squared cross-section. Hence, it cannot be opened exactly at the theoretical 35.3° but rather at 45°, which is however close enough to have a repulsive force of only about 20 N. It is specially made to investigate the stem or branches of plants at a field of 0.57 T with a homogeneity of about 50 ppm over a spherical volume of 5 mm in diameter. Since it was equipped with a gradient set, images of the stem and velocity measurements of its water transport and uptake could be achieved with it.[47]

While this prototype is still rather small and the forces scale with the square of its dimension (see eqn 5.4), it might be advantageous to seek arrangements with angles between magnets closer to 35.3°, as is the case

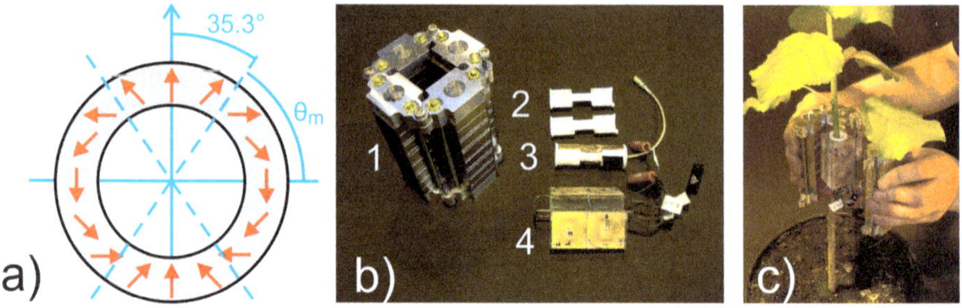

Figure 5.9 Principle and prototype of force-free opening Halbach cylinders. (a) Schematic representation of the planes (dashed blue lines) where an ideal Halbach cylinder can be opened without resulting magnetic forces; θ_m indicates the magic angle. (b) Photographs of a prototype (1) with four hinges and openings at 45°. It has a quadratic inner opening of 30 mm, produces a field of 0.57 T and weighs about 3 kg. The system is assembled by first clipping two PTFE templates (2) around the sample (here a plant stem), on which the solenoidal rf coils are wound (3). This is then enclosed by an open plane parallel gradient system (3). (c) Finally, the magnet is clasped around it. For more details see ref. 31.

for $n = 12$[48] or $n = 16$. Such hinged Halbach cylinders are ideally suited to investigate samples that as a whole are too large to fit into normal magnets, but have a smaller or slender section of interest that does, *e.g.* non-metallic tubes in experimental or industrial setups, extremities and necks in animals or humans, or stems and branches in plants.

5.3.3.2 *Halbach Spheres*

The Halbach sphere[49,50] (also known as "magic sphere", see Figure 5.1c) produces possibly the highest homogeneous flux per used mass of permanent magnet material. It is a third stronger than that of a Halbach cylinder of the same cross-section (*cf.* eqn (5.3))

$$B_{\text{sphere}} = \frac{4}{3} B_{\text{R}} \ln\frac{r_{\text{o}}}{r_{\text{i}}} = \frac{4}{3} B_{\text{cylinder}} \tag{5.5}$$

However, it differs from that of a cylinder in such a way that the flux is not zero outside. Halbach spheres are left with a residual magnetic moment, m, of magnitude:

$$m = \frac{VB_{\text{R}}}{3\mu_0} = \frac{4\pi B_{\text{R}}}{9\mu_0}\left(r_{\text{o}}^3 - r_{\text{i}}^3\right). \tag{5.6}$$

The stray field has roughly the form of that of a dipole. If this stray field is causing problems in the desired application, it can be cancelled by adding a thin uniformly magnetised shell on the outer surface.[15] This shell needs to be mounted with its magnetisation direction opposing that of the sphere. Its thickness must be optimised to just cancel the inner Halbach sphere and not alter the field inside it, since the cavities of homogeneously magnetised spheres are field free. Simulations[51] suggest that the Halbach sphere also has superior homogeneity when compared with Halbach cylinders, especially when considering truncated versions. Admittedly, the construction of such spheres appears to be extremely difficult, and only a few results are published. Bloch *et al.* report[52] an impressive flux density of 4.5 T inside a Halbach sphere over an inner volume of 20 mm^3, but we are not aware of anyone having ever constructed such a sphere for magnetic resonance purposes. This is probably due to the problem that once it is finished all access to the inner sample volume is blocked by tightly fixed magnets. Some concepts for polar tunnels and equatorial slots were suggested,[50,53] but this may distort the field homogeneity or only give access to a small fraction of the inner volume.

This problem can be solved by the same approach as shown for the Halbach cylinders in the previous section by looking for angles at which a Halbach sphere can be split open with no or minimal forces. Because we couldn't yet bring ourselves to construct such a device, the following ideas were published in a patent[54] only. In this particular case, the force calculations no longer give a single result but rather continuous solutions depending on two angles, as illustrated in Figure 5.10. Such hinged Halbach spheres could have applications in experiments, where the angle of the magnetic field must be varied in three dimensions, as for instance in magic angle turning[55] but with static samples and a rotating magnet.[17,56]

5.4 Applications

After the description of the hardware, we would like to highlight some magnetic resonance applications of magnets constructed by the above reviewed principles. Many studies have emerged in the last few years, probably due to better materials, designs and a general interest in mobile NMR.

5.4.1 Medical MRI of Living Organisms

Originally, the first envisioned application of Halbach magnets in magnetic resonance was whole body MRI of humans. They were compared with other designs and were typically found to be superior in performance but much harder to make.[24,57,58] Due to this and the extremely successful use of superconducting magnets in the field of clinical MRI not many were built to our knowledge. However, increasing prices for helium and generally lower maintenance costs might cause a renaissance, particularly in emerging nations.

Nevertheless, MRI systems for imaging extremities were successfully equipped with Halbach magnets[59] and similar systems were suggested to measure the bone composition in hands,[60] wrists[61] and heels[62] for diagnosing musculoskeletal diseases, such as rheumatic arthritis or osteoporosis. Much smaller systems were designed to perform whole body MRI of small

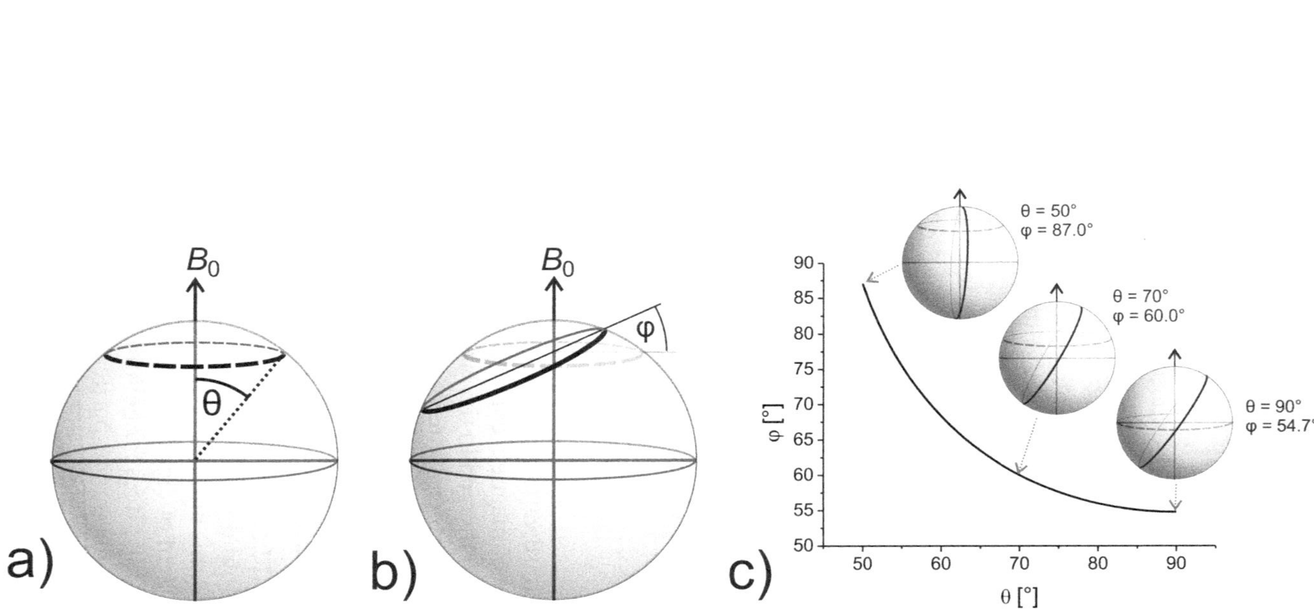

Figure 5.10 Solutions for force-free opening of Halbach spheres.[54] (a) and (b) illustrate schematically the involved angles, θ and φ, which are necessary to construct the solutions depicted in (c). First a cross-section at angle θ off the field direction (*cf.* Figure 5.1c) of the sphere is marked (a), which then must be additionally tilted by an angle φ (b) to give the cross-section (strong line) where the sphere can be opened without force. (c) Solutions to the problem as a function of both angles. Additionally three examples are depicted.

animals (mainly rodents) for pharmaceutical and biomedical research. Several companies are marketing such devices very successfully. An example of a portable MRI system using a Halbach design generating a field of 0.5 T with a homogeneity better than 1 ppm over a cylindrical volume of 40 mm in diameter is shown in Figure 5.11a. Equipped with a 3D active shielded gradient coil system and an optimized rf coil the system measures 3D images with high quality (see Figure 5.11b).

The easy access of the wide bore and the high sensitivity of the magnet make this desktop MRI system an ideal device for investigating small animals. Figure 5.11c shows 2D cross sections of a mouse taken from a full

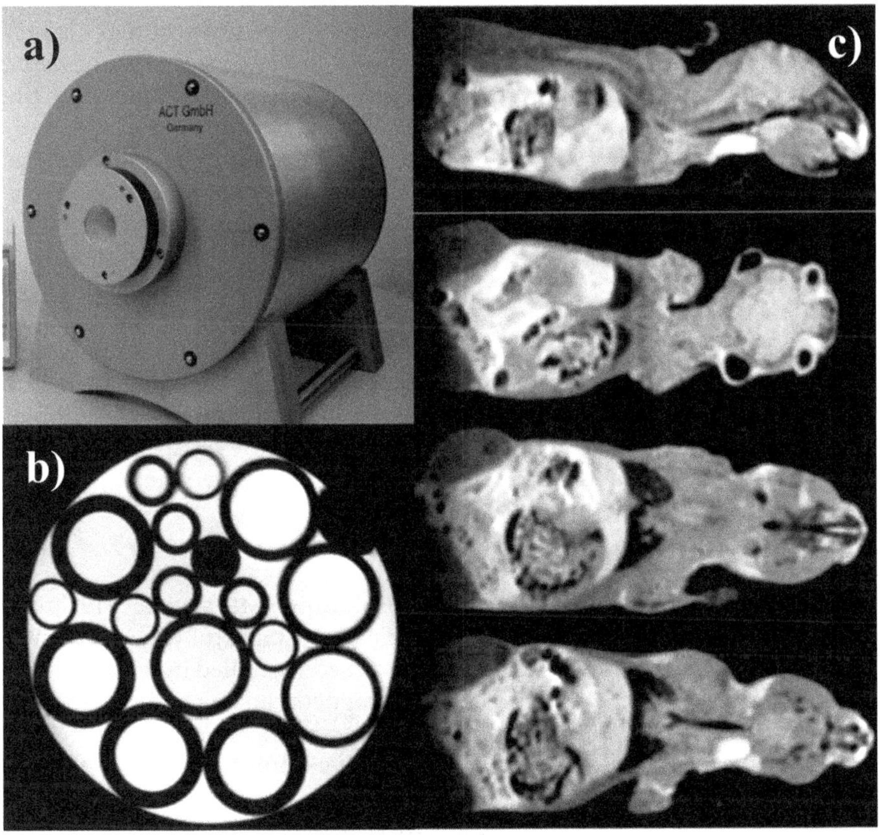

Figure 5.11 Desktop MRI system based on a Halbach magnet with 10 cm bore generating a magnetic field of 0.5 T. It is equipped with active shielded gradient coils and a solenoidal rf coil. The diameter of the samples fitting in the tomograph can be up to 4 cm. In this volume, the field homogeneity is better than 1 ppm. (b) Image of a phantom made of glass tubes immersed in water. No post-processing corrections were applied to the image, showing the uniformity of the gradients generated by the gradient coils in the mentioned volume. (c) Cross-section images extracted from a 3D image of a mouse acquired using a spin-echo sequence with an isotropic resolution of 390 μm.

3D image measured with spin echo sequence where the spatial resolution was set to be 390 μm along each direction. Another interesting application might be the investigation of plants or their fruit concerning their function and development following different genetic designs and/or cultivation.[63,64] Openable Halbach systems like the NMR-CUFF[31] offer unique insight to water transport and uptake and can in principle be scaled up to investigate even small trees.

5.4.2 Porous Media: Rocks, Soil, Construction Materials, Food

The investigation of the pore-size distributions and porosity of rocks is very important for geophysics and especially for the survey of reservoirs of water, gas or oil. The industrial importance of this field can be estimated by the numerous publications on single sided NMR used for *in situ* well logging by the oil companies, which will be shortly described in another chapter of this book, and for which Halbach magnets offer no particular advantage. However, the inspection of drill-cores on site is another application that would profit a lot from mobile NMR equipment. Anferova *et al.*[65] used a Mandhala to measure porosity and pore size distributions on water-saturated drill cores with 1D[65] and 2D $T_1 - T_2$ correlation experiments.[66] Similar experiments were also later used to characterise pores and water dynamics in natural soil,[67] cement pastes[68] and wood chips for the paper industry.[69] Using the same experimental approach also allows the determination of oil or water droplet size in food emulsions, the dynamics and spatial features of the rehydration and swelling of dried food[70] and the thawing of frozen vegetables.[71]

5.4.3 Chemistry and Process Control

With the possibility of measuring high-resolution NMR spectra with magnets built from permanent magnetic materials, compact, robust and maintenance-free NMR spectrometers can now be directly installed in the chemistry lab. As the magnetic stray field of Halbach magnets can be reduced to a minimum, they are ideal tools to be integrated in complex pre-existing technical facilities, where even multiple sensors can be incorporated without the need for much additional lab space. Moreover, as they do not require cryogenic cooling, safety issues associated with cryogenic liquids are eliminated.

Setting up the sensor directly at the place where the reaction proceeds would allow proper process monitoring and control of most of the relevant parameters governing the dynamical process, such as concentration, temperature, pressure, viscosity, mixing, impurities, *etc.* Without this information, the kinetics become difficult if not impossible to understand. Furthermore, for chemical reactions, the particular technology has to be chosen including the type of initialization, solvent, and catalyst to follow the optimum synthetic route. Except for very few simple or well-studied reactions, neither exact kinetic nor thermodynamic data are usually available.

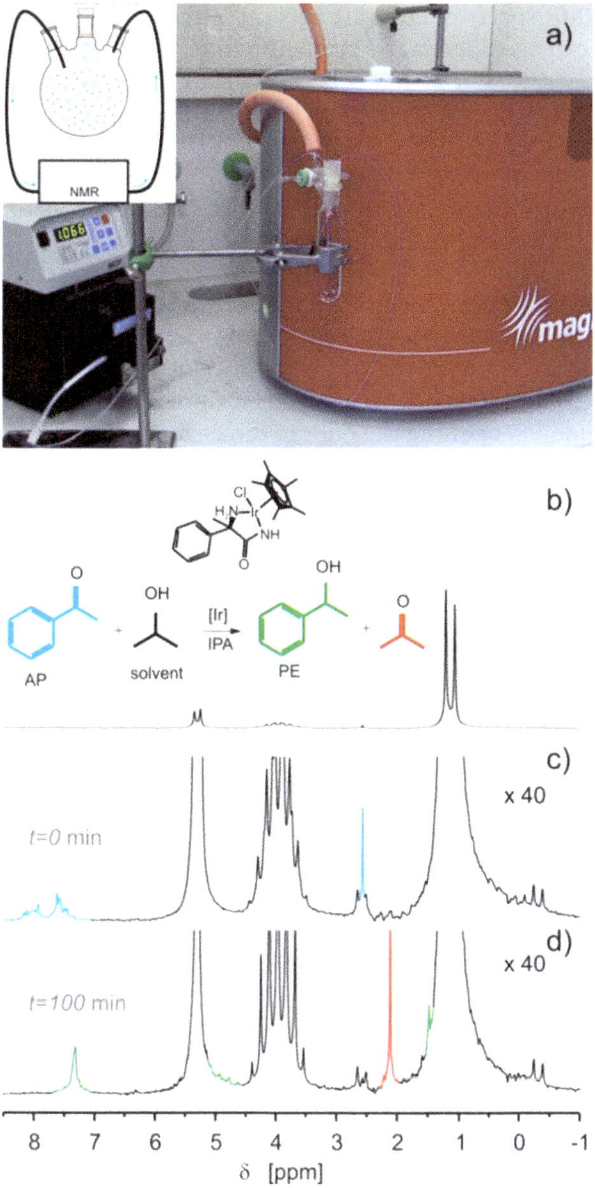

Figure 5.12 (a) Desktop high-resolution spectroscopy system installed in a chemistry fume hood. A peristaltic pump is used to continuously circulate the reaction mixture from the batch reactor to the magnet and back. (b) This setup was used to monitor the hydrogenation reaction that reduces acetophenone in phenylethanol using an iridium catalyst. The educt and product are dissolved in isopropanol at a concentration close to 1%, so at first sight a spectrum of the mixture shows only the characteristic lines of isopropanol. (c) Zoom of the spectrum measured at the beginning of the reaction, the signals of acetophenone are shown in blue. (d) Spectrum measured at the end of the reaction. It can be clearly observed how the signals of acetophenone disappeared and the signals of phenylethanol (green) and acetone (red) (byproduct) are resolved.[72]

Desktop NMR spectrometers are very attractive research tools for chemistry labs and can even be placed inside the fume hood to monitor hazardous chemical processes. For *in situ* process control in industrial environments, these new devices can be used to collect data regarding chemical composition or mixing efficiency where either optimization at the pilot-plant level or feedback control of full-scale reactors and mixers is needed. Figure 5.12 shows the setup mounted inside a chemistry fume hood to monitor the reduction of acetophenone in phenylethanol.

In the last few years, the first NMR spectra measured in real time inside a fume hood of a chemistry lab using a mobile NMR setup have been reported. Aided by a peristaltic pump, the reacting mixture is circulated through the magnet and returned to the reactor forming a closed loop. These experiments demonstrate that NMR spectroscopy can be made available to chemists for *in situ* measurements, opening the door to the design of setups where the NMR probes are integrated at specific points of interest of the process line.

5.5 Conclusions

In the last decade, the potential of Halbach arrays has been exploited for mobile and compact NMR. While the magnet development in mobile NMR was initially mainly driven by single-sided approaches, the idea to use enclosing structures like Halbach cylinders looked somehow like a step back. However, compact NMR had more to offer than investigating surfaces. It was shown that enclosing structures could be opened and nested to sweep fields. With the latest developments in shimming, high quality magnets are produced that can compete with superconducting magnet technology in terms of homogeneity and stability.[73] In the same time, the energy product, remanence and overall quality of the available permanent magnetic material have continuously increased,[7] so that flux densities of 1 T can be produced by handheld devices. Taken together, these developments allow the creation of permanent magnets that are comparable to the first generations of superconducting magnets, but fit in the fume hood, on the desk or even inside the palm.

Acknowledgements

Peter Blümler wants to thanks his former employers, H. W. Spiess (MPI for Polymer Research, Mainz) and U. Schurr (Forschungszentrum Jülich) for their continuous interest and support in conducting some of the presented work and W. Heil for agreeing to a research stay in Cambridge during which most of this paper was written. Furthermore, he wants to thank. Mike Mallett, Hanspeter Raich, Christian Bauer, Dagmar van Dusschoten, Helmut Soltner and Carel Windt for their stimulating ideas, help and friendship. Federico Casanova is thankful to Juan Perlo and Ernesto Danieli for their contributions, dedicated work and friendship. Finally, both authors are indebted to and thank Bernhard Blümich for his mentor- and friend-ship, and promotion of the entire field of compact NMR.

References

1. J. C. Mallinson, *IEEE Trans. Magn.*, 1973, **9**, 678–682.
2. K. Halbach, *IEEE Trans. Nucl. Sci.*, 1979, **NS-26**, 3882–3884.
3. K. Halbach, *Nucl. Instrum. Methods*, 1980, **169**, 1–10.
4. H. A. Shute and J. C. Mallinson, *IEEE Trans. Magn.*, 2000, **36**, 440–451.
5. H. Raich and P. Blümler, *Concepts Magn. Reson., Part B*, 2004, **23B**, 16–25.
6. Z. Q. Zhu and D. Howe, *IEEE Proceedings-Electric Power Applications*, 2001, **148**, 299–308.
7. J. M. D. Coey, *J. Magn. Magn. Mater.*, 2002, **248**, 441–456.
8. J. K. Lim, P. Frigola, G. Travish, J. B. Rosenzweig, S. G. Anderson, W. J. Brown, J. S. Jacob, C. L. Robbins and A. M. Tremaine, *Phys. Rev. Spec. Top.–Accel. Beams*, 2005, **8**, 072401.
9. K. K. H. Leung and O. Zimmer, *Nucl. Instrum. Methods Phys. Res.*, 2009, **611**, 181–185.
10. V. N. Samofalov, D. P. Belozorov and A. G. Ravlik, *Phys. Metals Metallogr.*, 2006, **102**, 494–505.
11. J. H. Jensen and M. G. Abele, *J. Appl. Phys.*, 1996, **79**, 1157–1163.
12. B. Blümich, F. Casanova and S. Appelt, *Chem. Phys. Lett.*, 2009, **477**, 231–240.
13. M. Kumada, Y. Iwashita, M. Aoki and E. Sugiyama, *Proc. Part. Accel. Conf., 20th*, 2003, 1993–1995.
14. S. Ebert, A. Amar, C. Bauer, M. Koelzer, P. Blümler, H. W. Spiess, D. Hinderberger and K. Münnemann, *Appl. Magn. Reson.*, 2012, **43**, 195–206.
15. H. A. Leupold, in *Rare-earth Iron permanent magnets*, ed. J. M. D. Coey, Clarendon Press, Oxford, 1996, pp. 381–429.
16. R. Haken and B. Blümich, *J. Magn. Reson.*, 2000, **144**, 195–199.
17. C. A. Meriles, D. Sakellariou, A. Moulé, M. Goldman, T. F. Budinger and A. Pines, *J. Magn. Reson.*, 2004, **169**, 13–18.
18. M. G. Abele and H. A. Leupold, *J. Appl. Phys.*, 1988, **64**, 5988–5990.
19. H. Soltner and P. Blümler, *Concepts Magn. Reson.*, 2010, **36A**, 211–222.
20. K. Turek and P. Liszkowski, *J. Magn. Reson.*, 2014, **238**, 52–62.
21. J. Z. Chen and C. Y. Xu, *J. Appl. Phys.*, 2007, **101**, 123926.
22. M. G. Abele, *J. Appl. Phys.*, 1994, **76**, 6247–6252.
23. M. G. Abele, R. Chandra, H. Rusinek, H. A. Leupold and E. Potenziani, *IEEE Trans. Magn.*, 1989, **25**, 3904–3906.
24. C. Li and M. Devine, *IEEE Trans. Magn.*, 2005, **41**, 3835–3837.
25. M. G. Abele and H. Rusinek, *J. Appl. Phys.*, 2008, **103**, 07E903.
26. B. P. Hills, K. M. Wright and D. G. Gillies, *J. Magn. Reson.*, 2005, **175**, 336–339.
27. G. Moresi and R. Magin, *Concepts Magn. Reson., Part B*, 2003, **19**, 35–43.
28. X. Zhang, V. Mahesh, D. Ng, R. Hubbard, A. Ailiani, B. O'Hare, A. Benesi and A. Webb, COMSOL Users Conference, Boston, 2005.
29. B. Manz, M. Benecke and F. Volke, *J. Magn. Reson.*, 2008, **192**, 131–138.
30. R. Bjork, *J. Appl. Phys.*, 2011, **109**, 013915.
31. C. W. Windt, H. Soltner, D. van Dusschoten and P. Blümler, *J. Magn. Reson.*, 2011, **208**, 27–33.

32. J. E. Hilton and S. M. McMurry, *IEEE Trans. Magn.*, 2007, **43**, 1898–1902.
33. J. Z. Chen and C. Y. Xu, *IEEE Trans. Magn.*, 2007, **43**, 3555–3557.
34. R. C. Jachmann, D. R. Trease, L. S. Bouchard, D. Sakellariou, R. W. Martin, R. D. Schlueter, T. F. Budinger and A. Pines, *Rev. Sci. Instrum.*, 2007, **78**, 035115.
35. E. Danieli, J. Mauler, J. Perlo, B. Blümich and F. Casanova, *J. Magn. Reson.*, 2009, **198**, 80–87.
36. C. Hugon, F. D'Amico, G. Aubert and D. Sakellariou, *J. Magn. Reson.*, 2010, **205**, 75–85.
37. J. Perlo, F. Casanova and B. Blümich, *Science*, 2007, **315**, 1110–1112.
38. S. K. Küster, E. Danieli, B. Blümich and F. Casanova, *Phys. Chem. Chem. Phys.*, 2011, **13**, 13172–13176.
39. E. Danieli, J. Perlo, B. Blümich and F. Casanova, *Angew. Chem., Int. Ed.*, 2010, **49**, 4133–4135.
40. H. A. Leupold, E. Potenziani II and M. G. Abele, *J. Appl. Phys.*, 1988, **64**, 5994–5996.
41. T. R. Ni Mhiochain, J. M. D. Coey, D. L. Weaire and S. M. McMurry, *IEEE Trans. Magn.*, 1999, **35**, 3968–3970.
42. T. R. Ni Mhiochain, D. Weaire, S. M. McMurry and J. M. D. Coey, *J. Appl. Phys.*, 1999, **86**, 6412–6424.
43. C. Bauer, H. Raich, G. Jeschke and P. Blümler, *J. Magn. Reson.*, 2009, **198**, 222–227.
44. B. D. Armstrong, M. D. Lingwood, E. R. McCarney, E. R. Brown, P. Blümler and S. Han, *J. Magn. Reson.*, 2008, **191**, 273–281.
45. O. Neudert, D. G. Zverev, C. Bauer, P. Blümler, H. W. Spiess, D. Hinderberger and K. Münnemann, *Applied Magnetic Resonance*, 2012, **43**, 149–165.
46. H. A. Leupold, A. S. Tilak and E. Potenziani II, *IEEE Trans. Magn.*, 1993, **29**, 2902–2904.
47. C. W. Windt and P. Blümler, *Acta Hortic.*, 2013, **991**, 115–122.
48. K. Menzel, C. W. Windt, J. A. Lindner, A. Michel and H. Nirschl, *Sep. Purif. Technol.*, 2013, **105**, 114–120.
49. H. A. Leupold and E. Potenziani II, *J. Appl. Phys.*, 1988, **63**, 3987–3988.
50. H. A. Leupold, A. S. Tilak and E. Potenziani II, *J. Appl. Phys.*, 2000, **87**, 4730–4734.
51. H. A. Leupold and E. Potenziani II, *J. Appl. Phys.*, 1991, **70**, 6621–6623.
52. F. Bloch, O. Cugart, G. Meunier and J. C. Toussaint, *IEEE Trans. Magn.*, 1998, **34**, 2465–2468.
53. H. A. Leupold, E. Potenziani II and A. S. Tilak, *J. Appl. Phys.*, 1994, **76**, 6856–6858.
54. P. Blümler and H. Soltner, Halbach magnet for magnetic resonance that can be opened and closed without effort, *Eur. Pat.*, EP2365353, 2011.
55. R. A. Wind, J. Z. Hu and D. N. Rommerereim, *Magn. Reson. Med.*, 2003, **50**, 1113–1119.
56. D. Sakellariou, C. Hugon, A. Guiga, G. Aubert, S. Cazaux and P. Hardy, *Magn. Reson. Chem.*, 2010, **48**, 903–908.
57. H. Zijlstra, *Philips J. Res.*, 1985, **40**, 259–288.
58. A. Podol'skii, *IEEE Trans. Magn.*, 1998, **34**, 248–252.

59. F. Bertora, E. Biglieri, G. C. Borgia, P. Fantazzini, P. Macini and A. Trequattrini, *Magn. Reson. Imaging*, 1994, **12**, 329–331.
60. H. Yoshioka, S. Ito, S. Handa, S. Tomiha, K. Kose, T. Haishi, A. Tsutsumi and T. Sumida, *J. Magn. Reson. Imaging*, 2006, **23**, 370–376.
61. S. Handa, S. Tomiha, T. Haishi and K. Kose, *Magn. Reson. Med.*, 2007, **58**, 225–229.
62. K. Kose, Y. Matsuda, T. Kurimoto, S. Hashimoto, Y. Yamazaki, T. Haishi, S. Utsuzawa, H. Yoshioka, S. Okada, M. Aoki and T. Tsuzaki, *Magn. Reson. Med.*, 2004, **52**, 440–444.
63. P. Blümler, C. W. Windt and D. van Dusschoten, *Nova Acta Leopold.*, 2009, **NF 96**, 17–30.
64. L. Borisjuk, H. Rolletscheck and T. Neuberger, *Plant J.*, 2012, **70**, 129–146.
65. S. Anferova, V. Anferov, D. G. Rata, B. Blümich, J. Arnold, C. Clauser, P. Blümler and H. Raich, *Concepts Magn. Reson., Part B*, 2004, **23B**, 26–32.
66. S. Anferova, V. Anferov, J. Arnold, E. Talnishnikh, M. A. Voda, K. Kupferschläger, P. Blümler, C. Clauser and B. Blümich, *Magn. Reson. Imaging*, 2007, **25**, 474–480.
67. A. Haber, S. Haber-Pohlmeier, F. Casanova and B. Blümich, *Vadose Zone J.*, 2010, **9**, 893–897.
68. E. Karakosta, G. Diamantopoulos, M. S. Katsiotis, M. Fardis, G. Papavassiliou, P. Pipilikaki, M. Protopapas and D. Panagiotaras, *Ind. Eng. Chem. Res.*, 2010, **49**, 613–622.
69. P. J. Barale, C. G. Fong, M. A. Green, P. A. Luft, A. D. McInturff, J. A. Reimer and M. Yahnke, *IEEE Trans. Appl. Supercond.*, 2002, **12**, 975–978.
70. A. Constantinesco, P. Choquet, G. Cauffet, J. M. Fournier, S. Ravier, J. M. Drillon and G. Aubert, *Magn. Reson. Chem.*, 1997, **35**, S69–S75.
71. M. Koizumi, S. Naito, T. Haishi, S. Utsuzawa, N. Ishida and H. Kano, *Magn. Reson. Imaging*, 2006, **24**, 1111–1119.
72. E. Danieli, J. Perlo, A. L. L. Duchateau, G. K. M. Verzij, V. M. Litvinov, B. Blümich and F. Casanova, *ChemPhysChem*, 2014, **15**, 3060–3066.
73. E. Danieli, J. Perio, B. Blümich and F. Casanova, *Phys. Rev. Lett.*, 2013, **110**, 180801.

CHAPTER 6

Hardware Developments: Handheld NMR Systems for Biomolecular Sensing

NAN SUN*[a] AND DONHEE HAM*[b]

[a]Department of Electrical and Computer Engineering, University of Texas at Austin, Austin, TX 78756, USA; [b]School of Engineering and Applied Sciences, Harvard University, Cambridge, MA 02138, USA
*E-mail: nansun@mail.utexas.edu, donhee@seas.harvard.edu

6.1 Introduction

NMR is the resonant energy exchange between RF magnetic fields and atomic nuclear spins subjected to static magnetic fields. It has served as a powerful means to examine the fundamental physics of atomic nuclei, to analyze molecular structures, and to probe properties of materials.[6–8] It has also had a prodigious impact in medicine and technology with such applications as medical imaging,[9] biomolecular sensing,[10,11] well logging,[12] and product quality control.[13]

One important limitation for existing NMR instruments is that they are bulky, heavy, and expensive. If they can be miniaturized, the scope of NMR applications can be broadened. For example, small, low-cost NMR biomolecular sensors would be relevant to medical diagnostics in the context of personalized medicine. Recently, we have developed such miniature NMR

New Developments in NMR No. 5
Mobile NMR and MRI: Developments and Applications
Edited by Michael L. Johns, Einar O. Fridjonsson, Sarah J. Vogt, and Agnes Haber

Published by the Royal Society of Chemistry, www.rsc.org

systems, aimed in particular at biomolecular sensing, which we originally reported in ref. 1–5. This book chapter seeks to review these developments.

An NMR system is comprised of three major components: a magnet, a coil, and an RF transceiver. The magnet produces strong static magnetic fields to polarize nuclear spins. The coil and RF transceiver are used to generate excitation RF magnetic fields and also to monitor the dynamics of the nuclear spins. By far the largest component in an NMR system is the magnet. A physically larger magnet (even for the same static field strength) is usually preferred, as it tends to produce a more homogeneous static field, which then can excite a larger volume of sample, yielding a stronger NMR signal. However, the large magnet leads to the traditionally bulky NMR instrumentation.

In creating our miniature systems, we used substantially smaller magnets, and to detect the NMR signal significantly weakened due to the use of the small magnets, we developed highly sensitive RF transceivers. In addition, these RF transceivers were implemented as CMOS integrated circuits (in conventional systems where large and expensive magnets dominate the system size and cost, such integrated transceiver would not be as meaningful from the cost and size points of view). Figure 6.1 shows our 1st, 2 kg, NMR prototype,[1,2] which is 60 times lighter and 40 times smaller than a typical 120 kg commercial benchtop NMR system. It uses a magnet the size of a fist (this small magnet and the large magnet of the commercial system produce similar static fields of ~0.5 T). The RF transceiver was partially integrated in a 0.18 μm CMOS chip. The in-house fabricated planar microcoil has a quality factor, Q, of 16.

Figure 6.2 shows our 2nd NMR prototype,[3,4] which integrates the planar coil and RF transceiver in the same CMOS 0.18 μm chip. The RF transceiver has a higher level of integration than the 1st prototype, and achieves a high-enough sensitivity to cope with the signal-to-noise reduction caused by the

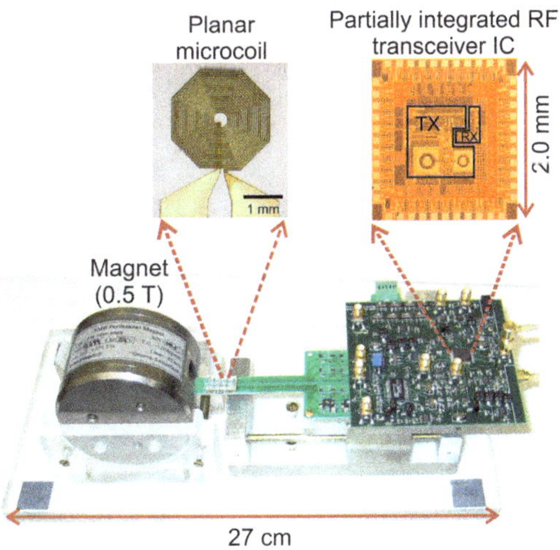

Figure 6.1 Our 1st miniature NMR system (figure adapted from ref. 1 and 2).

lossy on-chip coil with a Q of only 1.9. This '1-chip' NMR system, which operates with the same magnet as in the 1st prototype, manipulates nuclear spins on the CMOS chip in its coil region, enabling lab-on-a-chip experiments.

Our 3rd NMR prototype (Figure 6.3),[3,4] which uses an even smaller magnet, weighs only 0.1 kg; it is the smallest complete NMR system ever to the best of our knowledge. This system, which can be held in the palm of the hand, is 1200 times lighter and 1200 times smaller than the aforementioned commercial system. As the NMR signal is lowered due to the use of the small magnet, this 'palm' system uses a high-quality (Q = 28) solenoidal coil in order not to further weaken the signal, while employing the high-performance RF transceiver integrated circuits (IC), as in the 2nd prototype.

Figure 6.2 Our 2nd, '1-chip,' NMR system (figure adapted from ref. 3 and 4).

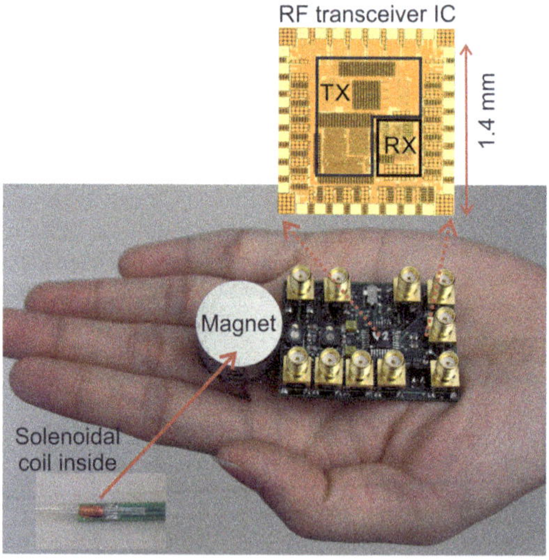

Figure 6.3 Our 3rd, 'palm,' NMR system (figure adapted from ref. 3 and 4).

In what follows, we will review the design of these miniature NMR systems and biomolecular detection experiments with them, but let us get started with the basic NMR review first.

6.2 Review of NMR Basics

Although NMR is a well-established subject,[6,7] we here review its basics to the extent pertinent to our work in hopes of making this chapter self-contained for readers' convenience. NMR experiments can be done with various atomic nuclei. We will consider protons, the nuclei of hydrogen atoms (^{1}H), as all our experiments concern ^{1}H NMR with water.

6.2.1 Protons in Static Magnetic Field

Due to spin, the proton has an intrinsic magnetic moment $\vec{m} = \gamma\vec{S}$ where $\gamma \approx 42.6$ MHz T^{-1} is the proton gyromagnetic ratio and $\vec{S}$ is the spin angular momentum. The single-proton Hamiltonian in an external static magnetic field $\vec{B}_0 = B_0\hat{z}$ ($\hat{z}$: unit vector along the z-axis) is its potential energy: $H = -\vec{m}\cdot\vec{B}_0 = -\gamma\vec{S}\cdot B_0\hat{z} = -\gamma B_0 S_z = -\omega_0 S_z$, where $\omega_0 \equiv \gamma B_0$ is the angular Larmor frequency. S_z has two eigenstates; the 'spin-up' state with eigenvalue $\hbar/2$ and the 'spin-down' state with eigenvalue $-\hbar/2$, where $\hbar \equiv h/2\pi$ (h: Planck constant). Therefore, the spin-up and -down states are the eigenstates of the Hamiltonian as well, with energies of $-\hbar\omega_0/2$ and $\hbar\omega_0/2$, respectively.

A large number of ^{1}H proton magnetic moments in a macroscopic sample, such as water, in $B_0\hat{z}$ and at ambient temperature T will reach thermal equilibrium, where the proton population ratio between the spin-up and -down states will be $e^{\hbar w_0/(2kT)}/e^{-\hbar\omega_0/(2kT)}$ (k: Boltzmann constant). Since $\hbar\omega_0 \ll kT$ at room temperature and with typical B_0 values, this ratio is only very slightly larger than 1. Nonetheless, with a typical macroscopic sample containing a large number of protons, the absolute difference in populations is still considerable, producing an appreciable net macroscopic magnetic moment $\vec{M}$ along the positive z-axis (Figure 6.4a). Specifically, $\vec{M} = M_0\hat{z}$ with:

$$M_0 = \gamma\sum_j S_{j,z} = N\gamma\frac{\hbar/2\cdot e^{\hbar\omega_0/2kT} - \hbar/2\cdot e^{-\hbar\omega_0/2kT}}{e^{\hbar\omega_0/2kT} + e^{-\hbar\omega_0/2kT}} \cong \frac{N\gamma^2\hbar^2 B_0}{4kT}, \tag{6.1}$$

where N is the total number of protons, the summation runs over all protons, and the last expression is an approximation using $\hbar\omega_0 \ll kT$. A stronger B_0 and a larger sample size (larger N) yield a stronger M_0.

6.2.2 Excitation Mode—NMR and Rabi Oscillation

Once the sample attains the equilibrium macroscopic magnetic moment $M_0\hat{z}$ in the static field $B_0\hat{z}$, let a perturbing RF magnetic field of angular frequency ω and amplitude B_1 be applied to the sample perpendicularly to $B_0\hat{z}$. Without

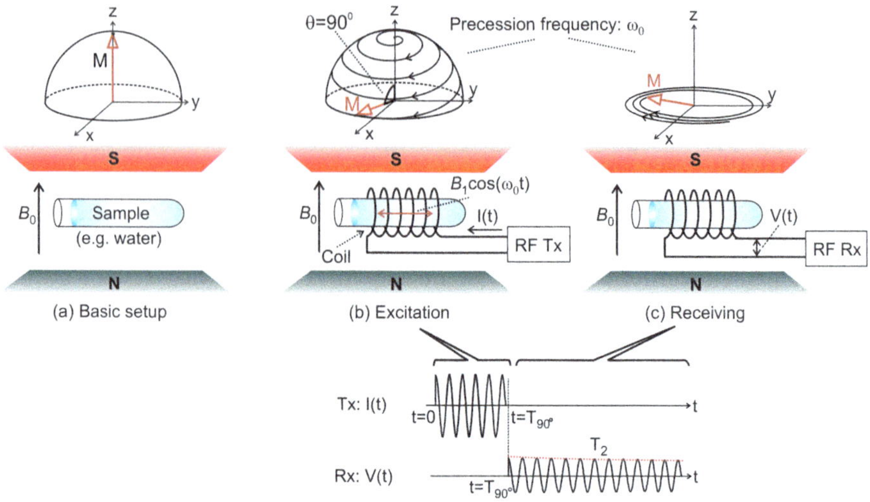

Figure 6.4 NMR basics.

loss of generality, the x-axis may be assigned to the direction of the RF field, which then can be written as $\vec{B}_1(t) = B_1\hat{x}\cos(\omega t)$. This RF field can be generated by wrapping a coil around the sample and transmitting an RF current into the coil (Figure 6.4b). The resulting time evolution of the macroscopic magnetic moment $\vec{M}(t)$ can be described by the following equation of motion:

$$\frac{\mathrm{d}\vec{M}(t)}{\mathrm{d}t} = \gamma\vec{M}(t) \times \left[B_0\hat{z} + B_1\hat{x}\cos(\omega t)\right], \tag{6.2}$$

which states that the torque $\vec{M}(t) \times [B_0\hat{z} + \vec{B}_1(t)]$ exerted on $\vec{M}(t)$ is equal to the time derivative of the macroscopic angular momentum $\vec{M}(t)/\gamma$. Solving eqn (6.2) is straightforward. We only discuss the result.

If the excitation frequency ω is different from the Larmor frequency ω_0, the perturbation is not effective, and $\vec{M}(t)$ remains to be $M_0\hat{z}$. If ω closes onto ω_0, $\vec{M}(t)$ absorbs energy from the RF magnetic field, and uses it to increase its potential energy by increasing the angle θ it makes with the static field $B_0\hat{z}$ (Figure 6.4b). If the RF perturbation at $\omega = \omega_0$ continues, θ keeps increasing to the maximum, π, with $\vec{M}(t)$ attaining the maximum potential energy, then, θ decreases back to 0 with $\vec{M}(t)$ releasing its potential energy back to the RF field, and then, θ grows again, and so on so forth. This resonant energy exchange between the RF field and $\vec{M}(t)$ is NMR. The oscillation of θ occurs at frequency $\omega_1 = \gamma B_1/2$ (Rabi oscillation).

In addition to the Rabi oscillation that periodically modulates the z-component of $\vec{M}(t)$ at frequency ω_1, $\vec{M}(t)$ also undergoes precession about the z-axis due to the torque the static field $B_0\hat{z}$ exerts, thus its x and y components are also periodically modulated. This precession motion, which does *not* alter the potential energy with respect to the static $B_0\hat{z}$ field, occurs at ω_0. Overall, $\vec{M}(t)$ will exhibit a spiral downward (or upward) motion (Figure 6.4b).

The downward/upward motion of $\vec{M}(t)$ (change of θ) is normally much slower than the spiral motion (precession), as typically $B_1 \ll B_0$ thus $\omega_1 \ll \omega_0$.

The resonant excitation frequency ω_0 corresponds to the energy splitting between the spin-up and -down states. In fact, the dynamics above can be obtained by using the standard quantum mechanical treatment instead of using the classical eqn (6.2); one can exactly solve the time-dependent Schrödinger equation, with the single-spin Hamiltonian $H(t) = -\gamma\vec{S}\cdot[B_0\hat{z} + B_1\hat{x}\cos(\omega t)] = -\omega_0 S_z - 2\omega_1\cos(\omega t)S_x$, and put together the solution with thermal distribution consideration.

6.2.3 Reception Mode

After a certain duration of the resonant RF excitation ($\omega = \omega_0$), which tips $\vec{M}$ away from the z-axis to a certain angle $\theta = \theta_0$, the coil is switched from the RF transmitter to the RF receiver (Figure 6.4c). This switchover terminates the resonant energy exchange process, and θ is now kept at θ_0 for a while ($\vec{M}$ will eventually relax back to the equilibrium position of $\theta = 0$, but this is a leisurely process taking several seconds, which is far slower than the faster dynamics considered here), but the ω_0-precession persists. The consequent variation of the magnetic flux across the coil induces a sinusoidal voltage $V(t)$ with frequency ω_0 across the coil, which is picked up by the RF receiver. We will call $V(t)$ NMR signal (although NMR actually occurs during the excitation mode). Typically, to maximize the amplitude of $V(t)$ in the reception mode, the excitation mode is terminated when $\theta = 90°$ with $\vec{M}$ on the xy plane; as the Rabi oscillation frequency is $\omega_1 = \gamma B_1/2$, the duration of the excitation mode to change θ from 0 to 90° is given by:

$$T_{90^\circ} = \frac{\pi}{\gamma B_1}. \tag{6.3}$$

$V(t)$ decays exponentially with a characteristic time called T_2 (Figure 6.4c). This damping is not due to, and occurs faster than, the energy relaxation parenthetically mentioned shortly before. It is caused by random interactions among the proton spins (spin–spin interactions), which perturb the ω_0-precession of each proton spin, causing its phase to undergo a random walk process.[14–16] Consequently, the precessions of the large number of protons grow out of phase with time, causing the vector sum $\vec{M}$ to exponentially decay, leading to the damped $V(t)$. For pure water at room temperature, $T_2 \sim 1$s. This damping is typically very slow as compared to the spin precession; for example, for $\omega_0 \approx 21$ MHz ($B_0 \approx 0.5$ T), 10^7 precession cycles occur before $V(t)$ decays appreciably.

6.2.4 NMR-Based Biomolecular Sensing

Consider a biological sample containing a large number of water molecules (thus a large number of ^{1}H protons). To detect particular proteins in the sample, magnetic nanoparticles coated with antibodies that can specifically

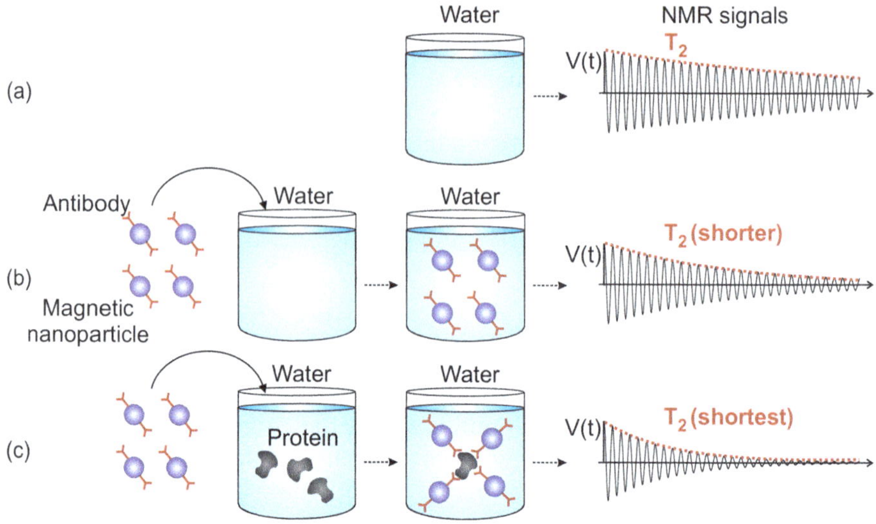

Figure 6.5 NMR-based biomolecular detection.[10]

bind to the target proteins are introduced into the sample (Figure 6.5).[10] In the absence of the target proteins (Figure 6.5b), the magnetic particles stay mono dispersed. These magnetic particles incessantly move around due to Brownian motion, producing fluctuating magnetic fields. These disturb precessions of the proton spins, increasing their phase noise beyond that due to the basic spin–spin interactions. Therefore, the phase coherence is lost at a higher rate, reducing T_2. In the presence of target proteins (Figure 6.5c), the magnetic particles assemble into local clusters, which are even more efficient in accelerating the dephasing of ^{1}H proton precessions, yielding an even smaller T_2. In summary, by monitoring T_2 of ^{1}H NMR signal, target proteins can be detected.[10]

6.2.5 Spin Echo Technique

Many important NMR applications rely on the accurate measurement of T_2. We have so far assumed a uniform static field B_0. In practice, magnetic fields of any magnet exhibit a spatial variation around the intended value B_0, with which proton spins at different locations precess at different Larmor frequencies, growing rapidly out of phase. Consequently, the damping of $V(t)$ is hastened with a new characteristic time T_2* that is smaller than T_2. This faster, spurious T_2*-damping overrides the T_2-damping we wish to observe.

The field inhomogeneity causing the T_2*-damping is spatially fixed and thus *deterministic*, while the T_2-damping is caused by the *random* perturbations, *i.e.*, spin–spin interactions and magnetic particle perturbations. This fundamental difference allows for techniques to recover T_2 despite the T_2*-damping. Figure 6.6 shows one such technique (Carr–Purcell technique).

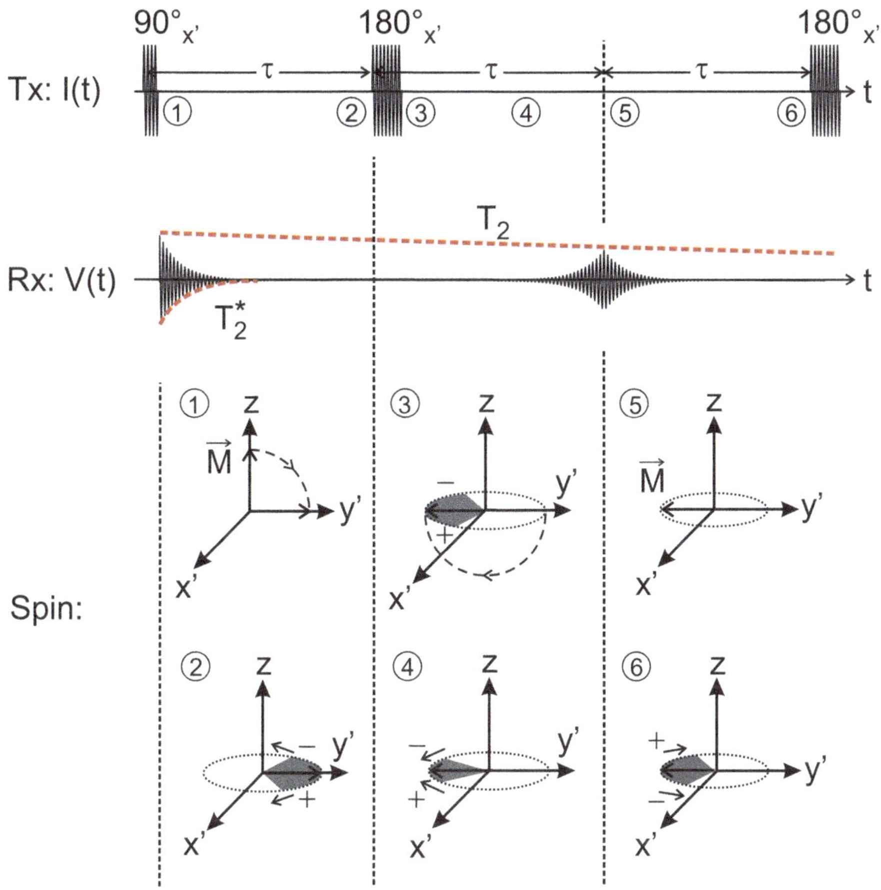

Figure 6.6 Carr–Purcell pulse sequence.

First, the RF magnetic field is applied for a duration of $T_{90°}$ (eqn (6.3)) so that $\vec{M}$ is tipped away from the z-axis onto the xy-plane ['1' in Figure 6.6]. The $x'y'$-frame of Figure 6.6 is a frame rotating about the z-axis in the same sense as the spin precessions and at the same rate as the nominal precession frequency, *i.e.*, the Larmor frequency at B_0 where the static field has spatial variations around B_0. This rotating frame removes the nominal-frequency precession. In the rotating frame, during the RF excitation for $T_{90°}$ ('90° pulse'), $\vec{M}$ exhibits a simple downward motion, instead of the spiraling downward motion of the laboratory frame; $\vec{M}$ is rotated around the x'-axis (this axis for the θ-changing rotation due to the RF excitation is set by the initial phase of the RF excitation) by 90° to the positive y'-direction. Due to the field inhomogeneity, precessions of individual spins with differing Larmor frequencies dephase rapidly, dampening $V(t)$ with T_2* time ['2' in Figure 6.6]; spins precessing faster (labeled '+') move clockwise (when looked at from the positive z-axis), while slower spins ('−') move counterclockwise. After a while ($t = \tau$), a 180°-pulse is applied to rotate the spins again around x'-axis by 180° ['3' in Figure 6.6], placing them now around the negative y'-axis. Faster (slower) spins still continue to move clockwise (counterclockwise), but due to the 180° flip, spins now re-phase ['4' in Figure 6.6], and $V(t)$ re-emerges, or 'echoes'. At $t = 2\tau$, the re-phasing becomes maximal, yielding $\vec{M}$ unaffected by the field inhomogeneity and a locally maximal $V(t)$ ['5' in Figure 6.6]. After this, faster (slower) spins continue to rotate clockwise (counterclockwise), again causing T_2*-damping. Applying another 180°-pulse will subsequently cause another echo to appear. Continued applications of 180°-pulses produce a sequence of echoes. As the random dephasing is not countered, T_2-damping continues to progress in the background, and the envelope of the echoes decays with T_2. In this way, T_2 can be measured.

The RF magnetic fields in excitation pulses can be inaccurate, causing errors in the rotation angles. As the angle errors accumulate, the envelopes of the echoes decay faster than T_2. This problem is solved (Figure 6.7; Carr–Purcell–Meiboom–Gill (CPMG) method) by executing the 180° rotations about the y'-axis, instead of the x'-axis (as mentioned shortly before in the footnote, the axis of rotation can be chosen by controlling the phase of the

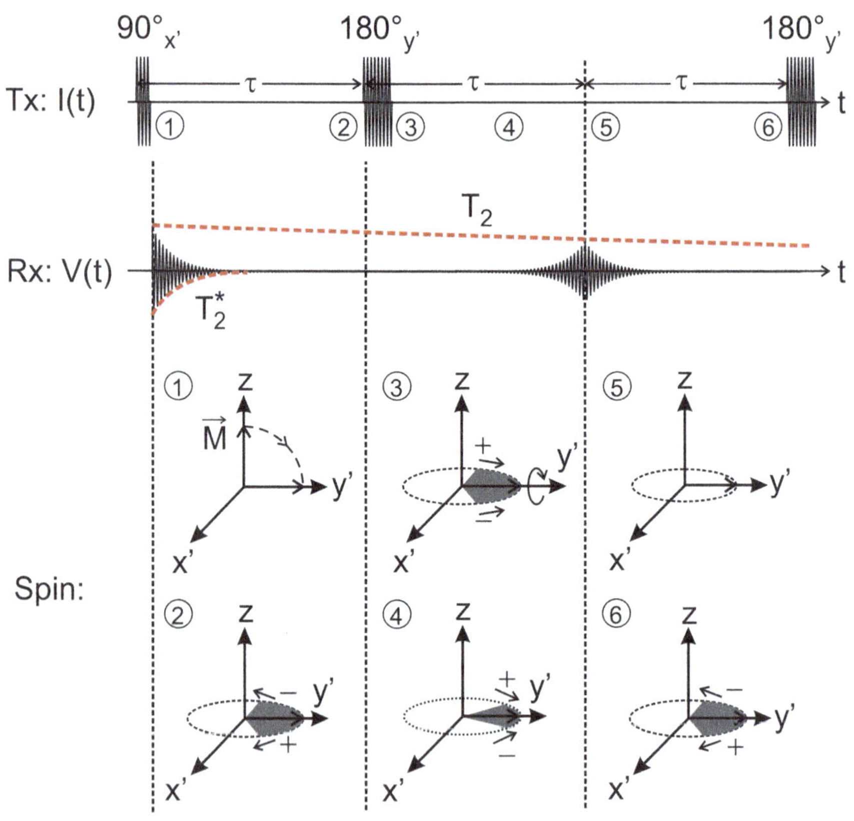

Figure 6.7 Carr–Purcell–Meiboom–Gill pulse sequence.

excitation signal). In this way, if an angle error is introduced by the first 180° rotation, it is removed by the second 180° rotation.

6.2.6 Signal-to-Noise Ratio

As in any measurement system, the signal-to-noise-ratio (SNR) is the most important parameter. Now let us qualitatively derive it. The amplitude of $V(t)$ electromagnetically induced by the precession of $\vec{M}$ is proportional to $\vec{M}$'s magnitude, $M_0 \sim NB_0$ (eqn (6.1)) and its precession frequency $\omega_0 = \gamma B_0$. In addition, the geometric configuration of the coil determines how effectively $V(t)$ is induced by the precession of $\vec{M}$; by resorting to reciprocity, this effectiveness can be measured by how much RF magnetic field magnitude B_1 is produced by an RF excitation current amplitude I_1, *i.e.*, B_1/I_1.[17,18] Overall, the amplitude of $V(t)$ is proportional to $N \cdot B_0^2 \cdot B_1/I_1$. The SNR at the output of the receiver is then proportional to:

$$\mathrm{SNR} \propto \frac{N \cdot B_0^2 \cdot B_1/I_1}{F\sqrt{4kTR\Delta f}} \tag{6.4}$$

where $4kTR\Delta f$ is the coil thermal noise (R: coil resistance) over the receiver bandwidth Δf, and F is the receiver noise figure in linear scale, which accounts for the receiver noise itself. This SNR expression plays the central role in the system-level design considerations, which will be presented in the next section.

6.3 System-Level Design Considerations

Before we present the detailed design of our miniaturized NMR systems, let us first discuss a few general system-level considerations.

6.3.1 Magnet Miniaturization

Eqn (6.4) shows that SNR is proportional to B_0^2. Thus, from the SNR point of view, it is desirable to increase the magnetic field B_0 as much as possible. Nowadays, magnetic fields as strong as 20 T have been used. In order to produce such strong magnetic field, the only viable option is to use a superconducting electromagnet. However, they are big, expensive, power hungry, and have high maintenance costs, as they require special cooling facilities. As a result, it is not suitable for a small low-cost NMR system. In order to significantly reduce magnet size and cost, the only feasible solution is to use a small permanent magnet. There are several challenges associated with this. First, a permanent magnet produces a relatively weak magnetic field B_0. Up to now, the strongest magnetic field for a permanent magnet made from $Nd_2Fe_{14}B$ alloy is 2 T.[19,20] For a typical small-sized permanent magnet, its magnetic field is usually around 1 T or below, which is much smaller than that for the superconducting electromagnet, leading to a smaller SNR. In addition to the low magnetic field strength, the other major challenge for using a small magnet is its magnetic field inhomogeneity. For an inhomogeneous magnetic field, nuclear spins at different locations in the sample precess at different Larmor frequencies. As a result, the spectral lines of the NMR signal are smeared, making it impossible to decipher the fine frequency shifts needed for high-resolution NMR spectroscopy applications. To solve this problem, a brute-force method is used before to increase the dimension of the permanent magnet. For instance, large permanent magnets that occupy 1 m^3 and weigh several hundreds of kilograms have been used; however, this method completely defeats the purpose of miniaturization.

To solve the field inhomogeneity problem for a small magnet, the key is to apply correction fields to cancel out inhomogeneities as large as several hundreds of ppm by devising accurate, low-power, and wide-correction-range magnetic field shimming techniques. One shimming technique is to use electrical shim coils that work like small electromagnets.[21] By tuning their currents, researchers can apply precise correction fields exactly opposite to B_0 inhomogeneities. The advantages of this method are simplicity and high accuracy as the current is easy to tune and the tuning step can be made very small. However, the drawback of this approach is its high current consumption and small correction range of less than 100 ppm, because a larger correction would require excessive currents that cause thermal and reliability problems. Recently, researchers have developed several mechanical shimming methods based on manually moving multiple small shim magnets.[20,22,23] The merit of mechanical shimming is its wide correction range (up to 20 000 ppm), but it has the following drawbacks: (1) it has limited accuracy (>1 ppm) due to low mechanical resolution; and (2) it is highly complex and time consuming, as it relies on manually and meticulously adjusting small magnet positions over many iterations, which is difficult to automate.[20,22,23] One ongoing research topic is to develop a hybrid mechanical–electrical shimming technique. During the shimming process, shim magnets are used first to correct the majority of first-order inhomogeneities to reduce the field inhomogeneity to be less 10 ppm. Once that is done, electrical shimming takes over and performs higher-order spherical harmonic correction until the field inhomogeneity decreases to 0.1 ppm. This hybrid method combines the merits of the two existing methods. It relaxes the precision requirement and complexity in mechanical shimming, and reduces the current required for electrical shimming, as its correction range is only 10 ppm.

Magnetic field shimming is necessary for high-resolution NMR spectroscopy, but it may not be needed for NMR relaxometry applications, whose parameters of interests are relaxation times T_1 and T_2. As reviewed in the previous section, relaxation times can be accurately measured in the presence of magnetic field inhomogeneity by using NMR spin echo techniques, such as the CPMG pulse sequence. This greatly reduces the burden on the magnet design. In our NMR miniaturization work, since our intended application is biomolecular sensing whose key information is T_2, we chose not to shim the magnet in order to reduce the magnet design complexity, and instead, rely on the use of spin-echo techniques to address the magnetic field inhomogeneity problem.

Note that even though the effect of the magnetic field inhomogeneity can be partially compensated by spin-echo techniques, there is still a limit on how inhomogeneous the magnetic field can be. The reasons are as follows. First, a very inhomogeneous magnetic field leads to a wide spread of Larmor frequencies. Thus, the signal bandwidth is enlarged, which requires the receiver bandwidth to increase proportionally. This leads to a larger noise bandwidth, which results in a smaller SNR, as shown in eqn (6.4). Second, the 90° RF excitation pulse has frequency components around $\omega_0 = \gamma B_0$ due to its finite duration of $T_{90°}$, with the bandwidth $1/T_{90°}$. The RF pulse then can excite only those spins in the region whose Larmor frequencies fall within the bandwidth. With significant static field inhomogeneity, the excitable region is reduced, decreasing the effective number of participant spins, N, in eqn (6.4), leading to a substantially decreased SNR. An excitation pulse with a shorter duration $T_{90°}$ can increase the excitable region, but this requires a larger B_1 (eqn (6.3)) or a larger transmitted power, which is typically limited by electronics. In practice, the upper limit for the magnetic field inhomogeneity is usually several hundreds of ppm. Beyond that, big SNR loss is inevitable even if NMR spin-echo techniques are used.

Although there are many challenges with using a small permanent magnet, it does bring the key benefits of being lightweight with small dimensions and low cost. Another advantage of using a permanent magnet is that it allows the use of solenoidal coils whose coil sensitivity B_1/i is three times higher than the saddle-shaped coils that are usually adopted for electromagnets.[18]

6.3.2 Coil Miniaturization

The size of the coil does not pose a challenge for NMR system miniaturization, as it is typically much smaller than the magnet and the transceiver box. The major motivation to develop a small coil is to increase the coil sensitivity B_1/I_1, as it directly affects SNR (see eqn (6.4)). For a solenoidal coil, it is easy to derive that its B_1/I_1 is inversely proportional to the coil length and diameter for the fixed number of turns. Thus, by shrinking the coil dimension, we can increase B_1/I_1 and the SNR. Another benefit is that an increased B_1/I_1 means that for the same B_1, the required coil current I_1 during NMR excitation is smaller, which lowers the requirement on the transmitter power.

So far, most research works have focused on solenoidal microcoils. The reason why solenoid is chosen over other coil geometries is that it offers higher B_1/I_1.[24–29] Nevertheless, because they are hand-wound, they cannot be batch fabricated and encounter severe fabrication difficulties at small dimensions (<1 mm).[30,31] Planar microcoils can be better candidates than solenoidal ones. Although the sensitivity of planar microcoils is lower, they have the following key advantages due to their standard photolithography based fabrication process: (1) they can be easily batch-fabricated into a large array with high (μm) resolution.[11,30–35] (2) Their planar structure allows them to be integrated with a microfluidic system.[32–35] As a result, the sample introduction can be greatly simplified and the sensitivity can also be improved.[30,31,34] (3) Because their fabrication process is compatible with the silicon IC process, it is possible to integrate the planar microcoil array with a multichannel transceiver on the same IC chip.[30,31,36,37] In this way, the wiring between the microcoil array and the transceiver becomes very simple.

6.3.3 Transceiver Miniaturization

Once the magnet size is reduced, the next biggest component to miniaturize is the transceiver. Existing commercial NMR transceivers are built using discrete electronics, leading to their bulky and expensive setup. Moreover, their performance is not optimal, especially in terms of sensitivity and power consumption. The conventional NMR transceiver, due to its bulky size, has to be placed far from the coil and connects to the coil through a long transmission line.[17] The use of this long wire not only causes signal loss, but also requires the transceiver to be matched to 50 Ω in order to avoid impedance mismatch. However, 50 Ω-matching provides neither efficient power delivery for the power amplifier (PA) nor noise matching for the receiver.

With the fast progress of RF integrated circuit (IC) technology over the past two decades, it is now feasible to integrate the entire RF transceiver onto a small millimeter-sized silicon chip, as demonstrated in ref. 1–4. The transceiver integration brings many advantages. It not only reduces the transceiver size and cost, but also increases SNR and reduces power consumption. With the significant size reduction, the transceiver IC can be placed in close proximity to the coil to completely obviate the need for the transmission line, which allows any desirable impedance to be used. Thus, during the RF transmission, the coil can be matched to a low impedance for efficient power delivery. In the receiving mode, receiver noise matching can also be achieved by having a different matching network to match the coil to a high impedance. This high impedance transformation can passively amplify the NMR signal, leading to a substantially reduced receiver noise figure (details are provided in later sections). For example, our research shows that a good noise figure of less than 1 dB can be achieved by using a low-power integrated transceiver. Compared to the discrete single-ended transceiver, the integrated transceiver can also be made fully differential, leading to reduced sensitivity to common-mode noise and perturbations from power supply, bias circuit, and silicon substrate. Also, the integration can greatly reduce parasitic inductances and capacitances that cause performance degradation.

In the conventional discrete NMR transmitter, the power amplifier (PA) usually operates under a large power supply (100 V) in order to deliver a large power of 100 W or beyond. Such a high-power PA is undesirable as it is big and expensive. It is hard to integrate on-chip because standard silicon CMOS technology cannot operate under such high voltage, as it can cause transistor gate breakdown. Fortunately, the increase in the microcoil sensitivity B_1/I_1 helps to lower the power. For the same B_1, a 10 times increase in B_1/I_1 means a 10 times reduction in I_1 and a 100 times reduction in the transmitter power.

If the sample has a relatively long relaxation time, we can also decrease B_1, which further reduces the power. For example, for the small NMR systems that we developed, since they are used to analyze fluidic samples that have long T_2 times (>10 ms), we can use relatively long excitation pulses (40 μs). Under this scenario, the required PA power can be as small as 1 mW. This makes it very easy to integrate the PA on-chip with other transceiver electronics.

6.4 Miniature NMR System—1st Prototype

Our 1st miniature NMR system (Figure 6.1),[1,2] which weighs 2 kg, employs a $B_0 \approx 0.5$ T, 1.25 kg magnet. The Larmor frequency $\omega_0/(2\pi)$ is ~21.3 MHz. The field inhomogeneity $\Delta B_0/B_0$ at the center of the magnet is about 50 ppm over a 5 μL volume. The 500 nH planar copper microcoil is in-house fabricated on a glass substrate,[11] where the electroplated copper is ~15 μm thick, with which $Q \approx 16$ at the Larmor frequency. A 5 μL sample is held on the microcoil with a 5 μm-thick passivation layer in between, inside a microfluidic container fabricated on top.[11]

6.4.1 RF Transceiver IC Architecture

Figure 6.8 shows the architecture of the RF transceiver used in this NMR system. The receiver path is in the lower portion. A weak spin-echo signal whose maximum available power is 0.5 fW appears at the front-end node ['2' in Figure 6.8]. In the frequency domain, the spin-echo signal is centered at the Larmor frequency with the bandwidth of $1/T_2^* \approx \gamma\Delta B_0/(2\pi) \approx 1.1$ kHz.

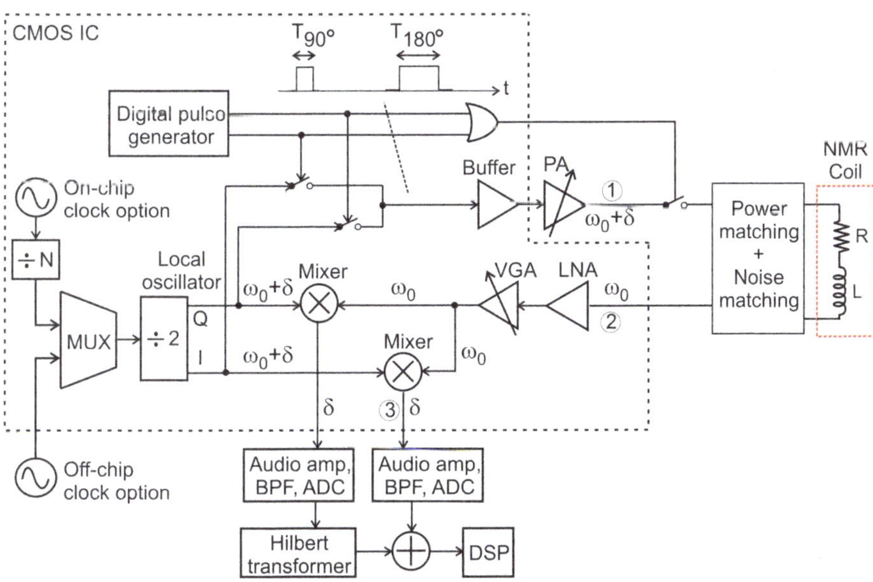

Figure 6.8 RF transceiver in the first miniature NMR prototype.

The signal is amplified by a low noise amplifier (LNA) and a variable-gain amplifier (VGA), and then is down-converted using mixers and quadrature local oscillator signals ('I' and 'Q') with frequency $\omega_0 + \delta$. We select $\delta/(2\pi)$ = 3 kHz, which is high enough to prevent swamping by $1/f$ noise, and is low enough to facilitate the rejection of the out-of-band noise with a band-pass filter with a moderate quality factor. The image noise is rejected by a digital-domain algorithm employing a Hilbert transformer.

The transmitter path is in the upper portion of Figure 6.8. The excitation RF magnetic field is produced by the same quadrature local oscillators used in the receiver. Their frequency, $\omega_0 + \delta$, deviates from ω_0, but the transmitted 90° and 180° pulses are windowed sinusoids with bandwidths of 15 kHz and 7.5 kHz, thus, they can still excite the entire sample, across which the Larmor frequency has a variation $\gamma\Delta B_0/(2\pi)$ = 1.1 kHz. By gating the quadrature oscillator signals with the digital pulse generator, we produce the CPMG pulse sequence. The power amplifier (PA) is implemented off chip.

6.4.2 Receiver LNA and Noise Matching

The LNA is of common-source configuration (Figure 6.9a). Since the Larmor frequency is smaller than the $1/f$ noise corner (~50 MHz) of NMOS transistors in the 0.18 μm CMOS technology used, PMOS devices M_1 and M_2 with smaller $1/f$ noise corner (~1 MHz) are used as input transistors. These PMOS transistors are built in separate *N*-wells, thus, they also help isolate the LNA from substrate noise produced by the transmitter. The coupling of the local oscillator signal into the LNA input is suppressed by cascoding

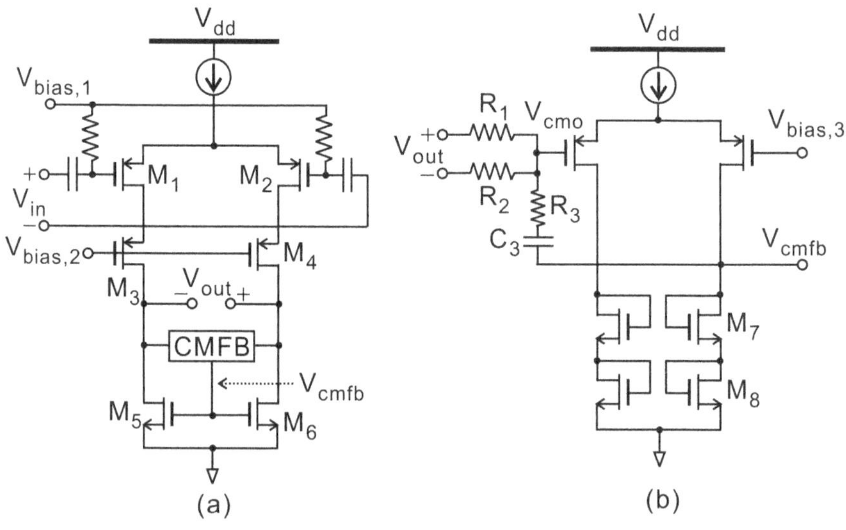

Figure 6.9 (a) Low noise amplifier (LNA). (b) Common-mode feedback circuit (CMFB).

(transistors M_3, M_4). With active load M_5 and M_6, the LNA achieves a high voltage gain of 41 dB. A common-mode feedback circuit (CMFB) (Figure 6.9b) ensures a correct output common mode in the LNA; the CMFB compares the output common-mode voltage V_{cmo} to $V_{bias,3}$, and its output V_{cmfb} is used to drive the gates of transistors M_5 and M_6. The Miller capacitor C_3 and resistor R_3 are for frequency compensation.

Channel thermal noise of transistors is the dominant noise source. The calculated input referred noise of the LNA is:

$$-v_{n,\mathrm{LNA}}^2 \Delta f \approx \frac{8kT\gamma_n}{g_{m1}}\left(1+\frac{g_{m5}}{g_{m1}}\right), \tag{6.5}$$

where g_m denotes transistor transconductance and γ_n denotes transistor channel thermal noise coefficient. Here we have ignored the noise of transistors M_3, M_4, and the CMFB circuit (and the input-referred current noise of the LNA is negligible at the Larmor frequency). To minimize the $1/g_{m1}$ factor, a large tail current (4 mA) and wide input transistors (900 μm) are used. To minimize the g_{m5}/g_{m1} term, transistors M_5 and M_6 are made much narrower than transistors M_1 and M_2. Subsequently, M_5 and M_6 need a large V_{GS}, which is provided by stacking transistors M_7 and M_8 in the CMFB (Figure 6.9b). In this way the input-referred noise of the LNA is minimized. The measured input-referred noise of the entire receiver is 1.8 nV/$\sqrt{Hz}$, which is primarily contributed by the LNA.

Ultimately, we seek to minimize the receiver noise figure F, for which both the input-referred noise minimization and the optimal LNA-coil matching play crucial roles. Consider a general passive network between the coil and the LNA, whose voltage transfer function from the coil side to the amplifier side assumes a value of α at the Larmor frequency. This is shown in Figure 10a along with the NMR signal $V(t)$'s rms value, V_{rms}, coil's thermal noise $\overline{v_{n,\mathrm{R}}^2} = 4kTR\Delta f$ over the signal bandwidth Δf, and the LNA's input referred voltage noise $\overline{v_{n,\mathrm{LNA}}^2}$ also over the signal bandwidth. With the voltage gain α in the matching network, noise figure F of the LNA is given by:

$$F = \frac{\mathrm{SNR}_{\mathrm{in}}}{\mathrm{SNR}_{\mathrm{out}}} \cdot \frac{V_{\mathrm{rms}} \Big/ \left(\overline{v_{n,\mathrm{R}}^2}\right)^{1/2}}{\alpha V_{\mathrm{rms}} \Big/ \left(\alpha^2 \overline{v_{n,\mathrm{R}}^2} + \overline{v_{n,\mathrm{LNA}}^2}\right)^{1/2}} = \left(1 + \frac{\overline{v_{n,\mathrm{LNA}}^2} \big/ \Delta f}{\alpha^2 \cdot 4kTR}\right)^{1/2}, \tag{6.6}$$

where we have neglected the noise of the passive network, which we will shortly justify. From eqn (6.6), it is clear that in addition to the small input-referred noise of the LNA, a larger voltage gain α helps lowering F. To attain a large value of α, we use a single shunt capacitor C as the passive network (Figure 6.10b), and the C value is chosen in such a way that it resonates with the coil inductor at the Larmor frequency. In this way, a large value of α is obtained, specifically: $\alpha = \sqrt{Q^2+1} \approx 16$. Note that this resonant noise matching leads to impedance mismatch between the LNA and the coil. At the Larmor frequency, capacitors are far less lossy than the coil, which justifies the

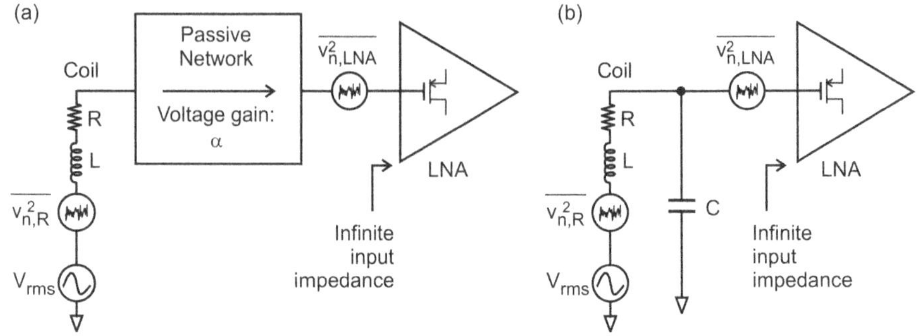

Figure 6.10 Coil–low noise amplifier (LNA) noise matching.

omission of the noise of the passive network in writing eqn (6.6). The large value of α due to the resonance noise matching together with the minimized input-referred noise of the LNA significantly reduces F, whose measured value is 0.7 dB.

6.4.3 Digital Pulse Generator

The CPMG pulse sequence is obtained by using one of the two quadrature oscillator signals for the 90° pulse, and the other quadrature oscillator signal for the 180° pulses. The axes of rotation for $\vec{M}$ in the rotating frame corresponding to the two quadrature signals are perpendicular, thus the CPMG sequence is obtained (Figure 6.7). The arrangement to gate the two quadrature oscillator signals in the transmitter to obtain the CPMG pulse sequence is shown in Figure 6.8, and is detailed in Figure 6.11. A finite-state-machine (FSM) and a counter are used. The counter output, t_S, is reset to 0 whenever the FSM transits between its internal states. Thus, t_S represents how long the FSM stays at its present state, and is used to control its state transition. The FSM's two outputs D_I and D_Q gate the quadrature signals 'I' and 'Q' to produce the CPMG pulse sequence. D_I and D_Q also feed an OR-gate, which controls the off-chip switch between the coil and the PA (Figure 6.8). The FSM deals with 4 states, '1', '2', '3', and '4' in Figure 6.11, which represent the time duration of the 90° pulse, the time interval between the 90° pulse and the first 180° pulse, the time duration corresponding to any 180° pulse, and the time duration between any two adjacent 180° pulses, respectively. The FSM starts from state '1' in Figure 6.11, and sequentially moves to states '2', '3', and '4' when t_S accumulates to the corresponding values. Then the FSM goes back and forth between states '3' and '4' to repeat the 180° pulses.

6.5 Miniature NMR Systems—2nd and 3rd Prototypes

The 2nd prototype (Figure 6.2)[3,4] uses the same magnet as the 1st prototype (Larmor frequency: ~21.3 MHz), but integrates the planar microcoil in the same RF transceiver IC for lab-on-a-chip operation. Due to thin

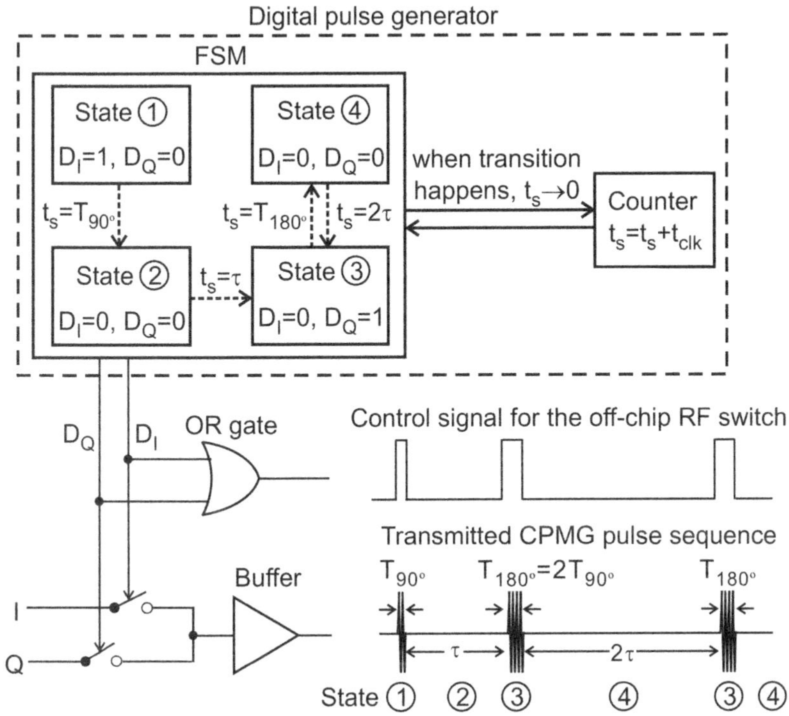

Figure 6.11 Block diagram of the digital pulse generator.

metals, the 430 nH integrated microcoil has a Q of 1.9, even after connecting 5 metal layers in parallel; this low Q is due mainly to the coil's dc resistance, while the substrate and skin effect are negligible at the Larmor frequency. The chip is packaged with the coil part exposed and the rest encapsulated. This open package provides an effective container on top of the coil to hold a 5 μL sample, while the sample is separated from the coil by the passivation layer native to the CMOS process. To cope with the signal-to-noise ratio reduction due to the lossy coil, the RF transceiver was redesigned. Its integration level was also increased, incorporating a power amplifier (PA) on chip. The transceiver architecture is shown in Figure 6.12. The dashed lines show the integration boundary, which includes the coil, capacitor C_m for resonant noise matching, and transceiver front-end except the local oscillator source. The mixer outputs feed the off-chip back-end signal processing unit.

The integrated PA is a differential chain of inverters that are quadrupled in size to amplify power and ensure the output drivability (Figure 6.12). The output amplitude of this class-D PA is fixed at V_{DD}, while tuning the output power is necessary to control the 90° and 180° pulse durations (*e.g.*, see eqn (6.3) for the 90° pulse). For power tuning, we vary the duty cycle of the transmitted signal: the signal power at the Larmor frequency is maximal for the duty cycle of 50%, and is 0 for the duty cycle of 0%. Other harmonic components of the output signal will be irrelevant, as they cannot excite the proton spins; that is, this scheme exploits the natural high-Q filtering ability of the proton spin in the static magnetic field. The duty cycle is tuned by using voltage-controlled delay lines (VCDLs) and logic gates. The receiver is a revision from the 1st prototype to further reduce the input-referred noise, *i.e.*: the active loads are replaced by resistive loads; the CMFB thus is removed; and two stages are cascaded to compensate for the reduced gain due to the resistive loads. The measured input-referred noise of the entire receiver is 1.26 nV/$\sqrt{Hz}$, which is 30% less than that of the first prototype. The measured noise figure is 2.2 dB, which is larger than that of the first prototype; this is because the voltage gain $\alpha = \sqrt{Q^2+1}$ of the noise matching network has been substantially reduced in this second prototype. The measured output impedance of the differential PA is 27 Ω. The duty cycle can be tuned from 0% to 45%, which translates to the output power tuning from 0 to 80 mW.

The 3rd prototype (Figure 6.3),[3,4] which employs an even smaller, 70 g, 0.56 T magnet (Larmor frequency: ~23.9 MHz), weighs only 0.1 kg. The magnet has a cylindrical shape with a height of 2.3 cm and a base diameter of 2.5 cm. This small magnet has a significant magnetic field inhomogeneity of 500 ppm over a 2 μL volume, inevitably reducing the NMR signal strength. To counter this, we use a 100 nH solenoidal microcoil with a Q of 28. The coil is wound around a capillary tube in which a 2 μL sample is placed. The RF transceiver used in this prototype shares almost the same design as that in

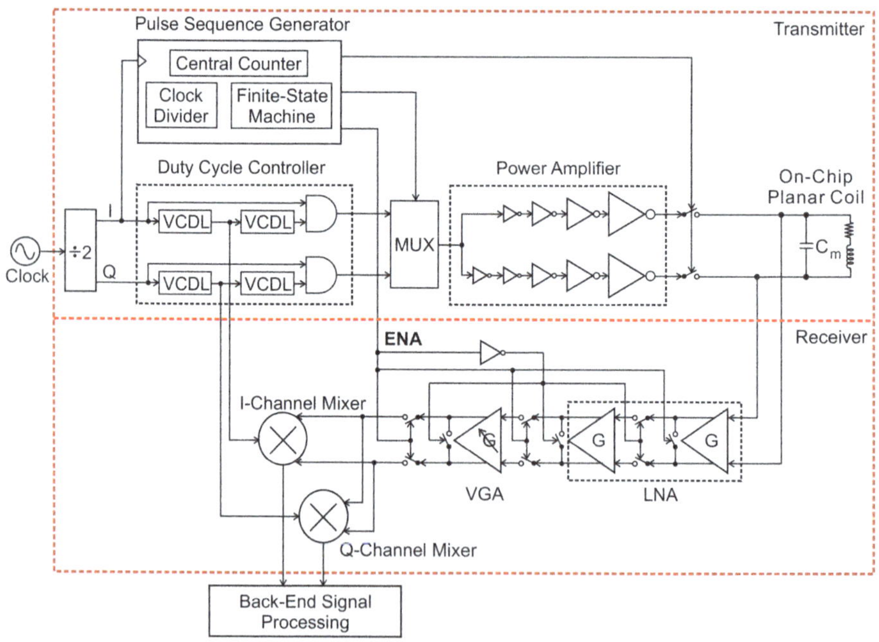

Figure 6.12 CMOS RF transceiver for the 2nd NMR prototype.

the 2nd prototype, except that the coil and the matching capacitor are not integrated. The measured receiver noise figure is 0.9 dB.

6.6 NMR and Biomolecular Sensing Experiments

Figure 6.13 shows the measurements of the down-converted ^1H NMR signal using the 3rd prototype with a water sample. The repeated ringings are spin echo responses to the CPMG pulse sequence. T_2 = 100 ms is extracted from the exponentially decaying envelope of the spin echoes. The repeated spikes between the echoes are due to the coupling of the large excitation pulses.

The miniature NMR systems detected a range of biological molecules and compounds.[1–4] For example, Figure 6.14 shows the detection of avidin using the 3rd prototype. Magnetic particles (diameter: 38 nm) coated with biotins are introduced into a water sample. In the absence of avidin, the particles stay mono dispersed, yielding a T_2 of 48 ms. In the presence of avidin, the biotinylated magnetic particles bind to avidin, forming clusters[10] and reducing T_2 to 40 ms. The 3rd prototype detected down to 1 avidin molecule in 600 million water molecules.

Yet another example, now with the 2nd prototype, is the detection of human bladder cancer cells (Figure 6.15). Magnetic particles (diameter: 40 nm) coated with monoclonal antibody to bladder cancer cell surface markers are introduced into a water sample. In the absence of cancer cells, the magnetic particles are mono dispersed, but in the presence of the cancer cells, magnetic particles bind to the cell surface. In the latter sample, centrifugation[38] separates the cells and unattached magnetic particles, and the unattached particles are washed away. The T_2 time difference is evident (Figure 6.15). The system detects down to ~18 cells per μL.

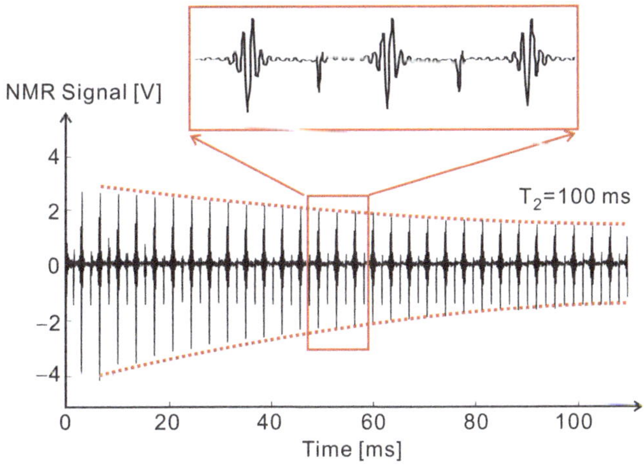

Figure 6.13 Down-converted ^1H NMR signal measured with the 3rd prototype (figure adapted from ref. 3 and 4).

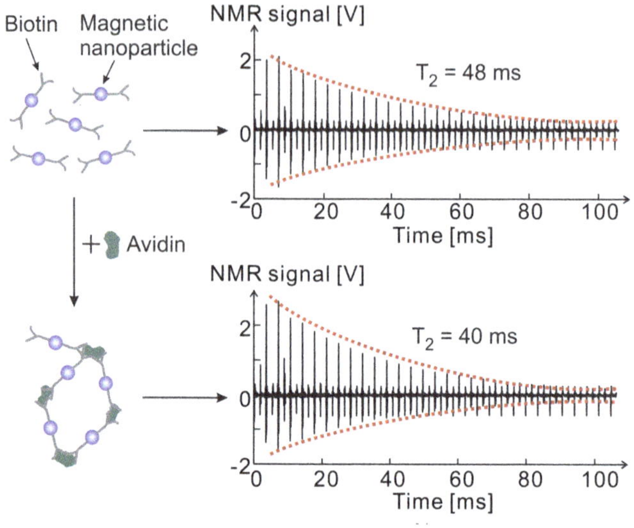

Figure 6.14 Avidin detection with the 3rd prototype (figure adapted from ref. 3 and 4).

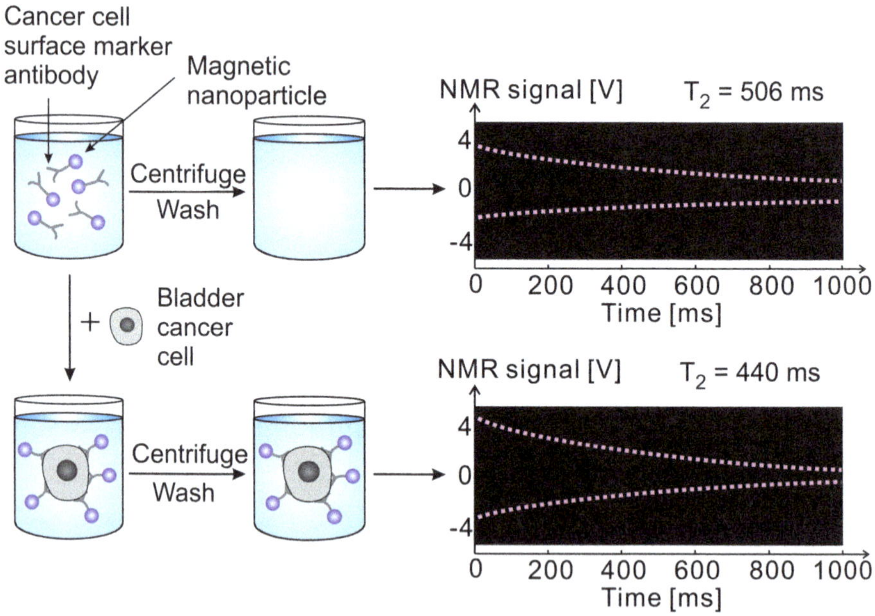

Figure 6.15 Human bladder cancer cell detection with the 2nd prototype (figure adapted from ref. 3 and 4).

6.7 Conclusion and Outlook

We reviewed our recent developments of small NMR biomolecular sensors.[1–4] The miniaturization was made possible by using small-sized, low-quality magnets, and by designing high-performance integrated RF transceivers that counter the effect of the low-quality magnets. The work involved a repertoire of technical areas, ranging from integrated circuits to biomolecular detection to the science and technology of NMR.

This work can be put in the general context of the research efforts to interface solid-state electronic chips with biological systems to provide bioanalytical capabilities.[39–50] The electronic and biological systems interfaces utilize spin-based magnetic interactions as in this reviewed work, or charge-based electric interactions as in the ion-sensitive field-effect transistors,[39–44] while more complex coupling modalities, such as electronics-regime plasmonic interactions,[51–53] may be made available in the future. While optics-based bioanalytical systems will continue to be used broadly into the foreseeable future, the solid-state electronic and bio systems interfaces have attractive features such as low cost, chip-scale operation, massive parallelism, and accessibility. There are thus continuing efforts to develop a host of such interfaces. Our NMR biomolecular sensors add to the list.

From the circuit engineering point of view, the solid-state electronic and bio systems interfaces create needs for a range of application-specific integrated circuits. For example, our NMR work reviewed here showcases how RF integrated circuits can be used not only for their prevalent wireless applications, but also for biomolecular detection aimed at medical diagnostics. For another example, integrated temperature sensors[54,55] will become important in systems interfaced with living organisms or *ex vivo* biological tissues, where monitoring and controlling temperature of the electronic chip is important not to damage these biological systems. For implanted, *in vivo*, or field-use applications, low-power designs of analog and mixed-signal circuits will be important; digitally-assisted design techniques,[56,57] which have been an important design trend, may prove especially useful for these applications.

The CMOS NMR systems and their biomolecular detection applications described in this chapter are focused on the measurements of relaxation times, limited by the field inhomogeneity of the specific small magnets used. The application scope of the small NMR systems will be significantly broadened if their capability can be expanded into the realm of high-resolution spectroscopy. This suggests another interesting future research path.

References

1. Y. Liu, N. Sun, H. Lee and D. Ham, 2008 IEEE International Solid-State Circuits Conference-Digest of Technical Papers, 2008.
2. N. Sun, Y. Liu, H. Lee, R. Weissleder and D. Ham, *IEEE J. Solid-State Circuits*, 2009, **44**, 1629–1643.

3. N. Sun, T.-J. Yoon, H. Lee and D. Ham, 2010 IEEE International Solid-State Circuits Conference-(ISSCC), 2010.
4. N. Sun, T.-J. Yoon, H. Lee, W. Andress, R. Weissleder and D. Ham, *IEEE J. Solid-State Circuits*, 2011, **46**, 342–352.
5. N. Sun, Y. Liu, L. Qin, H. Lee, R. Weissleder and D. Ham, *Solid-State Electron.*, 2013, **84**, 13–21.
6. C. P. Slichter, *Principles of magnetic resonance*, Springer, 1992.
7. D. Canet, *Nuclear magnetic resonance: concepts and methods*, Wiley, Chichester, 1996.
8. H. Günther, *NMR spectroscopy: basic principles, concepts and applications in chemistry*, John Wiley & Sons, 1995.
9. M. A. Brown and R. C. Semelka, *MRI: basic principles and applications*, John Wiley & Sons, 2010.
10. J. M. Perez, L. Josephson, T. O'Loughlin, D. Högemann and R. Weissleder, *Nat. Biotechnol.*, 2002, **20**, 816–820.
11. H. Lee, E. Sun, D. Ham and R. Weissleder, *Nat. Med.*, 2008, **14**, 869–874.
12. B. Sun and K.-J. Dunn, *J. Magn. Reson.*, 2005, **172**, 152–160.
13. A. J. Charlton, W. H. H. Farrington and P. Brereton, *J. Agric. Food Chem.*, 2002, **50**, 3098–3103.
14. D. Ham and A. Hajimiri, *IEEE J. Solid-State Circuits*, 2003, **38**, 407–418.
15. X. Li, O. Yildirim, W. Zhu and D. Ham, *IEEE Trans. Microwave Theory Tech.*, 2010, **58**, 2105–2117.
16. X. Li, W. Zhu and D. Ham, *arXiv preprint arXiv:0908.2214*, 2010.
17. D. I. Hoult, *Prog. Nucl. Magn. Reson. Spectrosc.*, 1978, **12**, 41–77.
18. D. I. Hoult and R. E. Richards, *J. Magn. Reson.*, *1969*, 1976, **24**, 71–85.
19. V. Demas, J. L. Herberg, V. Malba, A. Bernhardt, L. Evans, C. Harvey, S. C. Chinn, R. S. Maxwell and J. Reimer, *J. Magn. Reson.*, 2007, **189**, 121–129.
20. E. Danieli, J. Perlo, B. Blümich and F. Casanova, *Angew. Chem., Int. Ed.*, 2010, **49**, 4133–4135.
21. R. Gruetter, *Magn. Reson. Med.*, 1993, **29**, 804–811.
22. E. Danieli, J. Mauler, J. Perlo, B. Blümich and F. Casanova, *J. Magn. Reson.*, 2009, **198**, 80–87.
23. E. Danieli, J. Perlo, F. Casanova and B. Blümich, *Magn. Reson. Microsc.*, 2008, 485–499.
24. D. L. Olson, T. L. Peck, A. G. Webb, R. L. Magin and J. V. Sweedler, *Science*, 1995, **270**, 1967–1970.
25. B. Behnia and A. G. Webb, *Anal. Chem.*, 1998, **70**, 5326–5331.
26. A. P. M. Kentgens, J. Bart, P. J. M. Van Bentum, A. Brinkmann, E. R. H. Van Eck, J. G. E. Gardeniers, J. W. G. Janssen, P. Knijn, S. Vasa and M. H. W. Verkuijlen, *J. Chem. Phys.*, 2008, **128**, 052202.
27. F. D. Doty, in *Encyclopedia of Magnetic Resonance*, eds. R. K. Harris and R. Wasylishen, Wiley, New York, 2007.
28. K. R. Minard and R. A. Wind, *Concepts Magn. Reson.*, 2001, **13**, 128–142.
29. K. R. Minard and R. A. Wind, *Concepts Magn. Reson.*, 2001, **13**, 190–210.

30. C. J. Jones and C. K. Larive, *Anal. Bioanal. Chem.*, 2012, **402**, 61–68.
31. R. M. Fratila and A. H. Velders, *Annu. Rev. Anal. Chem.*, 2011, **4**, 227–249.
32. C. Massin, A. Daridon, F. Vincent, G. Boero, P.-A. Besse, E. Verpoorte, N. F. de Rooij and R. S. Popovic, *Micro Total Anal. Syst. 2001*, 2001, 438–440.
33. C. Massin, C. Azevedo, N. Beckmann, P. A. Besse and R. S. Popovic, International IEEE-EMB Special Topic Conference on Microtechnologies in Medicine & Biology 2nd Annual, 2002.
34. C. Massin, F. Vincent, A. Homsy, K. Ehrmann, G. Boero, P.-A. Besse, A. Daridon, E. Verpoorte, N. F. De Rooij and R. S. Popovic, *J. Magn. Reson.*, 2003, **164**, 242–255.
35. H. Ryan, S.-H. Song, A. Zaß, J. Korvink and M. Utz, *Anal. Chem.*, 2012, **84**, 3696–3702.
36. J. Anders, G. Chiaramonte, P. SanGiorgio and G. Boero, *J. Magn. Reson.*, 2009, **201**, 239–249.
37. V. Badilita, K. Kratt, N. Baxan, J. Anders, D. Elverfeldt, G. Boero, J. Hennig, J. G. Korvink and U. Wallrabe, 2011 IEEE 24th International Conference on Micro Electro Mechanical Systems (MEMS), 2011.
38. H. Lee, T.-J. Yoon, J.-L. Figueiredo, F. K. Swirski and R. Weissleder, *Proc. Natl. Acad. Sci. U. S. A.*, 2009, **106**, 12459–12464.
39. P. Bergveld, *Sens. Actuators, A*, 1996, **56**, 65–73.
40. J. M. Rothberg, W. Hinz, T. M. Rearick, J. Schultz, W. Mileski, M. Davey, J. H. Leamon, K. Johnson, M. J. Milgrew and M. Edwards, *Nature*, 2011, **475**, 348–352.
41. J. Bausells, J. Carrabina, A. Errachid and A. Merlos, *Sens. Actuators, B*, 1999, **57**, 56–62.
42. M. J. Milgrew, P. A. Hammond and D. R. S. Cumming, *Sens. Actuators, B*, 2004, **103**, 37–42.
43. F. Patolsky, B. P. Timko, G. Zheng and C. M. Lieber, *MRS Bull.*, 2007, **32**, 142–149.
44. F. Patolsky, G. Zheng, O. Hayden, M. Lakadamyali, X. Zhuang and C. M. Lieber, *Proc. Natl. Acad. Sci. U. S. A.*, 2004, **101**, 14017–14022.
45. N. Sun, Y. Liu, L. Qin, G. Xu and D. Ham, 2012 Proceedings of the ESSCIRC (ESSCIRC), 2012.
46. D. Ham and R. M. Westervelt, *Solid-State Circuits Soc. Newsl., IEEE*, 2007, **12**, 4–9.
47. H. Lee, D. Ham and R. M. Westervelt, *CMOS biotechnology*, Springer, 2007.
48. H. Lee, Y. Liu, E. Alsberg, D. E. Ingber, R. M. Westervelt and D. Ham, ISSCC. 2005 IEEE International Digest of Technical Papers. Solid-State Circuits Conference, 2005, 2005.
49. H. Lee, Y. Liu, D. Ham and R. M. Westervelt, *Lab Chip*, 2007, **7**, 331–337.
50. H. Lee, Y. Liu, R. M. Westervelt and D. Ham, *IEEE J. Solid-State Circuits*, 2006, **41**, 1471–1480.
51. H. Yoon, K. Y. M. Yeung, V. Umansky and D. Ham, *Nature*, 2012, **488**, 65–69.

52. W. F. Andress, H. Yoon, K. Y. M. Yeung, L. Qin, K. West, L. Pfeiffer and D. Ham, *Nano Lett.*, 2012, **12**, 2272–2277.
53. K. Y. M. Yeung, H. Yoon, W. Andress, K. West, L. Pfeiffer and D. Ham, *Appl. Phys. Lett.*, 2013, **102**, 021104.
54. D. Ha, K. Woo, S. Meninger, T. Xanthopoulos, E. Crain and D. Ham, *Very Large Scale Integr. (VLSI) Syst., IEEE Trans.*, 2012, **20**, 1590–1601.
55. K. Woo, S. Meninger, T. Xanthopoulos, E. Crain, D. Ha and D. Ham, Solid-State Circuits Conference-Digest of Technical Papers, 2009. ISSCC 2009, IEEE International, 2009.
56. N. Sun, H.-S. Lee and D. Ham, 2012 Proceedings of the ESSCIRC (ESSCIRC), 2012.
57. N. Sun, H.-S. Lee and D. Ham, *IEEE Trans. Circuits Syst. II: Express Briefs*, 2008, **55**, 877–881.

CHAPTER 7

Detection Using SQUIDs and Atomic Magnetometers

MICHELLE ESPY*[a,b], IGOR SAVUKOV[a], AND PETR VOLEGOV[a]

[a]Applied Modern Physics Group, Physics Division, Los Alamos National Laboratory, PO Box 1663, Los Alamos, New Mexico, USA 87545;
[b]Non-destructive Testing and Evaluation Group, Applied Engineering Technology Division, Los Alamos National Laboratory, PO Box 1663, Los Alamos, New Mexico, USA 87545
*E-mail: espy@lanl.gov

7.1 Introduction

In this chapter, by "mobile" MRI we are discussing systems that can operate outside of a traditional hospital or laboratory setting. For example, such a system might be one that can be deployed and set up in a field hospital or even at a bedside in an intensive care unit when a patient is too ill to be relocated. In principle, there is already "mobile" MRI for medical imaging in the sense that truck-deployed systems can and do exist to serve, for example, rural areas far from a hospital. These systems are, however, largely identical to what one would find in a hospital, taking the required hospital infrastructure for a large magnetic field (>1 T) and mounting it on wheels. In this chapter, we will take the approach of describing MRI in a rather different regime, which we call ultra-low field (ULF). Specifically, ULF MRI involves very low measurement magnetic fields from 10 μT to 10 mT, *i.e.* Larmor frequencies from ~400 Hz to 400 kHz. This approach often uses pre-polarization from tens to hundreds of mT to enhance the signal and very sensitive sensors.

New Developments in NMR No. 5
Mobile NMR and MRI: Developments and Applications
Edited by Michael L. Johns, Einar O. Fridjonsson, Sarah J. Vogt, and Agnes Haber

Published by the Royal Society of Chemistry, www.rsc.org

Both the Superconducting Quantum Interference Devices (SQUIDs) (see for example ref. 1–5) and atomic magnetometers (AM)[6] have been demonstrated as sensors that improve detection performance at the correspondingly low frequencies of ultra-low field MRI, with the vast majority of demonstrations being SQUID-based.[7] In addition to the fields being much lower than traditional MRI, the polarization field (B_p) and measurement field (B_m) are different. In this regard, ULF MRI resembles pulsed-field or field cycling MRI.

ULF systems pay a significant price in signal-to-noise (SNR) and thus speed and spatial resolution; however, they may provide advantages to mobile MRI applications for the following reasons: (1) ultra-low magnetic fields that do not require very high homogeneity[8] are easier to generate, and therefore lighter and less expensive systems may be realized; (2) ultra-low magnetic fields are safer for public places (*i.e.* MRI screening for security[9]) and are more tolerant and safer in the presence of metal;[10,11] (3) at very low fields unique pulse sequences such as those based on changing the orientation of the measurement field can be employed; (4) because T_1 changes as a function of the magnetic field, it may be possible to exploit different image contrast at ULF,[12,13] or vary image contrast by changing the magnetic field strength;[14] (5) ultra-low magnetic fields may be safer for certain populations, *e.g.* patients with medical implants, neonates; and (6) in some cases generation of high magnetic field is prohibitive, and there is no other option for MRI.

In this chapter, we will describe MRI systems based on the approach of ultra-low magnetic field and sensitive readout with SQUIDs or AMs. To provide the reader with an immediate idea of ULF images that have been produced to date, in Figure 7.1 we present images of the human brain based on sensitive detection with SQUIDs[3] and an AM.[6] Clearly, these images are

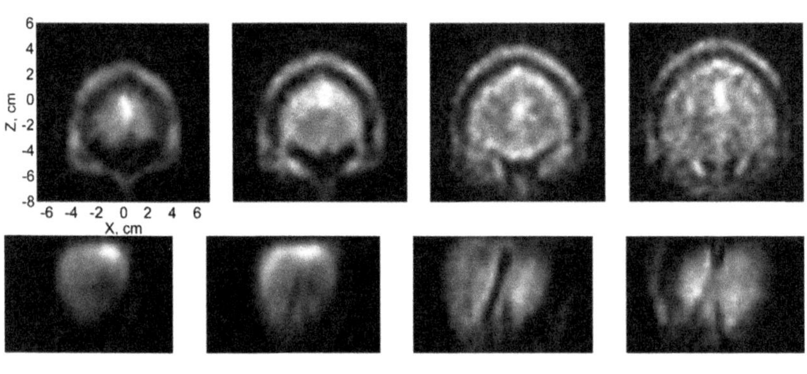

Figure 7.1 Upper Row: coronal slices of the human brain from the front of head, acquired with SQUID sensors. The slices are from the bottom of the cryostat (~18 mm above the head) in 6 mm depths. Pre-polarization was ~30 mT, with readout at 46 μT. An array of 7 SQUIDs was used to acquire the image, which consists of 5 averages and took ~45 min to acquire. Data from ref. 3. Lower: coronal slices of the human brain from the back of the head acquired with a single atomic magnetometer in ~5 mm depth slices. Pre-polarization was ~80 mT and readout ~4 mT. The total acquisition time was 13 min. Data from ref. 6.

lower in spatial resolution than what one is used to seeing in a clinical high field MRI system based on large superconducting magnets. While the fields used to generate these images are very low, and there is likely much room for improvement,[15] the question of trade-off between mobility and quality has to be carefully considered.

We will first introduce the basics of each sensor technology and the hardware considerations in performing MRI with them. Then we will describe several applications motivated by the advantages described above, discussing both the existing challenges and potential for improvement. Although at the time of this writing there are very few examples of ultra-low field MRI based on SQUIDs or AMs, let alone mobile ones, we will strive to illuminate applications to mobile MRI that might benefit from this approach. We will conclude with a discussion of how to model such systems, so that the reader can determine the "best case" images that can be expected.

7.2 Sensors

The signal in MRI with Faraday coil sensors, used as detectors in the vast majority of MRI systems, is proportional to the product of the polarization, B_p, and readout, B_m, magnetic field strengths. Since most traditional MRI scanners have a single fixed field for both, usually called B_0, the signal is said to scale as B_0^2. The B_p factor comes from the increase in signal with polarization. The factor proportional to the readout field, B_m, is due to the fact that coils perform better when there is more EMF, *i.e.* when the field is changing faster at higher frequencies. The behavior of noise is more complicated and depends on such factors as body conductivity and proximity effects, among others. For the Faraday coil when the intrinsic sensor noise dominates (below 0.065 T in a typical MRI geometry), SNR per voxel scales as $B_pB_m^n$ or B_0^{1+n}, where $0 < n < 1$ depending on the coil parameters, and at higher fields linearly with the field. Thus, as the field is lowered the SNR for MRI with a traditional coil rapidly decreases. Sensors like the SQUID and AM don't have the same dependence on measurement frequency (*e.g.* for an un-tuned SQUID, the magnetic field sensitivity is largely independent of frequency). Thus, it makes sense to consider using SQUIDs or AMs in ultra-low field MRI applications, below some frequency at which there will be a benefit to replacing the Faraday detectors. There is no hard "rule" as to what this frequency is. According to ref. 16, the trade-off might occur somewhere around 100 kHz for SQUID sensors. But of course, the issues of specific system configuration and practicality should also be factored in. Thus the importance of modeling any system to the fullest extent possible is reiterated. Here we describe the principle of operation and requirements of both SQUID and AM sensors.

7.2.1 SQUID Basics

A SQUID is a loop of superconductor interrupted by one or two weakly superconducting junctions (known as the Josephson junction). In this chapter we will be referring to applications of SQUIDs using two junctions, known as

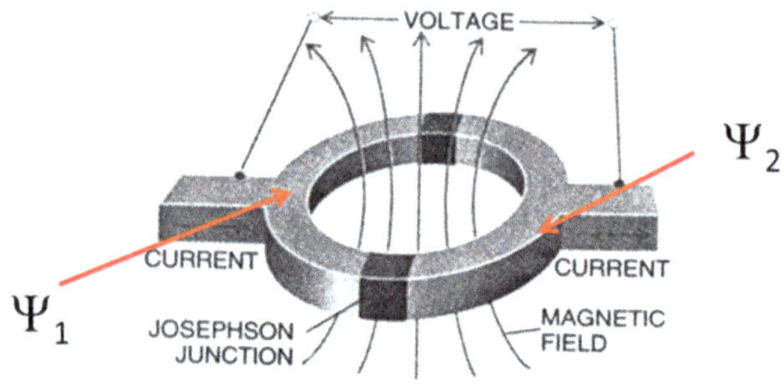

Figure 7.2 Schematic of a SQUID.

the DC SQUID. A schematic is shown in Figure 7.2. We will also be describing SQUIDs that are made of low temperature superconducting material (*e.g.* niobium), passively cooled to ~4 K by immersion in cryostats of liquid helium. An excellent description of the physics of SQUIDs can be found in ref. 17 and of practical SQUID design and applications in ref. 18.

During operation a bias current is applied to the SQUID above the critical current of the Josephson junctions, causing them to become resistive. The voltage, V, across the SQUID's junctions is oscillatory in the presence of changing flux (Figure 7.3a) Φ, with the period of oscillation being quantized as:

$$\Phi_0 = \frac{h}{2e} \cong 2.07 \times 10^{-15}\ T \cdot m^2 \tag{7.1}$$

where Φ_0 is known as the flux quantum. During typical operation, the SQUID is kept at a fixed working point where $dV/d\Phi$ is maximum, by application of current from a feedback coil. Typically the SQUID is connected to a gradiometer coil to reduce background from distant sources. A typical schematic for this is shown in Figure 7.3b.

The flux coupled to the SQUID from this configuration is:

$$\Phi_{\text{SQUID}} = \frac{M \cdot \Phi_p}{L_i + L_p} \tag{7.2}$$

where Φ_{SQUID} is the flux seen by the SQUID, Φ_p is the flux in the pickup coil, M is the mutual inductance between the SQUID and the input coil, L_i is the inductance of the input coil, and L_p is the inductance of the gradiometer pickup coil.

A typical low temperature SQUID has a noise floor of ~1 fT/√Hz (10^{-15} T/√Hz) and a dynamic range of 140 dB, with a magnetic field response that is flat from ~dC to MHz. As has been shown by ref. 19 SQUIDs typically outperform all other magnetic field sensors (including the tuned coils used for traditional MRI), at frequencies below 100 kHz. The relatively newer atomic magnetometer, which in principle can outperform a SQUID, will be discussed in Section 7.2.2.

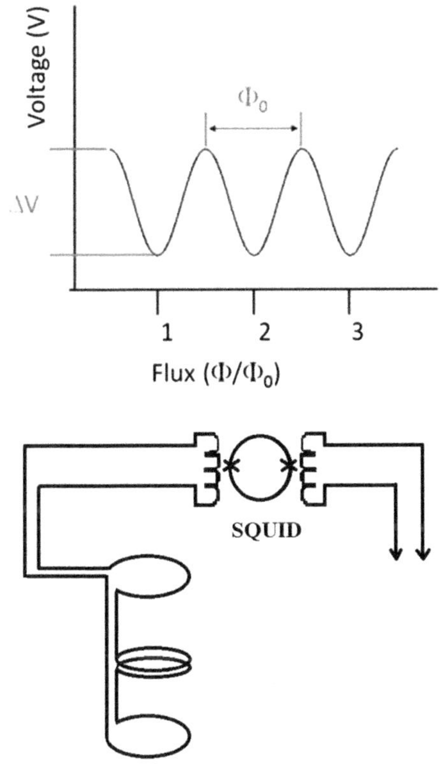

Figure 7.3 (Upper) Voltage *vs.* flux output, upper. (Lower) 2nd order axial gradiometer input to SQUID (left coil) and feedback coil (right coil).

Because of the exquisite sensitivity, SQUIDs are routinely used for extremely sensitive magnetic field measurements at low frequencies, such as magnetoencephalography (MEG),[20] which is the non-invasive and passive measurement of magnetic fields due to neuronal activity measured at some distance (~cm) from the head. The typical signal strength for MEG is from fT to pT depending on the activity at frequencies ~100 Hz and lower.

More recently, SQUIDs have been shown to be capable detectors for MRI at very low magnetic fields (and frequencies), see for example ref. 1 and 21 and references therein, where use of a Faraday coil would result in a severe loss in sensitivity. That very same sensitivity is also a practical downside to the use of a SQUID. For example, the magnetic noise associated with someone moving through the lab with a screwdriver at ~5 m is 10^{-9}–10^{-7} T, many orders of magnitude higher than what we are trying to measure. Thus, depending on the application, the issues of shielding from ambient noise can become quite challenging. For example, low-frequency noise from power-lines is a major issue for MEG. Most MEG applications take place inside large magnetically shielded rooms (MSRs). While SQUIDs certainly can operate outside an MSR for the higher frequency applications, the issues of keeping the SQUIDs within their dynamic range and keeping them protected from radiofrequency interference (to which they are highly susceptible) are major considerations in the design of any MRI system using SQUIDs.

7.2.2 Atomic Magnetometers

Atomic magnetometers have achieved fT sensitivity[22,23] and can potentially compete with SQUIDs in various applications, including that of magnetoencephalography[24,25] and ULF MRI.[6,26,27] Additionally, they have the advantage of non-cryogenic operation.

A typical AM is depicted in Figure 7.4. The central element is the atomic cell. It contains alkali-metal atoms, which have unpaired electron spins, nitrogen gas to quench excited states (30–60 mTorr), and optionally another buffer gas, usually helium, to reduce diffusion to the walls. The cell is heated to increase the density of alkali-metal vapor to the level of 10^{14} cm^{-3}. With this number density, the average spin $\boldsymbol{S}$ can be represented with a vector. Like nuclear spins, atomic spins react to a magnetic field and can be used for its detection. In AM, alkali-metal atomic spins are polarized and detected with light with high efficiency. Almost 100% polarization can be achieved with relatively low power, on the order of mW. The spin-pumping mechanism has been investigated in detail, especially in relation to production of hyperpolarized ^{3}He and ^{129}Xe gases.[28] The pump laser has to be tuned to the D1 line rather than the D2 line to achieve maximum polarization and allow the beam to propagate through optically thick vapor. In potassium AMs, the pump laser is tuned to 770 nm, and in Rb to 794 nm. With recent progress in laser technology, it is now possible for inexpensive high-quality diode lasers to be used for high-sensitivity AM.

The high sensitivity of AM, in addition to optical pumping enhancement, is due to very sensitive detection of spin orientation using the Faraday effect. The Faraday effect is the rotation of linear light polarization by atoms. Because of selection rules, the interaction of left and right circularly polarized light is unbalanced when the atomic spin projection is non-zero along the beam direction, and polarization rotation takes place. Light is polarized with a polarizer, and after the atomic cell, it is analyzed with a polarizing beam splitter (PBS), which is oriented at 45 degrees with respect to initial polarization.

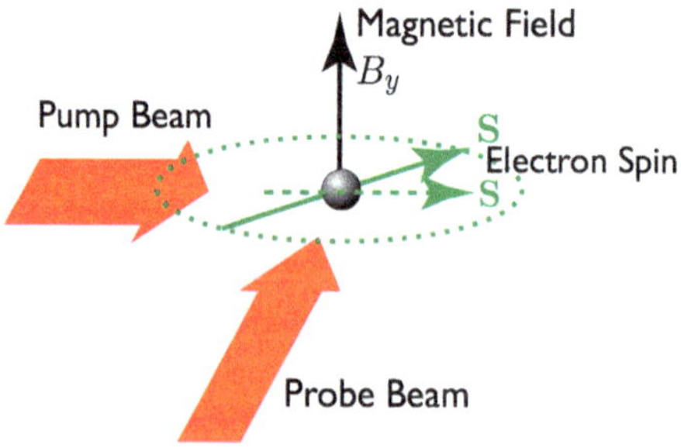

Figure 7.4 Principles of atomic magnetometer.

Any change in polarization orientation leads to differences in the outputs of the PBS. The difference amplifier output is directly proportional to the small polarization rotation. The sensitivity of optical detection of spins is limited by photon shot noise. Because linearly polarized light can be represented as the equal mixture of right and left circular polarization light, with each fluctuating according to a Poisson distribution, the noise of polarization angle is limited by $1/\sqrt{N_{Ph}}$, where N_{Ph} is the number of photons. This noise is quite low (an nrad range) even at a moderate laser power of a few mW.

Apart from this noise, spin-fluctuation noise can also limit the sensitivity.[29] This type of noise occurs due to quantum fluctuations of projections of the spin. Although as we mentioned above for a large number of spins the average spin behaves as a vector with 3 well-defined components, there are still small uncertainties arising from quantum fluctuations. The spin noise can be derived from the uncertainty principle. The spin noise scales as $1/\sqrt{N_{Spin}}$, where N_{Spin} is the number of spins. In a typical AM, the fundamental noise is much below 1 fT/Hz$^{1/2}$ and in practical systems, especially at a high frequency of operation, the sensitivity is not far off from the fundamental sensitivity. Considering that SQUID sensitivity is about 1 fT, AMs that demonstrate 0.2 fT/Hz$^{1/2}$ are quite promising.[30]

While the sensitivity of SQUIDs and AMs is roughly the same, they are not very similar in terms of other application properties. A key difference is that the AM is sensitive to static field and gradients, while SQUIDs are not, at least when precise gradiometers and feedback loops are implemented. In direct NMR/MRI detection, the MRI field and gradients have to be removed at the AM cell location. The field affects the AM frequency of maximum sensitivity, while the gradients broaden atomic magnetic resonances and reduce sensitivity. The difference in gyromangetic ratios between protons and K (Rb-87) spins is about 400, so the gradient applied in MRI will broaden the AM bandwidth (BW) 400 times beyond optimal operation. For these reasons, AMs in general are more problematic for MRI applications than SQUIDs, but solutions for these problems exist.

One strategy is remote MRI detection,[31] in which the liquid sample is arranged to flow from a pre-polarization region to an encoding region and to a detection region. All three regions can have different fields because they are far from each other. The prepolarization field can be produced by a strong non-uniform permanent magnet. This field determines the strength of the signal. The encoding region contains a uniform field and imaging gradients. Earth's field can be used to reduce the number of field-generating coils. The detection regions, placed in a shield, have a field close to zero and the AM operates at very low frequencies. MRI detection based on shuttling the sample[32] can be also called remote NMR. The disadvantage of remote MRI where the encoding region is outside the detection region is that one encoding step is done per excitation and time is lost by moving the spins from the encoding to the detection region. MRI as a rule takes a long time. It is also hard to implement anatomical imaging in this manner, and no demonstration of such imaging has been done.

The second strategy is based on ultra-low frequency detection (~kHz) with a NMR/MRI sample placed in a solenoid to separate the NMR and AM fields.[33,34] Because the field is quite low, field separation can be done with a solenoid that does not require a high level of perfection. Experimentally it was shown that this approach can be extended for NMR detection to frequencies of 60 kHz, but the effects of non-uniformity and field "leakage" from the solenoid become substantial. This approach also has the problem of the influence of MRI gradients on the AM sensitivity. No MRI has been demonstrated, and the problem of MRI gradients was not experimentally investigated. The confinement of the NMR sample into a solenoid is one disadvantage for imaging of large objects. Alternatively, the solenoid can be placed around the AM, but this has also not been demonstrated. In principle, this method can be useful for MRI, especially when kHz frequency operation is needed to penetrate into metallic objects. For high-frequency applications, with various precision winding techniques it can be possible to improve the uniformity of "the manual" method by at least an order of magnitude. Still, the read-out gradient has to be compensated at the AM to avoid loss of sensitivity.

The most currently feasible approach for anatomical imaging with an AM is detection mediated with a flux-transformer (FT) (Figure 7.5). FTs are widely used with SQUID magnetometers, where they are cryogenic and superconducting. The FT consists of two coils, an input coil that generates voltage from the MRI signal and an output coil electrically connected to the input coil that generates the magnetic field. Basically, an FT transfers the magnetic field

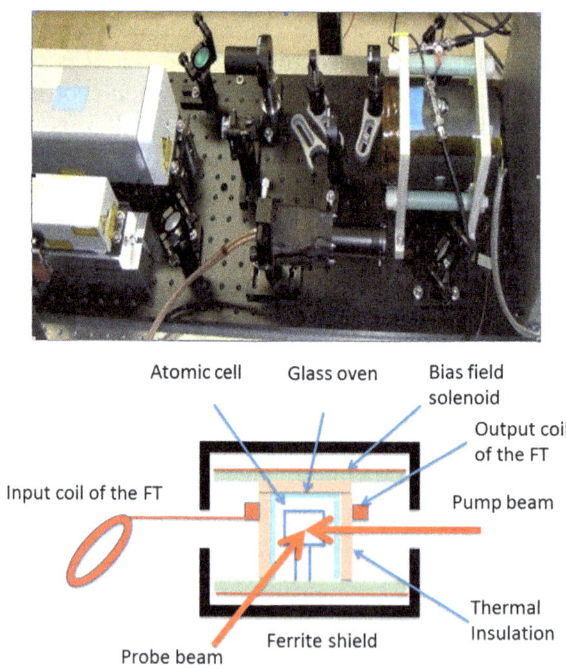

Figure 7.5 Atomic magnetometer and flux transformer.[6,27]

from the MRI region to the detection region. The input coil can be configured as a gradiometer to reduce noise. The output coil is placed near a magnetometer, SQUID or AM. AMs can in principle be used with a cryogenic FT, but the overall advantage of non-cryogenic operation would be lost. We have recently demonstrated MRI using a room temperature FT. The room temperature FT brings flexibility in the positioning and simplicity in operation. The room temperature FT removes the DC and attenuates low-frequency components of the field and gradients, thereby solving the problem of decoupling AM and MRI field systems. However, room temperature operation comes at the price of additional Johnson noise. Because with frequency the Johnson noise as well as the noise of surrounding metallic parts decreases, the increase in frequency can be a very efficient method for noise reduction.

Apart from DC offset and gradients, the BW of operation is another important issue for AMs if fast imaging and good resolution are needed. The following examples illustrate the required BW. If the acquisition window is set to 100 ms and the image has 100 pixels in the read-out direction, the MRI spectrum will be 1 kHz. Fast imaging sequences rely on multiple acquisitions per single prepolarization cycle, so the BW will increase proportionally to the number of acquisitions. With 10 acquisitions per excitation, BW becomes 10 kHz. This will be difficult to achieve at the 3 kHz frequency used in SQUID MRI experiments. The sensitivity of an AM and its BW are inversely related, so when an AM is configured to have the best sensitivity, its BW will be below 1 kHz. This BW might be sufficient in some cases, but for fast imaging methods it will definitely be a limiting factor. The FT + AM system allows broader BW when FT and AM are tuned to different frequencies (Figure 7.6), as we observed in actual experiments. Using multi-channel operation can further reduce the BW requirement, since each sensor would cover a smaller area of the object and the number of pixels would be reduced. The possibility of sensitive uncorrelated multi-channel operation is one of the significant advantages of the FT + AM method *vs.* the coil + capacitor + amplifier method.[6]

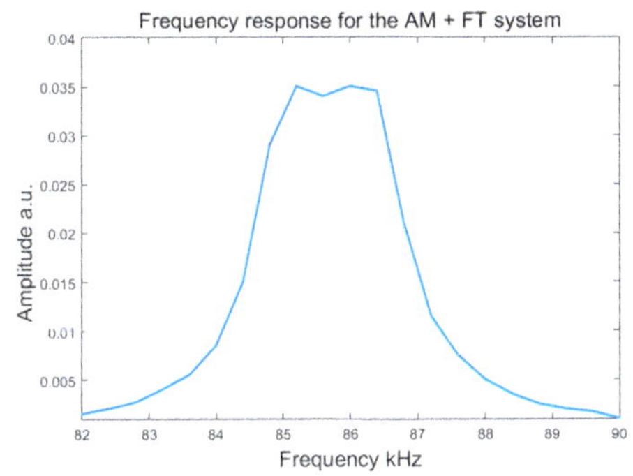

Figure 7.6 Response of AM + FT system, fairly broad and flat at the top, with input sensitivity about 1 fT/Hz$^{1/2}$.[6]

Other distinct features of AMs are their immunity to radiofrequency interferences and various potentially harmful "transients." AMs, unlike SQUIDs, do not require broad-band noise suppression and suppression of strong radio-frequency components. However, strong pulses can lead to temporary loss of sensitivity, and time of the order of 30 ms can be required for re-pumping spins, which are destroyed during pulses. Thus AMs used in ULF MRI will have some "dead time" issues. It is possible to reduce dead time by either detuning AM away from the strong excitation by applying a magnetic field[30] or by increasing the pumping rate.[34] In the AM + FT method, the input coil can be made orthogonal to the excitation coil, and transients can be zeroed with geometrical adjustments.

In terms of mechanical arrangement, further differences exist. SQUIDs are typically submerged at the bottom of a Dewar filled with liquid helium. AMs do not have this restriction. Although often AMs are constructed with a rigid optical configuration, recent trends have been to implement AMs with fiber-optics coupling, allowing flexible positioning.[25,35] Apart from this, when the FT + AM configuration is employed, the AM does not need to be located close to the imaging object.

7.2.3 Brief Overview of Hardware

A simple schematic of an ultra-low field MRI system based on SQUIDs is shown in Figure 7.7.

Examples of the field generation hardware used to acquire the images shown in Figure 7.1 are presented in Figure 7.8. As with traditional MRI, systems using SQUIDs or atomic magnetometers require the same basic components. These include: (1) magnetic fields for polarization, readout,

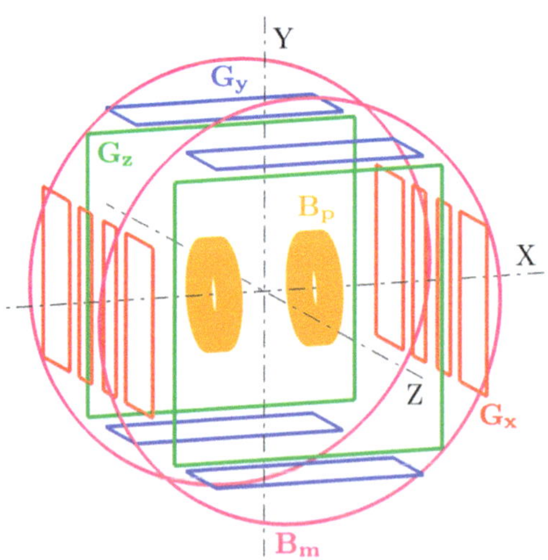

Figure 7.7 Schematic of the basic components of a SQUID-based ultra-low field MRI system.

spin-flip, and gradient encoding; (2) a detector; and (3) some method to generate pulse sequences and record them. One aspect that requires a bit more rigor than traditional MRI is shielding. This is because often the ambient magnetic fields and gradients (*e.g.* Earth's magnetic field, gradients associated with the presence of metal) are comparable in strength to the MRI fields. Here we will briefly discuss these components with special emphasis on differences associated with the use of lower magnetic fields and sensor approaches. More complete discussions of SQUID-based hardware can be found in ref. 21 and 36. In principle, both systems shown in Figure 7.8 could be mobile. What is not shown in the SQUID-based system, however, is the large magnetically shielded room (MSR) that encloses the system. This is, in some sense an artifact of the system also being designed for MEG, which demands such a magnetic shield. In MEG, these MSRs consist of one or more sandwiched layers of aluminum and mu-metal. SQUID-based systems that aren't required to perform MEG have been demonstrated (see for example ref. 1). However, such systems have still typically been located inside large room-like aluminum eddy-current and RF shields. If the issue of compact shielding can be resolved, the field generation is quite compatible with a mobile MRI concept. The AM-based system is designed with a small external shield that is rather compact, and thus demonstrates what might be possible.

7.2.3.1 *Magnetic Field Generation*

To improve SNR in ULF MRI the pre-polarization field approach is used as in pulsed low-field MRI[37] and field-cycling MRI.[38] Typically, such systems pre-polarize samples with a B_p of ~100 mT to achieve higher polarization, and then record spin-precession in a much weaker (~10 μT–3 mT) magnetic field B_m,[39] which we will refer to as the ultra-low field (ULF) regime. We note

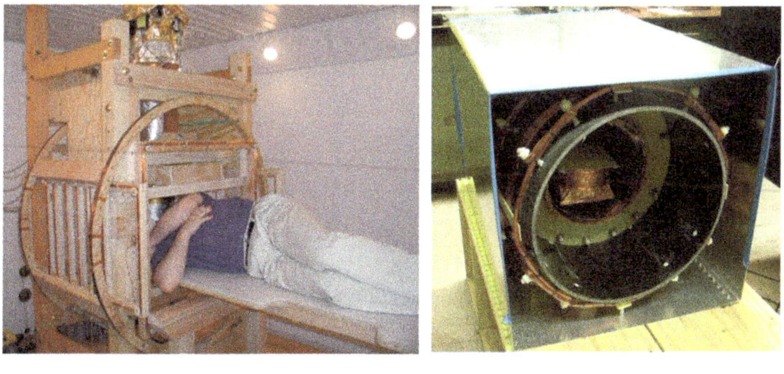

Figure 7.8 The SQUID (left) and AM (right) MRI systems used to acquire the images shown in Figure 7.1. The SQUID system is housed inside a large magnetically shielded room. The AM system (right) is located inside its own set of eddy-current shields.

that in high field MRI these fields would be the same, and are traditionally referred to as B_0. In ultra-low field MRI, B_p and B_m are generated by entirely different magnets, as is shown in Figure 7.4. Such an approach results in a much lower requirement for field homogeneity to achieve narrow line-widths than at high fields—since that is set by the absolute field strength.[8] A lower field readout (~kHz frequencies) also results in the ability to penetrate though metal and handle distortions[10,11] and reduced susceptibility to artifacts. Additionally, pulse sequences such as the measurement field echo, obtained by reversal of the B_m field, can be implemented, which is not possible at high fields. An example of such a pulse sequence used to generate the SQUID-based brain images in Figure 7.1 is shown in Figure 7.9. Perhaps most practically interesting, a low and/or variable B_m field strength may reveal unique contrast[12,13,40] because T_1 and T_2 are known to vary with magnetic field (see for example ref. 41).

However, there are some disadvantages to a pulsed field approach. The signal is often lost due to longitudinal polarization decay (T_1 is usually shorter at low fields and might approach T_2 due to the motional narrowing effect) during the ramp-down and delay necessary to avoid significant field transients. These transients are often induced in nearby conducting materials and can briefly exceed the dynamic range of the SQUIDs or distort the signal. The "dead time" can be as long as tens of milliseconds. Pre-polarization might also pose challenges for SQUID-based ULF MRI in that noise due to flux trapping in superconducting materials can be significant, as reported, and may put an upper limit on pre-polarization.[42] Atomic magnetometers can completely recover from pulsed B_p fields. However, as we noted previously, if the AM is inserted inside a ferromagnetic shield, the fields and gradients from remnant magnetization can negatively affect the performance of the AM. This problem is not present when FT-mediated detection is arranged.

As can be seen from Figure 7.8, ULF MRI systems look rather different from traditional MRI machines; however, the general principles of manipulation of the magnetization are similar. Traditionally in MRI, spin reorientation or the "spin flip" pulse used to tip magnetization off the field axis and

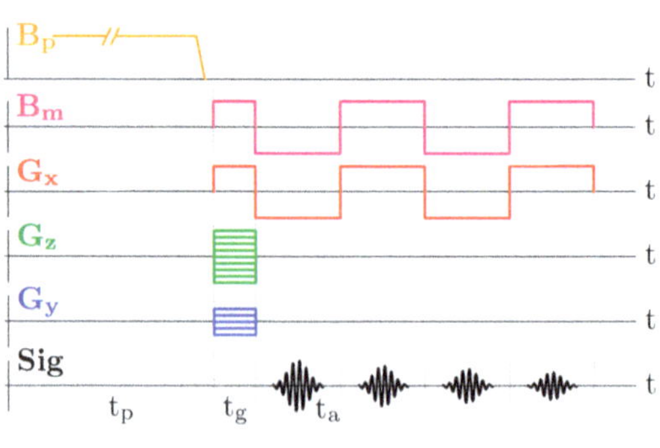

Figure 7.9 Example pulse sequence.

begin precession is done with a resonant field applied orthogonally to B_m for some time, usually to produce a 90° tip for maximum signal. This can also be the case in ultra-low field MRI: the magnetization is left aligned with B_m after pre-polarization and then a traditional spin-flip is used. However, because the fields can be reoriented or change amplitude in the ultra-low field technique, other methods for starting precession are also possible; see Chapter 4 in ref. 36 for more details. For example, in one approach, B_p and B_m can be mutually orthogonal, and precession begins when B_p is removed non-adiabatically such that the magnetization cannot follow the field, while B_m is turned on. In this case, the magnetization is left orthogonal to B_m and precession begins automatically. Practically speaking, when B_p and B_m are orthogonal and B_m is low, achieving a true adiabatic B_p ramp without exciting some spins is difficult. In this case, parallel B_p and B_m can be easier.

Encoding and readout gradients are largely used as in traditional MRI, although as mentioned the methods for producing echoes might be different.[21] Typical pulse sequences that have been employed so far are Fourier imaging,[39] as shown in Figure 7.9, and spin-warp (see for example ref. 1). However, because B_m can change orientation and field strength, one could envision novel pulse sequences such as projection imaging where the measurement field is rotated. This could be helpful for signal collection in the presence of surface coils oriented in many directions, for example like a helmet array for brain imaging. However, every time the magnetic field changes, remnant magnetic fields may be induced in any conducting material nearby (including magnetic shields) and thus considering how quickly fields can be reoriented in light of these effects and the dead time imposed by them is important.

Because B_m is low in strength and can in principle be varied easily, a unique opportunity for new contrast exists. As we mentioned, the variation in relaxation times as a function of frequency (B_m) and material can provide opportunities to enhance contrast. It is well known that T_1 differences exist between tissue types and between normal and cancerous tissue, and that contrast may be enhanced at fields ≤100 mT.[43] In some cases, this enhanced contrast may strongly outweigh the drawback of lower SNR compared to high field MRI, as has been shown by John Clarke's group at UC Berkeley in both phantoms[12] and excised prostate tissue.[13] Because one can vary B_m, the change in contrast as a function of magnetic field, T_1 dispersion, might also be accessible. Also, one can overlap the Larmor frequency with the msec–sec dynamics of molecular or physiological processes. However, probably most interesting for a mobile MRI application where we assume one of the largest challenges might be practical operation, it is also sometimes simply useful to be able to vary the Larmor frequency to avoid a particularly noisy part of the electromagnetic spectrum.

It is also important to mention that concomitant gradients are a more significant issue at ultra-low fields where SQUIDs and atomic magnetometers might be employed. There are numerous publications that explain and address this issue, for example ref. 44 and 45. We also note that transient fields or magnetization of nearby materials arising from B_p pulsing also produce artifacts that might need to be accounted for; see Chapter 4 of ref. 36 for more details. This appears especially important when performing imaging inside any sort of nearby magnetic shielding. We shall discuss this further in Section 7.2.3.3.

An interesting question to ask is: what are the optimal field and frequency for a ULF mobile system? The lowest end of ULF MRI, with fields of ~100 μT and Larmor frequencies of a few kHz, has been heavily explored with SQUID systems for applications such as combined MEG and MRI or unique contrast mechanisms. Here the advantages relevant to mobile MRI are extreme simplicity in generating uniform fields and a relaxed requirement on field stability. For example, for Helmholtz coils uniformity and stability in the order of 1000 ppm is sufficient. If one wants to construct a whole body scanner, this can be very easy. However, pre-polarization and a sensitive detector such as a SQUID or AM are typically required. Further, shielding at such low frequencies can become quite challenging.

At the other extreme of the ULF spectrum, field strengths of 3–10 mT, the systems for uniform field production become more complicated (Tetracoil, *etc.*) and the requirement for the current source stability goes beyond 100 ppm. Temperature instabilities either due to heating or the environment can become an issue. In our practical system,[26] heating of the coils led to field drifts that prevented long averaging to improve image quality. However, these problems with instabilities and uniformities can be solved with some additional effort. A more important advantage of higher frequencies is the simplification in shielding. In fact, if we only consider the shielding coefficient, environmental noise, and shield noise (*e.g.* Johnson noise), a high frequency of operation would win over any other advantages of low frequency operation. For example, the shield noise of copper foil drops to the level of 1 fT at 100 kHz frequency, while it is in the order of 10 fT at 1 kHz frequency. To achieve a 100-fold shielding factor at 2 kHz requires about 10 mm aluminum, while at 100 kHz just 1 mm copper foil is sufficient. In addition to this, transients become shorter at higher frequency as well as available bandwidth. If one intends to have multi-echo sequences, high frequency operation will be necessary. A multi-echo approach can remove effects of slow field and gradient transients without a complicated system for compensating them. With respect to AM applications with a flux-transformer, Johnson noise of the coil decreases with increasing frequency. Thus the sensitivity of AM + FT detection, based on our experience, becomes sufficient for anatomical imaging around ~60 kHz. Increasing the frequency from 120 kHz to 420 kHz might bring a factor of 2 increase in sensitivity, and hence would considerably improve image quality or reduce scan time. However, this again brings up the question of at what frequency a SQUID or an AM is really better than an inductive coil? To answer this question, it is very important to consider the application and the relative benefits to image quality. Appropriate modeling as discussed in Section 7.4 can greatly assist in answering these questions.

7.2.3.2 Sensor Arrays

In the case of a SQUID or AM due to the nature of the sensor needing to be cooled or heated, respectively, a volume coil configuration for detection can be quite challenging. Typically for SQUID- or AM-based MRI, the sensors are implemented as surface coils. And often, an array of such sensors is used to implement parallel imaging (see ref. 36 chapter 4). And while typical high field MRI relies on volume coils, there are many examples of parallel imaging methods that use arrays of surface coils, for example SENSE.[46]

With regard to AM applications in parallel imaging, two cases should be distinguished: direct detection and detection *via* FT. In direct detection, an advantage of realizing a multi-channel system with a single atomic cell is that it can make the system inexpensive and simple. AM have sensitivity on the order of 1 fT for 1 cm volume, so the direct detection with AM can be ideal for imaging flat objects such as a hand or small animals. The FT + AM method can also be realized in multi-channel modality. Here the advantage of using AM is to decouple input coils, since otherwise they would require resonance operation and decoupling would be extremely difficult, if not impossible for large arrays. An additional advantage of the FT + AM approach is that gradiometers can be arranged easily for FT input coils and the input coils can be arbitrarily positioned to cover the object.

With SQUID sensors, parallel imaging may be easier to implement at ULF as opposed to HF MRI because the noise at ULF is largely independent of the sample and because SQUID sensors are not highly inductively coupled like tuned coils. The array in general can help to solve the BW issue and accelerate the imaging. In some cases, fast imaging might be needed without high-resolution requirements, and so arrays can be the only possible solution.

In the SENSE method of parallel imaging, the number of phase-encoding steps, N, are decreased by a factor of R but the end values of the gradients are kept the same. As long as the end values of the gradients are the same, the spatial resolution will not change. However, taking N/R steps (instead of N steps) will reduce the field-of-view (FOV) by a factor of R. The reduction in FOV, using normal reconstruction methods, will produce an aliased image. The SENSE reconstruction described by Pruessmann *et al.*[46] uses linear algebra and sensitivity maps for each pickup coil to reconstruct the full FOV images from the aliased ones. This method has been successfully implemented with SQUID-based ULF MRI[47] as well. While image acceleration can be a big win for ULF MRI methods, there is some penalty in SNR when using an array to perform image acceleration. The SNR is reduced by a factor of $\sqrt{R}$ because the SNR in MRI is related to the number of steps, which is reduced by R-fold. In addition, there is an additional SNR penalty that depends on the underlying geometry of the coil array.

7.2.3.3 Shielding and Noise

Background electromagnetic noise, which will surely be found in almost any environment, affects ULF MR measurements in a few ways: (1) the magnetic background is superposed on top of the ULF MR signal of interest; (2) the background signal directly interferes with the sensor (*e.g.* electromagnetic noise in the RF regime directly interferes with the performance of a SQUID sensor, and fluctuating magnetic fields will interfere with the operation of the AM); and (3) the background electromagnetic noise (if large enough) may interfere with the fields we are trying to use to perform the ULF MRI itself. In fact, for mobile MRI it may be that shielding is the real practical limitation. There is likely no one solution to this issue, and a variety of requirements and tradeoffs must be examined, including the power spectral density (PSD) of the noise in the environment, details of the ULF MRI methods being employed (*e.g.* amplitude of pulsed fields and range of frequencies being measured, to name a couple) and the goals of the measurement. While it is pretty straightforward to shield fields >10 kHz it can be very challenging to shield static (DC) and low-frequency fields (<10 kHz). In principle, there are three categories of shielding approaches for ULF MRI: (1) passive shielding of both high- and low-frequency fields inside high-permeability MSRs; (2) passive high-frequency shielding by a highly conducting shield combined with static compensation of the local fields and gradients; and (3) active shielding of high- and low-frequency fields. The first method of passive shielding has been demonstrated for combined MEG and ULF MRI. Smaller versions of these shields have been shown in our deployed liquid screening systems.[9,48] Such shields are rather heavy and expensive when one considers a size large enough to accommodate an anatomical MRI application. Thus they are probably not practical for mobile MRI as we envision it (*i.e.* something that can be easily set up in a field hospital or moved around the intensive care unit). While various groups have used combined passive and active shielding for MEG, it has not yet been demonstrated in combination with ULF MRI, where the pulsed fields used to produce the MRI will present challenges for the active shield sensors. Groups are now starting to investigate the use of open purely "compensation"-based systems operated by using both static[49] and active generation of magnetic fields to control the measurement environment for the ULF MRI.[50] As noted above, the pulsed fields used to produce the MRI present challenges for the active shield sensors. Nevertheless, this may be a path towards expedient and lightweight shielding, and advances in this area are likely critical to achieve an effective mobile MRI system. For a more detailed discussion in the context of SQUID sensors, we refer the reader to Chapter 2 of ref. 36.

7.3 Current Applications of ULF MRI

There have been very limited examples of what would be considered "mobile" MRI using SQUIDs and AMs. In part this has been because one large driver for the ultra-low field MRI methods where these sensors are used is the combination of MEG with MRI, which requires the use of a heavy MSR, typically not portable. Here we very briefly mention some ongoing applications of SQUIDs- and AMs-based ultra-low field MRI that have been demonstrated, especially in the context of practicality for mobile applications.

7.3.1 MEG and MRI

The motivation for pursuit of this application is that MEG[20] provides millisecond temporal resolution of the neuronal activity, as compared to much slower (~1 s) functional MRI which measures the blood flow or oxygenation changes related to neuronal activity. However, to obtain the location of neuronal activity derived from a typically large (>300) SQUID array requires the solution of a fundamentally ill-posed inverse problem. An anatomical image, from high field MRI, is typically required in addition to the MEG to provide anatomical constraints. At present, MEG and an anatomical MRI acquired in a separate system are the state of the art. The MEG demands a sensor with exquisite sensitivity and full head coverage. Such measurements are routinely only achieved thus far by SQUIDs (although promising demonstrations with AM sensors are being made with partial head coverage[24,25]). To capture the small (fT to pT) signals from the brain requires sensitivity that leaves SQUID and AM sensors entirely incompatible with conventional MRI hardware (which operates at typically > 1T strengths). Thus at present, MEG requires two complex and expensive machines, one for MEG and one for MRI, as well as a resolution-degrading data co-registration procedure.

To resolve these issues, the proposal to combine MEG with ULF MRI has been quite attractive since one could imagine a single detector system based on a common sensor platform to perform both. In this case, the magnetic fields for the MRI are simply reduced to the point where the sensor has enough dynamic range to still operate (typically μT). This reduces the system to one set of hardware with all data co-registered. This is a clear application where ULF MRI based on ultra-sensitive detectors provides an obvious advantage over conventional MRI.

The ability to combine MEG and ULF MRI has been reported with SQUID hardware by LANL twice[3,4] and a group based in Aalto University.[5] Systems have revolved around traditional MEG hardware, including the large MSR. To be mobile, such a system would have to resolve the challenge of low-frequency shielding (the MEG signal is typically < 100 Hz). A SQUID-based system for MEG has been demonstrated that uses a compact superconducting outer shield;[51] however, this hasn't been shown to be "mobile" in the sense that we defined at the beginning of this chapter. Further complications arise when combining MEG and MRI because dynamically varying magnetic fields (required for MRI) in the presence of the MSR can introduce transient magnetic fields that can exceed the dynamic range of the sensors and introduce artifacts. A self-shielded B_p coil design[52] or the use of field compensation coils and/or an optimized MSR design[53] can somewhat mitigate these transients. In addition, some of the transient artifacts that arise can be corrected (see for example ref. 36, Chapter 3). One true challenge to simultaneous or interleaved MEG and ULF MRI is that the transients in the shielding associated with the MRI are at low frequencies and virtually impossible to de-convolve from the MEG signal.[4] However, for many potential applications, simultaneous or interleaved MEG and ULF MRI might not be required. For example, the measurements could be made sequentially in the same instrument. In this case, removal of artifacts in the MRI associated with the MSR is still an issue, but may be resolvable.

In terms of reduction of weight of the shield and making the system portable, it has been shown that MEG and MRI can be performed (for now each separately) without a large shielded room. There have been demonstrations of MEG acquired with AM sensors[24] inside a rather compact cylindrical 3-layer mu-metal shield. ULF MRI has been performed inside a mu-metal cylinder without noticeable artifacts (unpublished result by Savukov). However, combining MRI with MEG in a single system inside a small shield will encounter similar difficulties if not more severe: (1) AM performance in the presence or pulsed fields can be impaired; and (2) the MRI pulsed magnetic fields will induce transients and noise affecting both MRI and MEG regardless of sensor. Challenging as these difficulties might be, solutions do exist through a compromise: to perform MRI and MEG separately but inside the same system, and perhaps with separately optimized sensors for the MEG and MRI.

7.3.2 Unique Contrast and Anatomical Imaging

One of the most important drivers for ULF MRI, more generally, is the unique T_1 contrast not available in conventional high-field MRI. That T_1 changes as a function of field is well known.[41,54] The work of John Clarke, for example, has shown that the contrast between benign *vs.* cancerous prostate tissue at ULF is marked[13] and that the improved contrast-to-noise ratio (CNR) can make up for the lower signal-to-noise ratio (SNR) at ULF.[12] Clarke's group has also demonstrated that appropriate pulse sequences taking advantage of variable magnetic fields can further highlight the available contrast,[1] and other groups have shown that ULF enables unique methods to access contrast, such as T_1 encoding by the polarization field.[55] Additionally, the work of the German group at PTB has shown that even for well-studied materials, such as water, unexpected contrast mechanisms can arise at very low, <1 kHz, Larmor frequencies.[56] However, moving to very low Larmor frequencies again raises the issue of shielding and whether or not such a system provides any benefits by being mobile.

We began this chapter with a discussion of ULF MRI with AM or SQUID sensors as being perhaps more suitable for mobile MRI because in such situations it might be challenging to produce or use high magnetic fields. We note that thus far there has not been a demonstration of ULF MRI in this class of applications, but work is in progress on the underlying technologies that would be required in terms of sensors, imaging methods, and shielding. We also state again that appropriate modeling of any such system, as described in Section 7.4, is vital for understanding what such a mobile MRI system for anatomical imaging will be able to do.

7.3.3 Materials

The issue of screening of liquids inside closed containers is one where ULF NMR/MRI with a SQUID sensor has truly been made portable. One implementation of ULF NMR/MRI methods was to develop a SQUID-based relaxometer for the inspection of liquids in a security checkpoint. The ULF approach used pre-polarization followed by readout at very low magnetic fields, so signals can penetrate most packaging (including aluminum cans) owing to large skin depths. By moving to ULF, the magnetic fields are low, safe for the public, and conducive to screening through most kinds of streams-of-commerce (SOC) packaging, including aluminum cans. An imaging relaxometer known as MagViz was developed with the support of the Department of Homeland Security Science and Technology Directorate and was used to demonstrate the inspection of liquids by relaxometry.[9,57,58] By measuring T_1 in the pre-polarization field (~2 MHz) using different polarization times, while measuring T_2 at the much lower measurement field (~2 kHz; at 2 kHz $T_1 \cong T_2$), we are able to discriminate between many different liquid types using T_1 dispersion and comparison to a database. MagViz was shown to be able to screen liquids in an airport with SQUID-based ULF MRI. The MagViz unit was also used for some limited anatomical imaging.[57] However, we note that in subsequent incarnations of the device SQUID sensors were not used[59] to keep the size (~2 m on a side driven by shielding), scan time, and cost down. This was done at the expense of multiple bottle screening.[48]

7.4 How to Model System Performance

In this section we cover the topics of how to model a system, choose reference model points (performance at several plausible configurations), use the model to compare sensor performance at various field strengths, and optimize the system. Then we show how such a model might be used to determine the best system for the task at hand: *e.g.*, does a constant field system with coils (B_0 = 20 mT) work better than a pulsed field system with SQUIDs (B_p = 20 mT)? Our main motivation is to try and convince the reader that they ought to model to the fullest extent possible. This point of view was very eloquently laid out by Myers.[16] Everything is (in principle) known except the contrast in a ULF MRI system, and there is no point in building a device without a clear notion of what the resulting image quality will be.

7.4.1 How to Model an MRI System

We start our description of the simulation framework by introducing a simplified equation for a detected MRI signal. Assuming that everywhere inside a sample the magnetic field is aligned along the *z*-axis, the precession of the magnetic moment at any arbitrary position $\boldsymbol{r}$ of the sample can be written as:

$$\boldsymbol{m}(\boldsymbol{r},t)\mathrm{d}^3 r = e^{-\frac{t}{T_2(\boldsymbol{r})}} \cdot e^{\mathrm{i}\int_0^t \omega(\boldsymbol{r},t')\mathrm{d}t'} \cdot \hat{\boldsymbol{e}}_+ \cdot (\hat{\boldsymbol{e}}_+^\dagger \cdot \boldsymbol{m}_0\, \mathrm{d}^3 r) + \mathrm{c.c.} \tag{7.3}$$

where $\boldsymbol{m}(\boldsymbol{r},t)\mathrm{d}^3r$ is the magnetic moment of a voxel with volume d^3r around position $\boldsymbol{r}$ at time t, $\boldsymbol{m}_0\mathrm{d}^3r \equiv \boldsymbol{m}(\boldsymbol{r},0)\mathrm{d}^3r$ is the initial magnetic moment of the voxel, $T_2(\boldsymbol{r})$ is transverse relaxation time of the sample at position $\boldsymbol{r}$, $\omega(\boldsymbol{r},t) \equiv \gamma|\boldsymbol{B}(\boldsymbol{r},t)|$ is the instantaneous Larmor frequency at position $\boldsymbol{r}$ and time t, $\hat{e}_{\pm} = \hat{e}_x \pm \mathrm{i}\hat{e}_y$, and $\hat{e}_x$, $\hat{e}_y$, $\hat{e}_z$ are unit vectors aligned with x-, y-, z-axis respectively, and c.c. denotes the complex conjugate. Here we also assumed that the initial magnetization of the sample is orthogonal to the magnetic field.

The signal detected by a sensor is proportional to the magnetic field flux generated by the magnetic moment:

$$\mathrm{d}S(\boldsymbol{r},t) = V(\omega)\cdot e^{-\frac{t}{T_2(\boldsymbol{r})}}\cdot e^{i\int_0^t \omega(\boldsymbol{r},t')\mathrm{d}t'}\cdot\left(\boldsymbol{F}^{\dagger}(\boldsymbol{r})\cdot\hat{\boldsymbol{e}}_+\right)\cdot\left(\hat{\boldsymbol{e}}_+^{\dagger}\cdot\boldsymbol{m}_0\right)\mathrm{d}^3r + \text{c.c.} \quad (7.4)$$

Here $\mathrm{d}S(\boldsymbol{r},t)$ is the MRI signal generated, $V(\omega)$ is a coefficient connecting magnetic field flux through the sensor pick-up coil to the sensor output, *i.e.* signal of the sensor per unit flux, and a vector function $\boldsymbol{F}(\boldsymbol{r})$, sometimes called "lead field", describes the response of a sensor to a unit magnetic dipole at the position $\boldsymbol{r}$. Specifically $F_x(\boldsymbol{r})$ is the magnetic field flux through the pick-up coil of a sensor generated by a unit magnetic dipole aligned along the x-axis, correspondingly $F_y(\boldsymbol{r})$ and $F_z(\boldsymbol{r})$ are the magnetic field flux due to dipoles aligned along the y- and z-axis. Due to the reciprocity principle $\boldsymbol{F}(\boldsymbol{r})$ is equal to the magnetic field at the position $\boldsymbol{r}$ generated by unit current in a sensor pick-up coil.[60] Here it is convenient to remind the reader that if the sensor in question is an untuned SQUID with a superconducting pick-up coil the coefficient $V(\omega)$ does not depend on the frequency and is simply constant. However, in the case of the inductive pick-up sensor this coefficient is equal to the angular frequency of spin precession: $V(\omega) = \omega$. In the terms of ordinary frequency, $\nu \equiv \omega/2\pi$: $V(\nu) = 2\pi\nu$.

Spatial encoding of the MRI signal is usually implemented by applying a magnetic field gradient $\boldsymbol{G}(t)$ that causes the local magnetic field and thus Larmor frequency to vary across a sample in a known way, such that:

$$\omega(\boldsymbol{r},t) = \omega_0 + \gamma\boldsymbol{G}(t)\cdot\boldsymbol{r} \quad (7.5)$$

Here we introduce the concept of the reciprocal space vector:[61,62]

$$\boldsymbol{k}(t) \equiv (2\pi)^{-1}\gamma\int_0^t \boldsymbol{G}(t')\mathrm{d}t' \quad (7.6)$$

and following the outline in Callaghan[63] (and references therein) we re-write eqn (7.4) as

$$\mathrm{d}S(\boldsymbol{r},t) = V(\omega_0)e^{\mathrm{i}\omega_0 t}\cdot e^{i(2\pi\boldsymbol{k}(t)\cdot\boldsymbol{r})}\cdot e^{-\frac{t}{T_2(\boldsymbol{r})}}\cdot\left(\boldsymbol{F}^{\dagger}(\boldsymbol{r})\cdot\hat{\boldsymbol{e}}_+\right)\cdot\left(\hat{\boldsymbol{e}}_+^{\dagger}\cdot\boldsymbol{m}_0\right)\mathrm{d}^3r + \text{c.c.} \quad (7.7)$$

Here we assumed that $V(\omega) \approx V(\omega_0)$ in the bandwidth of the MRI signal. The initial magnetic moment of the voxel $\boldsymbol{m}_0\mathrm{d}^3r$ is defined by following equation:

$$\boldsymbol{m}_0\mathrm{d}^3r = \chi_0\boldsymbol{R}_{\mathrm{sf}}(\boldsymbol{r})\boldsymbol{B}_{\mathrm{p}}(\boldsymbol{r})(1 - e^{-T_{\mathrm{p}}/T_1(\boldsymbol{r})})\cdot e^{-T_{\mathrm{e}}/T_2(\boldsymbol{r})}\cdot N(\boldsymbol{r})\mathrm{d}^3r \quad (7.8)$$

Here $\boldsymbol{B}_\mathrm{p}(\boldsymbol{r})$ is the applied magnetic field during polarization, $\boldsymbol{R}_\mathrm{sf}(\boldsymbol{r})$ is a rotation matrix describing rotation of magnetization due to polarization field ramp down (adiabatic or non-adiabatic) and/or spin flip pulses (if any), T_p is duration of the polarization, $T_1(\boldsymbol{r})$ is longitudinal relaxation time of the sample at position $\boldsymbol{r}$, T_e is the duration of the encoding step (if any), *i.e.* time between end of the polarization period and start of the readout, and $N(\boldsymbol{r})$ is spin density, *i.e.* number of nuclei in unit volume. Specific nuclear susceptibility χ_0, *i.e.* magnetic moment induced by unit magnetic field per nuclei, is defined by:

$$\chi_0 = \frac{\gamma^2 \hbar^2 I(I+1)}{3k_\mathrm{B} T} \tag{7.9}$$

where I is the spin of nuclei (for protons $I = 1/2$), T is sample temperature, $\hbar \equiv h/2\pi$ is the reduced Plank's constant (h here denotes the Planks constant) and k_B is the Boltzmann constant. As a reference: $h \approx 6.62606957 \times 10^{-34}$ J s, $k_\mathrm{B} \approx 1.3806488 \times 10^{-23}$ J K^{-1}. For protons at 300 K $\mu_0\chi_0 \approx 6.0744 \times 10^{-34}$ and $\mu_0 \equiv 4\pi \times 10^{-7}$ T^2 J^{-1} m^3 is the permeability of vacuum.

Finally, substituting eqn (7.8) into (7.7) and integrating the last equation over the sample volume V, we will get the equation for the detected MRI signal:

$$S(t) = e^{\mathrm{i}\omega_0 t} \cdot \int_V e^{\mathrm{i}(2\pi \boldsymbol{k}(t)\cdot \boldsymbol{r})} \cdot e^{-\frac{t}{T_2(\boldsymbol{r})}} \cdot M(\boldsymbol{r}) \cdot q(\boldsymbol{r}) \mathrm{d}^3 r + \mathrm{c.c.} \tag{7.10}$$

Here the function $M(\boldsymbol{r})$, often referenced to as a spatial sensitivity map, is defined only by parameters of a MRI system, *i.e.* hardware:

$$M(\boldsymbol{r}) \equiv V(\omega_0)\chi_0 N_0 (\boldsymbol{F}^\dagger(\boldsymbol{r}) \cdot \hat{\boldsymbol{e}}_+) \cdot (\hat{\boldsymbol{e}}_+^\dagger \cdot \boldsymbol{R}_\mathrm{sf}(\boldsymbol{r}) \boldsymbol{B}_\mathrm{p}(\boldsymbol{r})) \tag{7.11}$$

where N_0 is the spin density of water, introduced here for the convenient normalization. As a reference $N_0 \approx 6.65654 \times 10^{28}$ m^{-3}. Specifically, $2|M(\boldsymbol{r})|$ is the amplitude of the sensor signal produced by a water sample at position $\boldsymbol{r}$ per unit volume.

Function $q(\boldsymbol{r})$, called the contrasted spin density, is defined only by local properties of the sample and imaging sequence:

$$q(\boldsymbol{r}) \equiv (1 - e^{-T_\mathrm{p}/T_1(\boldsymbol{r})}) \cdot e^{-T_\mathrm{e}/T_2(\boldsymbol{r})} \cdot \rho(\boldsymbol{r}) \tag{7.12}$$

where $\rho(\boldsymbol{r}) \equiv N(\boldsymbol{r})/N_0$ is relative spin (proton) density, usually referred to as PD in the context of MR imaging.

The contrasted spin density $q(\boldsymbol{r})$ is primarily proportional to the local spin density $\rho(\boldsymbol{r})$ of the observed nuclides. When a suitable preparatory sequence is used, $q(\boldsymbol{r})$ can also be made to depend on the relaxation times $T_1(\boldsymbol{r})$ and $T_2(\boldsymbol{r})$ and on the diffusion coefficient $D(\boldsymbol{r})$; all local properties at point $\boldsymbol{r}$.

It should be mentioned that in the approach detailed in eqn (7.10) we are assuming that the signal is obtained from the entire sample. Another way to extract information is through the excitation of a sub-space of the sample, for example by slice selection. In this approach, a "slice select" gradient is applied in conjunction with a specific RF pulse to excite only a slab of the sample. In this case, the information from eqn (7.10) would simply involve a 2D integration over the slab.

Eqn (7.10) suggests that the starting point to simulate a MRI system is to compute its sensitivity map according to eqn (7.11). As follows from eqn (7.11), to compute the sensitivity map of an MRI system we need to compute magnetic fields $\boldsymbol{B}_{\mathrm{p}}(\boldsymbol{r})$ and $\boldsymbol{F}(\boldsymbol{r})$ generated by the polarization and the sensor pick-up coils over the field of view of the system. Expressions for the magnetic field in free space are well known, the most basic of which is the Biot–Savart Law, which gives the magnetic field due to a line current source in free space:

$$\boldsymbol{B}(\boldsymbol{r}) = \frac{\mu_0 I}{4\pi} \int_C \mathrm{d}\boldsymbol{l}' \times \frac{\boldsymbol{r} - \boldsymbol{r}'}{|\boldsymbol{r} - \boldsymbol{r}'|^3} \tag{7.13}$$

where I is the current, $\boldsymbol{r}$ is the observer location and the source $\boldsymbol{r}'$ is evaluated along the contour C, and μ_0 is permeability of vacuum. Note that eqn (7.13) is a near field approximation, *i.e.* it is valid only when the distance scale of the experiment is much less than the wavelength of the electromagnetic field. If this approximation is not true, one needs to use a generalized form of the Biot–Savart Law.[60]

In practice, a polarization coil is typically a solenoid, *i.e.* a combination of circular loops. A SQUID pick-up coil is typically an axial gradiometer consisting of several circular loops. To compute magnetic fields from such coils we need to integrate eqn (7.13) over a circular contour as shown in Figure 7.10. Doing this exercise we will get the following well-known formula for magnetic field generated by a circular loop. (The derivation is based on the following integral: $\int_0^{2\pi} \frac{\cos\varphi \mathrm{d}\varphi}{\sqrt{1 \pm b\cos\varphi}} = \frac{2}{b\sqrt{1+b}}\{K(k) - (1+b)E(k)\}$, where K and E

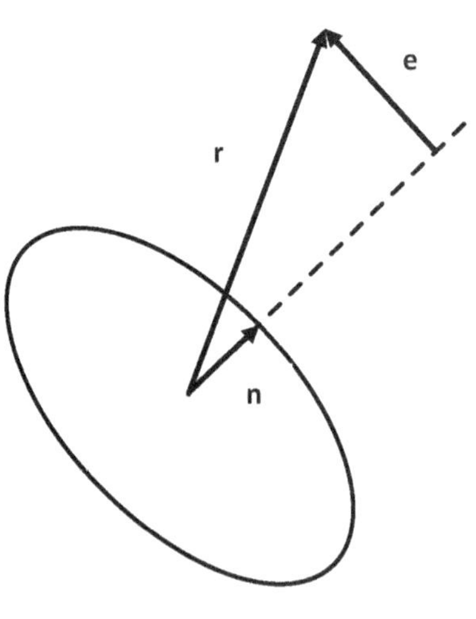

Figure 7.10 A circular coil.

are complete elliptical integrals of the 1st and 2nd kind. The argument is modulus k: $k = \sqrt{2b/(1+b)}$.):

$$\boldsymbol{B}_{\text{loop}}(\boldsymbol{r}) = \frac{\mu_0 I}{2\pi R\sqrt{Q}}\left\{\left(E(k)\frac{1-\alpha^2-\beta^2}{Q-4\alpha}+K(k)\right)\hat{\boldsymbol{n}} + \gamma\left(E(k)\frac{1+\alpha^2+\beta^2}{Q-4\alpha}-K(k)\right)\hat{\boldsymbol{e}}\right\} \quad (7.14)$$

Here the geometry of the problem (see Figure 7.10) is defined as follows: R is the loop radius, $\hat{\boldsymbol{n}}$ is the unit vector orthogonal to the loop plane, $\hat{\boldsymbol{e}}$ is the unit vector orthogonal to the loop axis laying in the plane defined by the loop axis and observation point, $\boldsymbol{r}$ is the vector from the loop center to the observation point, $h = (\boldsymbol{r}\cdot\hat{\boldsymbol{n}})$ is the distance, on the axis from the center of the current loop to the observation point, and $d = |\boldsymbol{r} - (\boldsymbol{r}\cdot\hat{\boldsymbol{n}})\hat{\boldsymbol{n}}|$ is the distance from the axis of the current loop to the observation point. Coefficients α, β, γ, Q, and k are computed from the geometry parameters: $\alpha = d/R$, $\beta = h/R$, $\gamma = h/d$, $Q = (1+\alpha)^2 + \beta^2$, and $k = \sqrt{4\alpha/Q}$. Functions $K(k)$ and $E(k)$ are complete elliptical integrals of the 1st and 2nd kind. (Note that here the argument is "modulus k". For a reference to complete elliptical integrals see M. Abramowitz and I. A. Stegun, *Handbook of Mathematical Functions*, Dover Publications, 1965. For useful polynomial approximation of complete elliptic integrals of the first and second kind, see Sections 17.3.34 and 17.2.36. The approximation error is less than 2×10^{-8}).

Another type of coil very often used in practice is a rectangular coil, which is a combination of straight line segments. Evaluating the integral in eqn (7.13) for a straight segment shown in Figure 7.11 we will get the following formula:

$$\boldsymbol{B}_{\text{seg}}(\boldsymbol{r}) = \frac{\mu_0 I}{4\pi}\cdot\left[\boldsymbol{r}_1 \times \boldsymbol{r}_2\right]\cdot\frac{r_1 + r_2}{r_1 r_2(r_1 r_2 + (\boldsymbol{r}_1\cdot\boldsymbol{r}_2))} \quad (7.15)$$

where $\boldsymbol{r}_1$ and $\boldsymbol{r}_2$ are vectors from the segment ends center to the observation point (see Figure 7.11).

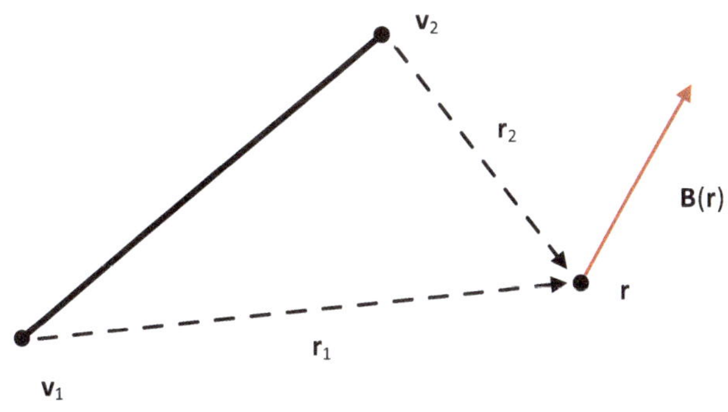

Figure 7.11 A straight line.

In general, representing an arbitrary current carrying wire as a combination of n straight segments $\{(\boldsymbol{v}_i, \boldsymbol{v}_{i+1})\}_1^n$, we can compute the magnetic field generated by such a wire as a sum of the fields generated by those segments:

$$\boldsymbol{B}_{\text{line}}(\boldsymbol{r}) = \frac{\mu_0 I}{4\pi} \cdot \sum_{i=1}^{n} \left[\boldsymbol{r}_i \times \boldsymbol{r}_{i+1}\right] \cdot \frac{r_i + r_{i+1}}{r_i r_{i+1}\left(r_i r_{i+1} + \left(\boldsymbol{r}_i \cdot \boldsymbol{r}_{i+1}\right)\right)} \tag{7.16}$$

where $\boldsymbol{r}_i = \boldsymbol{r} - \boldsymbol{v}_i$, are vectors from the points $\boldsymbol{v}_i$, which break a wire into straight segments, to the observation point.

Knowing the spatial sensitivity map of a MRI system $M(\boldsymbol{r})$, physical parameters of a sample $\{\rho(\boldsymbol{r}), T_1(\boldsymbol{r}), T_2(\boldsymbol{r})\}$, and imaging sequence times T_{p}, T_{e} it is straightforward to compute the MRI signal using eqn (7.10). Direct evaluation of the volume integral in eqn (7.10), however plausible, could be computationally expensive. One approach often used for efficient evaluation of eqn (7.10) is to split a sample into several objects assuming that each object has the same NMR relaxation parameters over the object. In this case we can write the contrasted spin density as a sum:

$$q(\boldsymbol{r}) = \sum_{i=1}^{N_0} C^{(i)} \cdot \rho^{(i)}(\boldsymbol{r}) \tag{7.17}$$

Here N_0 is number of the objects in the sample, $\rho^{(i)}(\boldsymbol{r})$ is spin density distribution of the i-th object, and constant $C^{(i)}$ contains all of aspects of contrast information for the i-th object. In the simplest case:

$$C^{(i)} = (1 - e^{-T_{\text{p}}/T_1^{(i)}}) \cdot e^{-T_{\text{e}}/T_2^{(i)}} \tag{7.18}$$

In this case the MRI signal can be evaluated efficiently using 3D Fourier transform:

$$S(t) = e^{\mathrm{i}\omega_0 t} \cdot \sum_{i=1}^{N_0} e^{-t/T_2^{(i)}} C^{(i)} \cdot \mathcal{F}\left\{M(\boldsymbol{r})\rho^{(i)}(\boldsymbol{r})\right\}\Big|_{\boldsymbol{k}(t)} + \text{c.c.} \tag{7.19}$$

Here $\mathcal{F}\{\ldots\}\big|_{\boldsymbol{k}(t)} \equiv \int_V e^{\mathrm{i}(2\pi \boldsymbol{k}(t)\cdot \boldsymbol{r})} \cdot \ldots \mathrm{d}^3 r$ denotes a Fourier transform, which can be computed very efficiently using an FFT algorithm.

It should be noted that eqn (7.10) and (7.19) give the MRI signal recorded by an ideal sensor. To make the simulations realistic we need to introduce distortions and noise caused by a measurement system, as described in Section 7.2.

Finishing our description of the simulation framework we would like to note that here we considered ideal measurement field and gradients. Effects of non-ideal gradients, such as concomitant gradients, and non-perfect spin-flip pulses or field inversion, *i.e.* "ghost" artifacts are considered in detail in Chapter 4 of ref. 36.

We illustrate the introduced MRI simulation framework using the example of the 7-channel ULF MRI system ULF MRI developed at LANL that was used to acquire interleaved MEG and brain images.[4] The detailed properties of the system have been described elsewhere.[39] Here we summarize the main parameters of the system used in the simulations:

- Polarization coil: a horizontal nitrogen cooled coil placed directly under a sample producing ~75 mT at the sample center.
- Sensors: seven 2nd order axial gradiometers with 120 mm baseline.
- Gradiometer pick-up loop: diameter 36 mm, the input inductance is 0.7 μH.
- SQUID: input inductance is 0.5 μH, the mutual inductance between the SQUID and pick-up loop is 8 nH, intrinsic noise $5\mu\Phi_0/\sqrt{\text{Hz}} \geq$ input flux noise $0.75\ \text{m}\Phi_0/\sqrt{\text{Hz}}$.
- Equivalent input field noise referred to one turn of the pick-up loop ~1 fT/√Hz.
- Signal conversion factor 1.3×10^{-13} Wb V^{-1} (referred to the flux in the pick-up coil), or 0.13 fT μV^{-1} (referred to one turn of the pick-up loop).
- Total noise of the system, *i.e.* "SQUID noise" + "ambient field noise" + "electronics noise" + *etc.*, at the output 10–30 μV/√Hz, or about 1.5–3 fT/√Hz (referred to one turn of the pick-up loop).

To characterize the performance of the system, *i.e.* resolution and signal-to-noise ratio, we used an acrylic phantom, as shown in Figure 7.12. The holes are 3 mm in diameter, 6 mm deep, and spaced 5 mm apart. During the experiments, the holes were filled with de-ionized water doped with $CuSO_4$ to obtain a transversal relaxation time $T_2 \approx 150$ ms.

Using these carefully measured parameters, we create a realistic computational model of the system under study, as shown in Figure 7.13.

The first step to simulate the system is to compute the sensitivity map of each sensor of the system according to eqn (7.11). Computing magnetic fields

Figure 7.12 Test object.

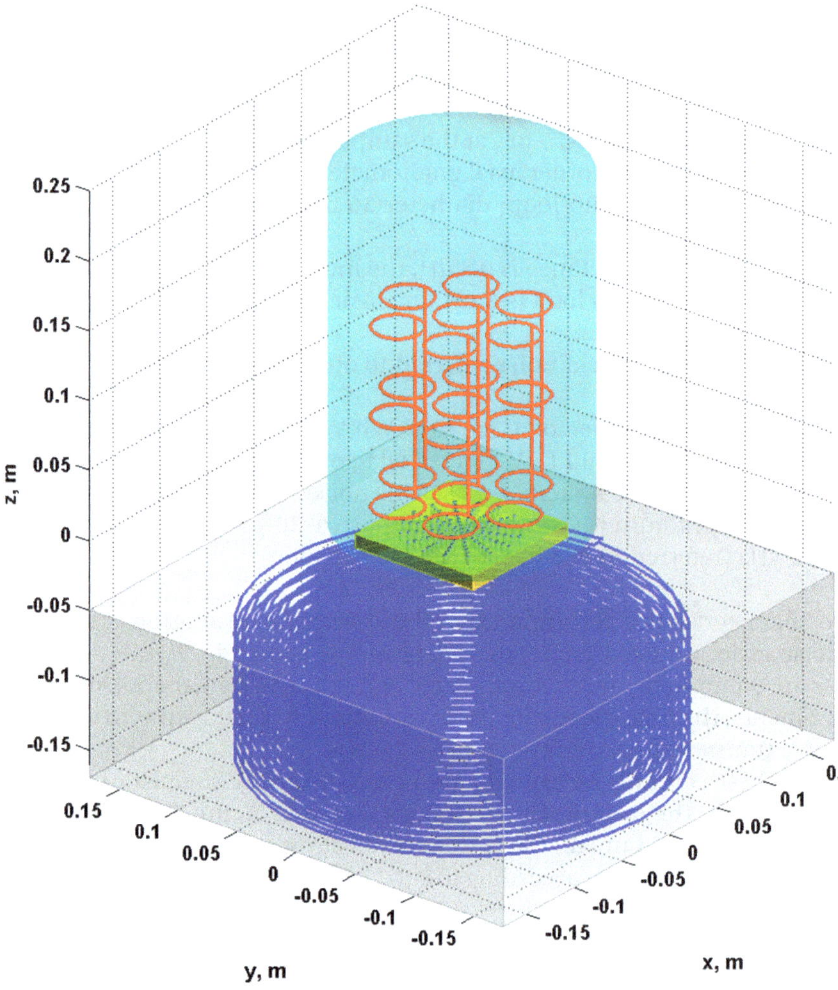

Figure 7.13 System simulation configuration showing gradiometer coils (red) in the Dewar (cyan cylinder), polarization coil (blue) inside a reservoir with liquid nitrogen (grey box), and the phantom (yellow rectangle with holes). Figure 7.12 shows the picture of the actual acrylic phantom.

inside the sample volume generated by a unit current through the polarization coil (shown in blue in Figure 7.13) and by a unit current in the sensor pick-up coils (shown in red in Figure 7.13) we will obtain a sensitivity map of the considered ULF MRI system, as shown in Figure 7.14.

Interpretation of this sensitivity map is simple. For example, from Figure 7.14 we can easily see the sensitivity associated with being located directly under any one of the seven sensors, and immediately deduce that if we place 1 cc water sample in the center of the field of view at the top of the phantom, polarize it long enough in the field generated by 1 A current in

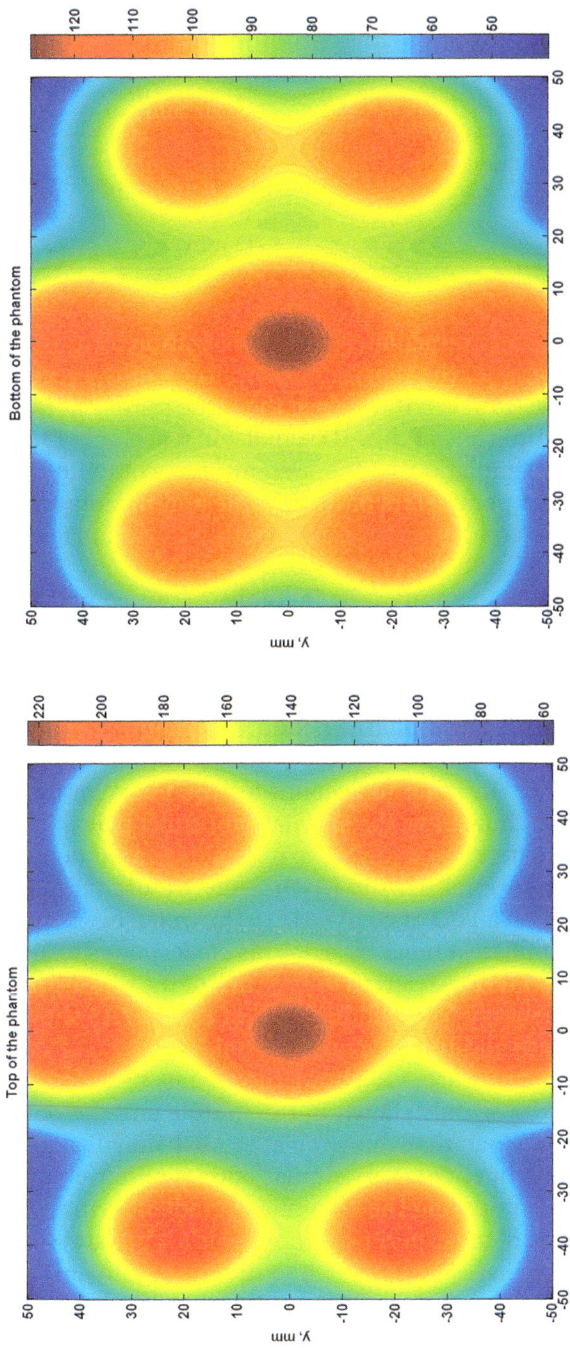

Figure 7.14 Sensitivity map of 7 ch ULF-MRI system. Color map: Amplitude of the signal, *i.e.* $2|M(r)|$, in μV per 1 cubic cm of a water sample polarized by 1 A current.

Table 7.1 Parameters of the imaging sequence.

Polarization time, T_p	2000 ms
Polarization field, B_p	75 mT
Encoding time, T_e	25 ms
Readout time, T_{acq}	50 ms
Readout gradient, G_x	22.0 Hz mm^{-1}
Number of phase encoding steps, N_y	81
Phase encoding gradient range, G_y	22.0 Hz mm^{-1}
Voxel size, dV	0.9 × 0.9 × 6 mm^3
Number of averages, N_a	10
Total imaging time, T_{tot}	~34 min

the polarization coil, and immediately after polarization record an FID signal generated by this sample, the maximum amplitude of the recorded signal will be about 220 μV. However, if we place the same sample at the bottom of the phantom, the amplitude of the FID signal will reach only 130 μV.

The next step in simulation of the system is to compute the expected MRI signals given the parameters of the sample and the imaging sequence. Parameters of the imaging sequence used both in experiments and for the simulation are shown in Table 7.1.

Using the parameters summarized in Table 7.1 we compute, according to eqn (7.19), the MRI signals. These signals are shown in the top panel in Figure 7.15. Here each line represents the signal $S_i(t)$ where subscript i = 1, ..., N_y numerates the phase encoding step. The thick red line corresponds to the step with a zero phase encoding gradient.

Note that the signals shown on the top panel of the figure are ideal, *i.e.* do not contain any noise. In order to account for the noise invariably present in a MRI system and thus make the simulation results plausible we need to add a noise signal. There are many ways to generate a random signal with the required power spectral density (PSD). Here we used a random phase periodic process[64] with a PSD of 20 μV/√Hz in the frequency band from 3000 Hz to 6000 Hz to simulate a noise signal. This corresponds to an equivalent field noise PSD of ~3 fT/√Hz (referred to one turn of the pick-up loop) such as we observe experimentally. The middle panel of Figure 7.15 shows the simulated MRI signals with addition of a noise signal.

The bottom panel of Figure 7.15 shows the signals recorded in the experiment. Comparing the signals shown in the middle and the bottom panels of the figure, it is obvious that the simulated MRI signals and the signals measured in the experiment are very similar both qualitatively and quantitatively. For example, comparing the zero phase gradient signals we have the maximum amplitude of the experimental signal is 8.4 mV and standard deviation (RMS) of the signal is 0.76 mV, while the maximum amplitude of the simulated signal is 7.8 mV and the RMS is 0.71 mV.

Finally, we apply an inverse Fourier transform both to the simulated and to the measured signals to reconstruct the phantom image. Figure 7.16 shows reconstructions of the phantom image. Once again it is obvious that the

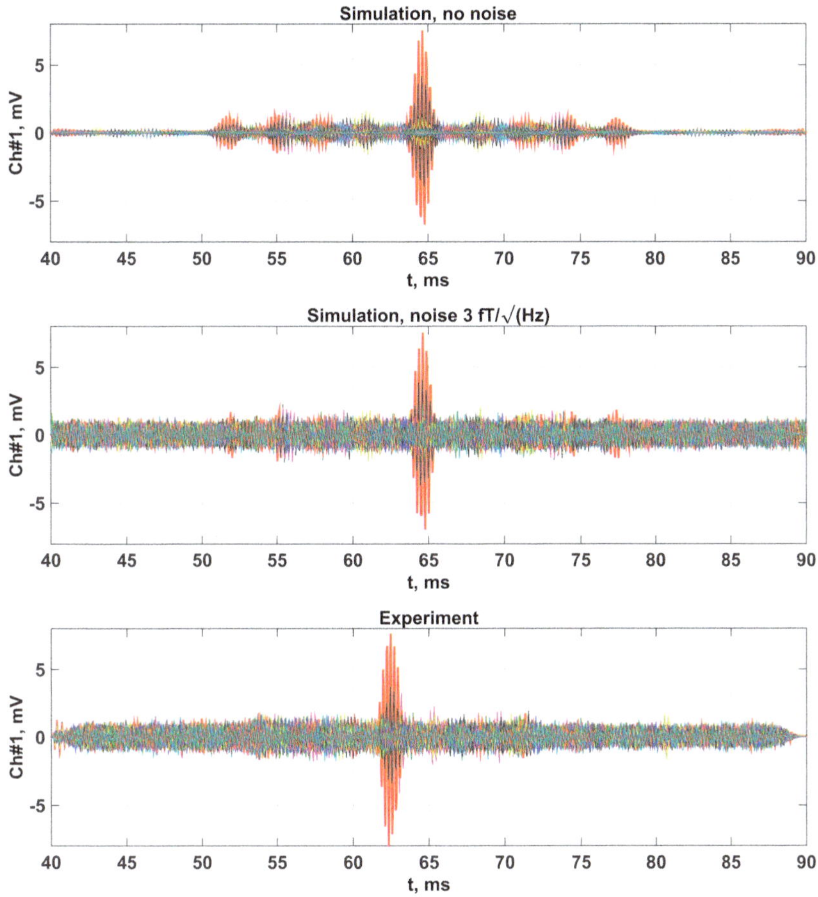

Figure 7.15 Simulated and measured signals.

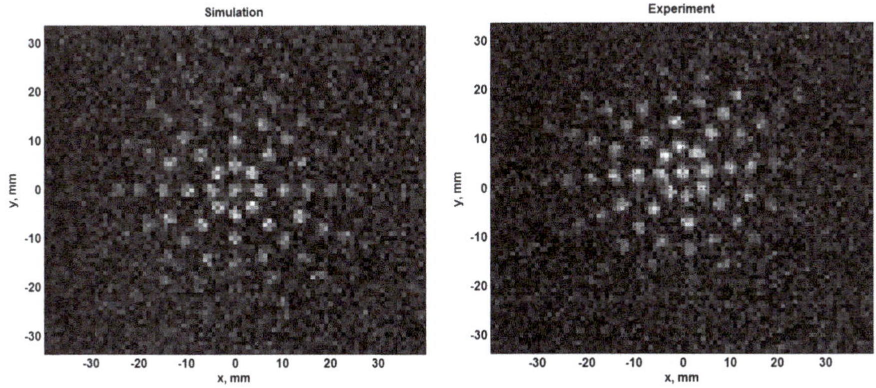

Figure 7.16 Reconstructed images. Left: simulation, right: experiment.

simulated MRI signals and the signals measured in the experiment produce qualitatively and quantitatively similar images.

Concluding our description of the simulation framework, it is very important to emphasize that to compute the simulated signals we did not use any "normalization" coefficients. The factor converting the magnetic field flux (Wb) in a sensor pick-up loop to the signal (V) was obtained independently, as described in Section 7.2, without any NMR signal measurement being involved. This shows that the described model has a predictive power and can be used to design and optimize ULF-MRI systems.

Having established the validity of the described simulation framework we can use it to simulate a novel instrument being considered or developed. These simulations are crucial (1) to initially demonstrate that the instrument in question is theoretically capable of generating the contrast and resolution at these novel ultra-low signal strengths at achievable acquisition times, and (2) to iterate the design trade-offs of the operating instrument to adjust to observed contaminants, constraints, and other limitations in order to still obtain a diagnostically useful image.

In order to achieve these goals one needs to employ a realistic test object for the simulations. Nowadays, due to the progress in computer-assisted tomography and magnetic resonance imaging, a plethora of computational phantoms exists. These phantoms range from individual organ phantoms, such as, for example, a brain phantom, to a whole body phantom.[5]

Here we will consider a human brain phantom developed by researchers at The McConnell Brain Imaging Center. This phantom known as the Montreal Neurological Institute (MNI) phantom[65] is standard in the neuroimaging research community, including research software funded by the National Institutes of Health (NIH). The phantom is derived from T_1, T_2, and PD-weighted images formed from the average of several scans of the same normal subject. The images are then segmented into eleven volumes that define the spatial distribution for different tissues (*e.g.* gray matter, white matter, CSF, vessels, bone, scalp, *etc.*) where voxel intensity is proportional to the fraction of tissue within the voxel. The volumes are defined at a 0.5 mm isotropic voxel grid. In Figure 7.17, sagittal slices of several representative volumes are show as an example.

To complete the phantom we need to attribute physical properties relevant to the simulation under study to each tissue class. In the case of MRI simulations we need to attribute at least three parameters to each tissue: a relative proton density (PD), a longitudinal relaxation time (T_1), and a transversal relaxation time (T_2). These parameters will adequately describe the majority of MRI sequences. However, if we want to simulate more complicated MRI sequences, for example, a pulsed-gradient-spin-echo (PGSE) sequence designed to extract the diffusion properties of tissue, we also need to assign a diffusion coefficient (D) to the tissue classes.

The physical properties of tissue that we will use in the following examples are summarized in Table 7.2.

Note that a i-th volume that defines the spatial distribution of the i-th tissue multiplied by the relative proton density of the tissue (parameter

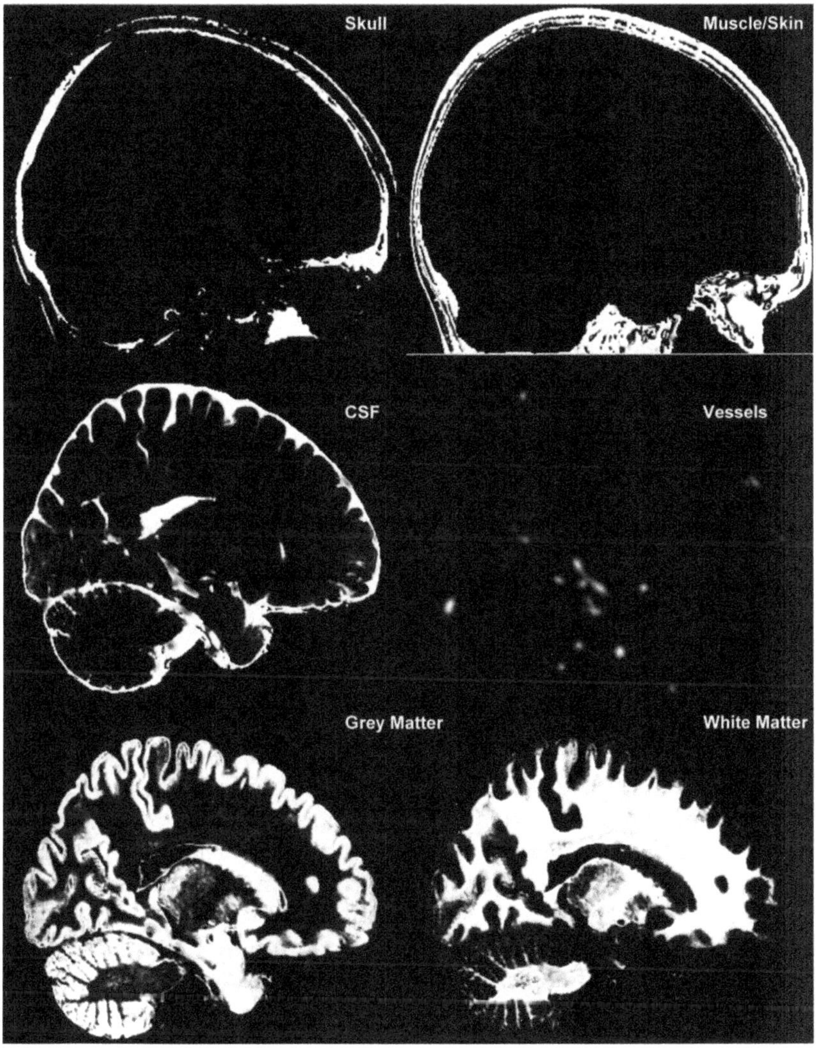

Figure 7.17 The Montreal Neurological Institute (MNI) phantom.

PD in Table 7.2) constitutes the function $\rho^{(i)}(\boldsymbol{r})$, *i.e.* absolute spin density, in eqn (7.17).

To start our discussion of the potential for mobile MRI, we will use this phantom to predict the performance of a mobile ULF-MRI device being considered at Los Alamos National Laboratory for the detection of brain injuries. Tentative parameters of the prototype of the system are listed below:

- Polarization coil: a water-cooled coil producing ~250 mT at the center of the field of view (FOV).
- Sensors: seven 2nd order axial gradiometers with 120 mm baseline.

Table 7.2 Physical properties of tissue used in examples.

Volume#	Tissue name	PD	T_1	T_2
1	CSF	1	2569	329
2	Grey matter	0.86	635	83
3	White matter	0.77	360	70
4	Fat	1	350	70
5	Muscle	1	120	47
6	Muscle/skin	1	120	47
7	Skull	0	0	0
8	Vessels	0	0	0
9	Connective	0.77	500	61
10	Dura matter	1	2569	329
11	Bone marrow	0.77	500	70

- Gradiometer pick-up loop: diameter 36 mm, the input inductance is 0.7 μH.
- Total noise of the system, *i.e.* "SQUID noise" + "ambient field noise" + "electronics noise" + *etc.*, is projected to be about 1 fT/√Hz (referred to one turn of the pick-up loop).

The computational model of the system is shown in Figure 7.18.

We can use our simulation with the goal of evaluating the proposed system for detection of brain abnormalities such as hemorrhage or hydrocephalus. To address this goal we added artificial inclusions to the MNI brain phantom. The first inclusion is a sphere of 10 mm in diameter containing 50% blood and 50% tissue placed at position (25 mm, −50 mm, 60 mm) in the Talairach coordinate system to simulate hemorrhage. (The Talairach coordinate system is defined by making two points, the anterior commissure and posterior commissure, lie on a straight horizontal line. Since these two points lie on the midsagittal plane, the coordinate system is completely defined by requiring this plane to be vertical. Distances in Talairach coordinates are measured from the anterior commissure as origin. Talairach coordinates are sometimes also known as stereotaxic coordinates. Talairach coordinates are defined such that the right hemisphere has positive X values, the anterior part has positive Y values, and the superior part has positive Z values; with AC being at coordinate (0,0,0). The second inclusion is a sphere of 25 mm in diameter containing 50% CSF and 50% tissue placed at position (17 mm, 17 mm, 20 mm) in the Talairach coordinate system to simulate hydrocephalus. In addition to those two inclusions, we also added a water cylinder of 20 mm in diameter and 200 mm in length placed vertically in the simulation FOV to provide a reference point to validate signal strength, SNR, *etc.*

To highlight the inclusions, we employed an inversion recovery sequence, which consisted of polarization of the sample for time T_{inv}, and then an inversion of the direction of the polarization field and polarization of the sample for time T_p. Parameters of the imaging sequences used in simulations are summarized in Table 7.3.

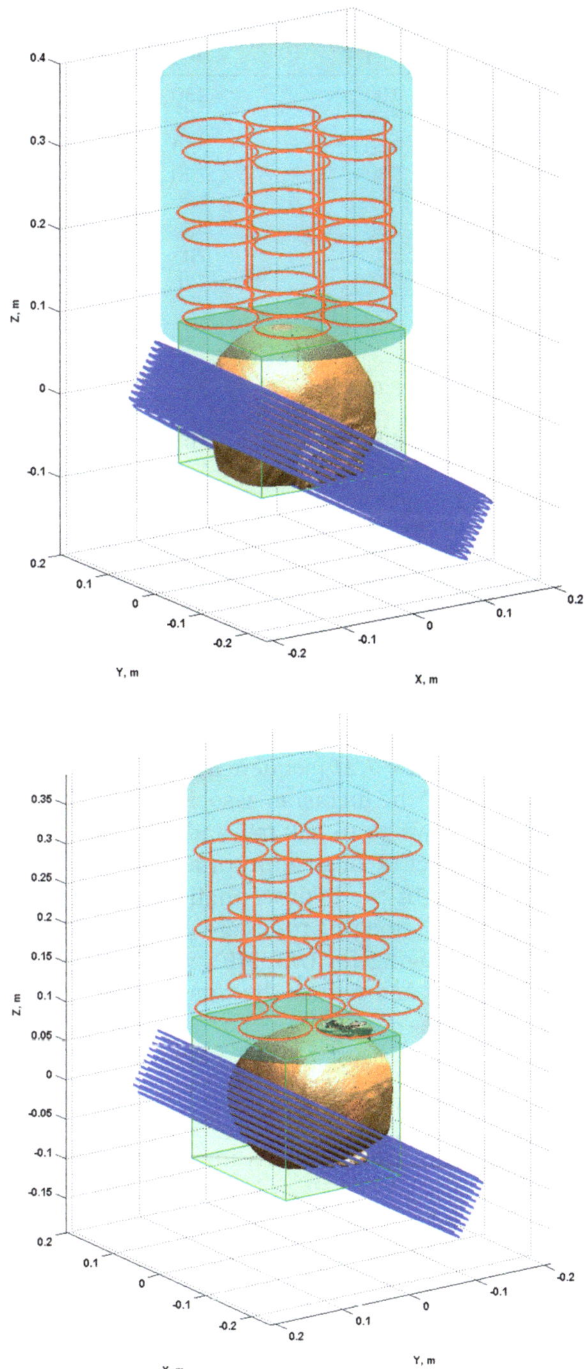

Figure 7.18 Left: coronal slices, right: sagittal slices.

Table 7.3 Parameters of imaging sequences used in simulation examples.

	Seq. #1	Seq. #2	Seq. #3
Polarization field at the phantom center B_p, mT	100	250	250
Polarization time T_p, ms	750	750	2750
Polarization inversion time T_{inv}, ms	750	750	100
Encoding time T_e, ms	35	35	35
Readout time T_{acq}, ms	70	70	70
Readout gradient G_x, Hz mm^{-1}	7.0	7.0	7.0
Phase encoding gradient range G_y, Hz mm^{-1}	7.0	7.0	7.0
Phase encoding gradient range G_z, Hz mm^{-1}	3.0	3.0	3.0
Number of phase encoding steps, N_y	103	103	103
Number of phase encoding steps, N_z	41	41	41
Voxel size dV, mm^3	2.0 × 2.0 × 4.8	2.0 × 2.0 × 4.8	2.0 × 2.0 × 4.8
System noise, fT/√Hz	1.8	0.9	0.9
Number of averages, N_a	1	1	1
Total imaging time T_{tot}, min	106.5	106.5	201.5

Results of the simulations are shown in Figure 7.19 (sequence #1), Figure 7.20 (sequence #2), and Figure 7.21 (sequence #3).

In Figure 7.19 we can see the inclusion associated with blood, but not that for hydrocephalus. When we go to 250 mT and 0.9 fT/√Hz we can see both (Figure 7.20), and with selection of an appropriate contrast weighting (Figure 7.21) we can really highlight the hydrocephalic inclusion. One might also note from the table that the imaging times are very long. This highlights the importance of parallel imaging for ULF MRI, which offers the potential to skip phase encoding steps and reduce the imaging time by factors of ten[66] or more.[65]

7.4.2 What if We Just Used Coils?

While we have seen that diagnostic quality images may be obtained (at least with $B_p > 100$ mT and noise ~1 fT/√Hz, it is a completely fair (and necessary) question to ask how things really compare to more traditional "low-field" MRI systems with more mainstream coil-based readout. Direct attempt to replace SQUIDs or AM sensors with an inductive detection is destined to fail at ULF-MRI. As an illustrative example, let us consider noise properties of a 7-channel inductive magnetometer developed for a liquid explosives detection device based on ULF-MRI technique.[59] This magnetometer, which is presently state-of-the-art, consists of seven identical induction coils and has the same footprint as the SQUID gradiometers described above. Each coil has

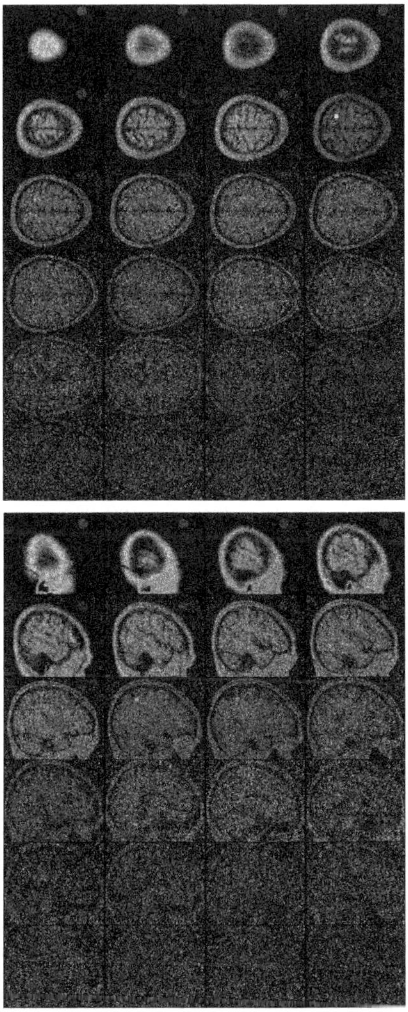

Figure 7.19 Sequence #1, B_p = 100 mT, noise 1.8 fT/√Hz.

90 mm outer diameter, 20 mm inner diameter and 14 mm height. It consists of 1400 turns of AWG24 (ø0.51 mm) copper wire. Its inductance is 70 mH and its resistance is 20 Ohm. The field transfer coefficient is about 70 V mT^{-1} at 3.3 kHz. Such induction magnetometers have a magnetic field resolution of about 20 fT/√Hz at 3.3 kHz.

As was illustrated in the previous section, in order to obtain clinically relevant images the equivalent noise of the system should be less than 1 fT/√Hz. Comparing that with the reported 20 fT/√Hz noise level of the described induction magnetometer it becomes obvious that such magnetometers are not suitable for brain imaging using the ULF-MRI technique.

The obvious way to increase the sensitivity of induction magnetometers is to increase the Larmor frequency, *i.e.* the strength of the measurement field.

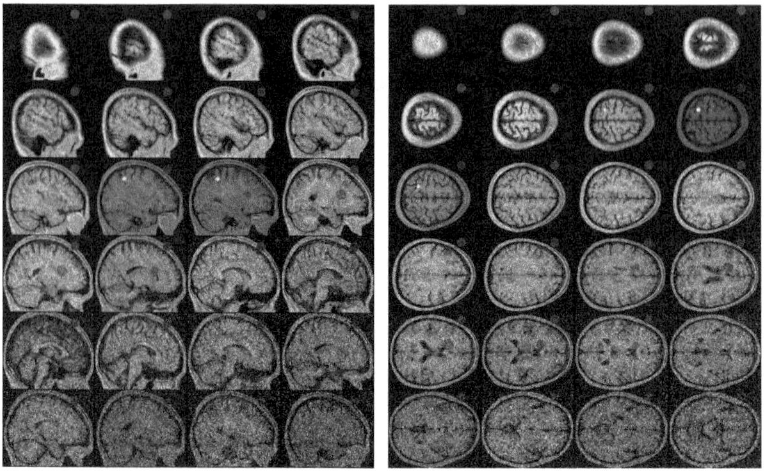

Figure 7.20 Sequence #2, $B_p = 250$ mT, noise 0.9 fT/√Hz.

As was mentioned before, the induction magnetometers reach the sensitivity of a SQUID-based magnetometer at frequencies of around 100 kHz.[19]

To illustrate this we will consider a coil designed to work at 100 kHz. The coil has the following parameters:[67]

- 70 mm inner diameter;
- 40 turns of LITZ 25/38 SPNSN multi-stranded copper wire;
- Field transfer coefficient is about 96.7 V mT^{-1} at 100 kHz;
- Resistance ~1.35 Ohm;
- Inductance ~276 μH at 100 kHz;
- Q-factor ~90 at 100 kHz;
- 0.18 nV/√Hz intrinsic (Johnson) noise level at room temperature.

Note that parameters of the coil, such as resistance, inductance, Q-factor and the intrinsic noise, were experimentally verified. Given those parameters, we can estimate that the equivalent magnetic field resolution is about 1.9 fT/√Hz at a Larmor frequency 100 kHz, *i.e.* at ~2.3487 mT measurement field.

To evaluate the performance of such coils we replaced the SQUID sensors in the ULF MRI system considered in the previous section with inductive gradiometers. Because induction sensors are more robust with respect to performance in the presence of RF interference, in this case it is possible to use a first order gradiometer. The computational model of such a system with 7 induction gradiometers is shown in Figure 7.22. We assumed that the 7 coils in the model do not interact and produce additional noise, which is realizable with the FT–AM idea. Resonant operation would strongly couple the coils and lead to significant deviation from the prediction.

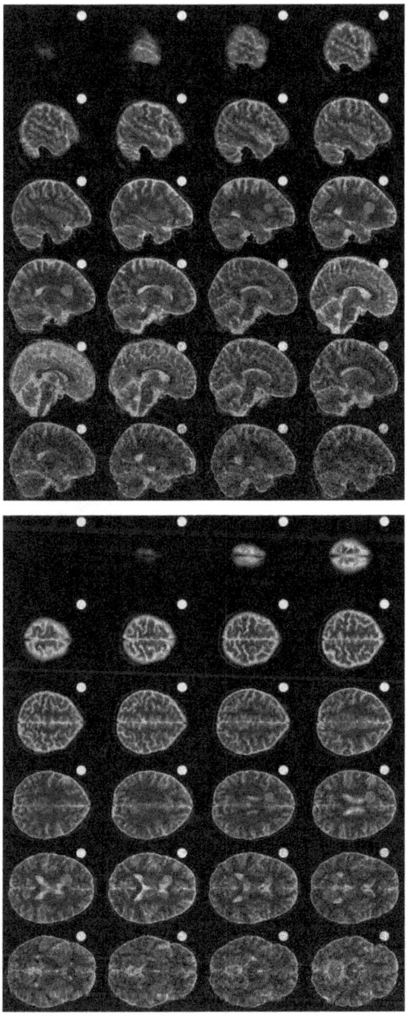

Figure 7.21 Sequence #1, B_p = 250 mT, noise 0.9 fT/√Hz.

The simulation results of this system using the imaging sequence #2 (see Table 7.3) and assuming that the MRI signals are detected at the Larmor frequency of 100 kHz, *i.e.* in the measurement field of 2.3487 mT, are shown in Figure 7.23.

As expected, the quality of the reconstructed images is very similar to the quality of the images simulated from a SQUID-based system considered in the previous section. With this knowledge in hand, onc can then begin to make the determination with regard to whether lower frequency readout (a few kHz) using a SQUID is easier than a higher frequency readout using a regular coil.

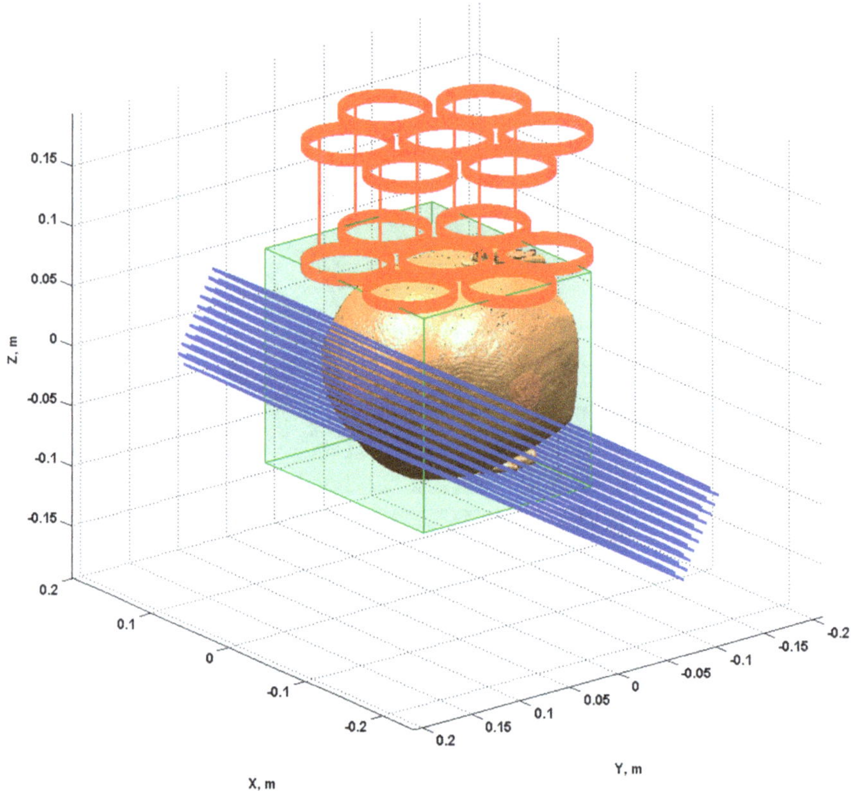

Figure 7.22 Induction gradiometers.

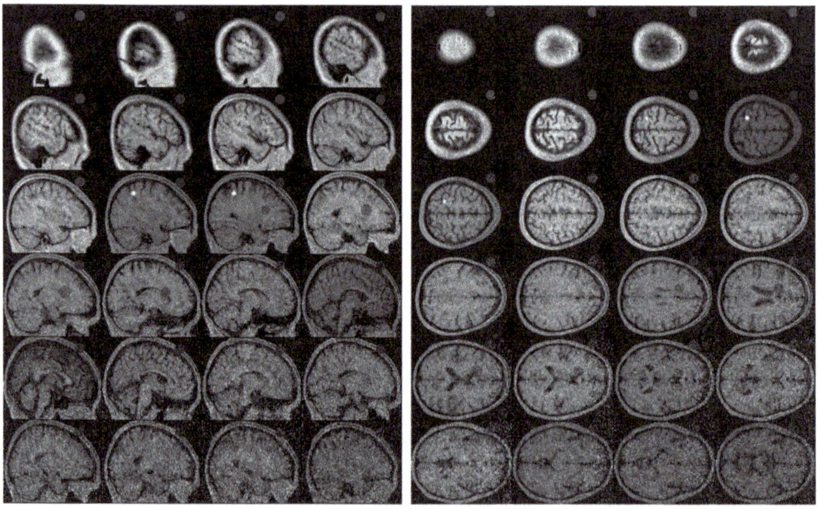

Figure 7.23 Sequence #1, B_p = 250 mT, noise 0.18 nV/√Hz (equivalent 1.9 fT/√Hz at a Larmor frequency of 100 kHz).

7.5 Conclusions

We have introduced the idea of pre-polarized ultra-low field MRI using an atomic magnetometer or SQUID including the potential benefits of contrast and simplified field generation, and then introduced the basic hardware components and considerations. Most importantly, we have provided the reader with a feel for what image quality might arise and how to predict it ahead of time to make an informed decision regarding all the trade-offs for mobile MRI.

Nomenclature

AM	Atomic magnetometer
B_m	Measurement or readout magnetic field
B_p	Pre-polarization magnetic field
BW	Bandwidth
FT	Flux transformer
MEG	Magnetoencephalography
MSR	Magnetically shielded room
PBS	Polarizing beam splitter
SNR	Signal to noise ratio
SQUID	Superconducting quantum interference device
ULF	Ultra-low field

References

1. B. Inglis, K. Buckenmaier, P. SanGiorgio, A. F. Pedersen, M. A. Nichols and J. Clarke, *Proc. Natl. Acad. Sci. U. S. A.*, 2013, **110**, 19194–19201.
2. H. C. Seton, J. M. S. Hutchison and D. M. Bussell, *Meas. Sci. Technol.*, 1997, **8**, 198–207.
3. V. S. Zotev, A. N. Matlashov, P. L. Volegov, I. M. Savukov, M. A. Espy, J. C. Mosher, J. J. Gomez and R. H. Kraus Jr, *J. Magn. Reson.*, 2008, **194**, 115–120.
4. P. E. Magnelind, J. J. Gomez, A. N. Matlashov, T. Owens, J. H. Sandin, P. L. Volegov and M. A. Espy, *IEEE Trans. Appl. Supercond.*, 2011, **21**, 456–460.
5. P. T. Vesanen, J. O. Nieminen, K. C. J. Zevenhoven, J. Dabek, L. T. Parkkonen, A. V. Zhdanov, J. Luomahaara, J. Hassel, J. Penttilä, J. Simola, A. I. Ahonen, J. P. Mäkelä and R. J. Ilmoniemi, *Magn. Reson. Med.*, 2013, **69**, 1795–1804.
6. I. M. Savukov, V. S. Zotev, P. L. Volegov, M. A. Espy, A. N. Matlashov, J. J. Gomez and R. H. Kraus, *J. Magn. Reson.*, 2009, **199**, 188–191.
7. J. Clarke, M. Hatridge and M. Mößle, *Annu. Rev. Biomed. Eng.*, 2007, **9**, 389–413.
8. R. McDermott, A. H. Trabesinger, M. Muck, E. L. Hahn, A. Pines and J. Clarke, *Science*, 2002, **295**, 2247–2249.

9. M. Espy, M. Flynn, J. Gomez, C. Hanson, R. Kraus, P. Magnelind, K. Maskaly, A. Matlashov, S. Newman, T. Owens, M. Peters, H. Sandin, I. Savukov, L. Schultz, A. Urbaitis, P. Volegov and V. Zotev, *Supercond. Sci. Technol.*, 2010, **23**, 034023.
10. M. Mößle, S.-I. Han, W. R. Myers, S.-K. Lee, N. Kelso, M. Hatridge, A. Pines and J. Clarke, *J. Magn. Reson.*, 2006, **179**, 146–151.
11. A. N. Matlachov, P. L. Volegov, M. A. Espy, J. S. George and R. H. Kraus, *J. Magn. Reson.*, 2004, **170**, 1–7.
12. S. K. Lee, M. Mößle, W. Myers, N. Kelso, A. H. Trabesinger, A. Pines and J. Clarke, *Magn. Reson. Med.*, 2005, **53**, 9–14.
13. S. Busch, M. Hatridge, M. Mößle, W. Myers, T. Wong, M. Muck, K. Chew, K. Kuchinsky, J. Simko and J. Clarke, *Magn. Reson. Med.*, 2012, **67**, 1138–1145.
14. S. E. Ungersma, N. I. Matter, J. W. Hardy, R. D. Venook, A. Macovski, S. M. Conolly and G. C. Scott, *Magn. Reson. Med.*, 2006, **55**, 1362–1371.
15. M. Espy, P. Magnelind, A. Matlashov, S. Newman, A. Urbaitis and P. Volegov, *IEEE Trans. Appl. Supercond.*, 2013, **23**, 1603107.
16. W. R. Myers, Potential Applications of Microtesla Magnetic Resonance Imaging Detected Using a Superconducting Quantum Interference Device, PhD thesis, University of California Berkeley, 2006.
17. R. P. Feynman, R. B. Leighton and M. L. Sands, *The Feynman lectures on physics*, Addison Wesley Longman, 1963–1965.
18. J. Clarke and A. I. Braginski, *The SQUID Handbook*, Wiley, 2005.
19. W. Myers, D. Slichter, M. Hatridge, S. Busch, M. Mossle, R. McDermott, A. Trabesinger and J. Clarke, *J. Magn. Reson.*, 2007, **186**, 182–192.
20. M. Hamalainen, R. Hari, R. J. Ilmoniemi, J. Knuutila and O. V. Lounasmaa, *Rev. Mod. Phys.*, 1993, **65**, 413–497.
21. M. Espy, A. Matlashov and P. Volegov, *J. Magn. Reson.*, 2013, **229**, 1–15.
22. J. C. Allred, R. N. Lyman, T. W. Kornack and M. V. Romalis, *Phys. Rev. Lett.*, 2002, **89**, 130801.
23. I. K. Kominis, T. W. Kornack, J. C. Allred and M. V. Romalis, *Nature*, 2003, **422**, 596–599.
24. H. Xia, A. B. A. Baranga, D. Hoffman and M. V. Romalis, *Appl. Phys. Lett.*, 2006, **89**, 211104.
25. C. Johnson, P. D. D. Schwindt and M. Weisend, *Appl. Phys. Lett.*, 2007, **90**, 81102.
26. I. Savukov and T. Karaulanov, *Appl. Phys. Lett.*, 2013, **103**, 43703.
27. I. Savukov and T. Karaulanov, *J. Magn. Reson.*, 2013, **231**, 39–45.
28. T. G. Walker and W. Happer, *Rev. Mod. Phys.*, 1997, **69**, 629–642.
29. I. M. Savukov, S. J. Seltzer, M. V. Romalis and K. L. Sauer, *Phys. Rev. Lett.*, 2005, **95**, 063004.
30. S. K. Lee, K. L. Sauer, S. J. Seltzer, O. Alem and M. V. Romalis, *Appl. Phys. Lett.*, 2006, **89**, 214106.
31. S. J. Xu, V. V. Yashchuk, M. H. Donaldson, S. M. Rochester, D. Budker and A. Pines, *Proc. Natl. Acad. Sci. U. S. A.*, 2006, **103**, 12668–12671.
32. V. Biancalana, Y. Dancheva and L. Stiaccini, www.http://arxiv.org/pdf/1401.6454v1.pdf, 2014.

33. I. M. Savukov and M. V. Romalis, *Phys. Rev. Lett.*, 2005, **94**, 123001.
34. I. M. Savukov, S. J. Seltzer and M. V. Romalis, *J. Magn. Reson.*, 2007, **185**, 214–220.
35. I. Savukov, T. Karaulanov and M. G. Boshier, *Appl. Phys. Lett.*, 2014, **104**, 023504.
36. R. H. J. Kraus, M. A. Espy, P. E. Magnelind and P. L. Volegov, *Ultra-Low Field Nuclear Magnetic Resonance: A New MRI Regime*, Oxford University Press, Oxford, 2014.
37. A. Macovski and S. Conolly, *Magn. Reson. Med.*, 1993, **30**, 221–230.
38. D. J. Lurie, S. Aime, S. Baroni, N. A. Booth, L. M. Broche, C. H. Choi, G. R. Davies, S. Ismail, D. O. HOgain and K. J. Pine, *C. R. Phys.*, 2010, **11**, 136–148.
39. V. S. Zotev, A. N. Matlashov, P. L. Volegov, A. V. Urbaitis, M. A. Espy and R. H. J. Kraus, *Supercond. Sci. Technol.*, 2007, **20**, S367–S373.
40. V. S. Zotev, A. N. Matlashov, I. M. Savukov, T. Owens, P. L. Volegov, J. J. Gomez and M. A. Espy, *IEEE Trans. Appl. Supercond.*, 2009, **19**, 823–826.
41. H. W. Fischer, P. A. Rinck, Y. Van Haverbeke and R. N. Muller, *Magn. Reson. Med.*, 1990, **16**, 317–334.
42. J. O. Nieminen, Ultra-low-field MRI: techniques and instrumentation for hybrid MEG-MRI, PhD thesis, Aalto University, 2012. https://aaltodoc.aalto.fi/handle/123456789/7250.
43. S. H. Koenig and R. D. Brown, in *NMR Spectroscopy of Cells and Organisms*, ed. R. K. Gupta, CRC Press, 1987, vol. 2.
44. P. L. Volegov, J. C. Mosher, M. A. Espy and R. H. Kraus, *J. Magn. Reson.*, 2005, **175**, 103–113.
45. J. O. Nieminen and R. J. Ilmoniemi, *J. Magn. Reson.*, 2010, **207**, 213–219.
46. K. P. Pruessmann, M. Weiger, M. B. Scheidegger and P. Boesiger, *Magn. Reson. Med.*, 1999, **42**, 952–962.
47. V. S. Zotev, P. L. Volegov, A. N. Matlashov, M. A. Espy, J. C. Mosher and R. H. Kraus, *J. Magn. Reson.*, 2008, **192**, 197–208.
48. R. Austin, M. Espy, A. Matlashov, H. Sandin, L. Schultz, A. Urbaitis and P. Volegv, in *NATO Sciencc for Pcacc and Sccurity Scrics B: Physics and Biophysics*, ed. B. R. Tomaž Apih, Georgy Mozzhukhin and Jamie Barras, Springer Science+Business Media Dordrecht, 2014.
49. H. Dong, L. Q. Qiu, W. Shi, B. L. Chang, Y. Qiu, L. Xu, C. Liu, Y. Zhang, H. J. Krause, A. Offenhausser and X. M. Xie, *Appl. Phys. Lett.*, 2013, **102**, 102602–102604.
50. L. Q. Qiu, G. F. Zhang, Y. L. Wang and X. M. Xie, *J. Phys.: Conf. Ser.*, 2010, **234**, 042028.
51. H. Ohta, T. Matsui and Y. Uchikawa, *IEEE Trans. Appl. Supercond.*, 2007, **17**, 730–733.
52. J. O. Nieminen, P. T. Vesanen, K. C. J. Zevenhoven, J. Dabek, J. Hassel, J. Luomahaara, J. S. Penttila and R. J. Ilmoniemi, *J. Magn. Reson.*, 2011, **212**, 154–160.
53. K. C. J. Zevenhoven, S. Busch, M. Hatridge, F. Oisjoen, R. J. Ilmoniemi and J. Clarke, *J. Appl. Phys.*, 2014, **115**, 103902.
54. S. H. Koenig and R. D. Brown, *Magn. Reson. Med.*, 1984, **1**, 437–449.

55. J. O. Nieminen, J. Voigt, S. Hartwig, H. J. Scheer, M. Burghoff, L. Trahms and R. J. Ilmoniemi, *Metrol. Meas. Syst.*, 2013, **20**, 327–336.
56. S. Hartwig, J. Voigt, H. J. Scheer, H. H. Albrecht, M. Burghoff and L. Trahms, *J. Chem. Phys.*, 2011, **135**, 054201.
57. M. Espy, M. Flynn, J. Gomez, C. Hanson, R. Kraus, P. Magnelind, K. Maskaly, A. Matlashov, S. Newman, M. Peters, H. Sandin, I. Savukov, L. Schultz, A. Urbaitis, P. Volegov and V. Zotev, *IEEE Trans. Appl. Supercond.*, 2009, **19**, 835–838.
58. M. Espy, S. Baguisa, D. Dunkerley, P. Magnelind, A. Matlashov, T. Owens, H. Sandin, I. Savukov, L. Schultz, A. Urbaitis and P. Volegov, *IEEE Trans. Appl. Supercond.*, 2011, **21**, 530–533.
59. A. N. Matlashov, L. J. Schultz, M. A. Espy, R. H. Kraus, I. M. Savukov, P. L. Volegov and C. J. Wurden, *IEEE Trans. Appl. Supercond.*, 2011, **21**, 465–468.
60. E. K. Insko, M. A. Elliott, J. C. Schotland and J. S. Leigh, *J. Magn. Reson.*, 1998, **131**, 111–117.
61. S. Ljunggren, *J. Magn. Reson.*, 1983, **54**, 338–343.
62. D. B. Twieg, *Med. Phys.*, 1983, **10**, 610–621.
63. P. T. Callaghan, *Principles of Nuclear Magnetic Resonance Microscopy*, Clarendon Press, Oxford, 1991.
64. G. A. Korn and T. M. Korn, *Mathematical handbook for scientists and engineers: definitions, theorems, and formulas for reference and review*, Courier Dover Publications, New York, 2000.
65. B. Aubert-Broche, A. C. Evans and D. L. Collins, *NeuroImage*, 2006, **32**, 138–145.
66. Y. D. Zhu, C. J. Hardy, D. K. Sodickson, R. O. Giaquinto, C. L. Dumoulin, G. Kenwood, T. Niendorf, H. Lejay, C. A. McKenzie, M. A. Ohliger and N. M. Rofsky, *Magn. Reson. Med.*, 2004, **52**, 869–877.
67. I. M. Savukov, private communication, Los Alamos, 2014.

CHAPTER 8

Software Developments: Improvements in Data Analysis

DANIEL J. HOLLAND*[a] AND LYNN F. GLADDEN[a]

[a]Department of Chemical Engineering and Biotechnology University of Cambridge, Pembroke Street, Cambridge CB2 3RA, United Kingdom
*E-mail: daniel.holland@canterbury.ac.nz

8.1 Introduction

There are three major challenges when performing Nuclear Magnetic Resonance (NMR) using permanent magnets or magnets with a low magnetic field strength that could be improved through advanced data analysis techniques. The first problem is that NMR is an insensitive technique owing to the small gap between energy states. This problem is exacerbated when using low magnetic field strengths as the energy gap is proportional to the field strength. Thus, low field instruments will generally suffer from a low signal-to-noise ratio (SNR). The second problem relates to the sensitivity of NMR to molecular structure. One of the most useful properties of NMR is that molecules have a resonant frequency (or frequencies) that is determined by the local molecular structure—giving rise to the phenomena of chemical shift. However, the chemical shift is proportional to the magnetic field strength. Therefore, when using low field strength magnets, spectra compress leading to significant overlap of nearby peaks. The third major problem of low field instruments is that the homogeneity of low field magnets is typically relatively poor. Poor magnetic field homogeneity leads to broader

New Developments in NMR No. 5
Mobile NMR and MRI: Developments and Applications
Edited by Michael L. Johns, Einar O. Fridjonsson, Sarah J. Vogt, and Agnes Haber

Published by the Royal Society of Chemistry, www.rsc.org

line widths and hence even more overlap between nearby peaks. The development of low field instruments has focused primarily on improvements to the hardware. Significant progress has been made in this regard, for example it is now possible to obtain chemical spectra with astonishingly narrow line widths,[1] and improvements are continuously being made in this area. However, an alternative method of improving the quality of information that can be obtained from NMR is to use advanced data processing techniques. The advantage of improving our data analysis techniques is that the same, or at least a similar, approach can be applied regardless of the hardware used. Further, as the hardware improves, the benefits of improved data processing will remain. This chapter will outline the basic approach to data processing that is used in conventional NMR and Magnetic Resonance Imaging (MRI) and then discuss Compressed Sensing and Bayesian analysis, two techniques that are increasingly being used to improve the resolution, SNR and/or structural information that can be obtained.

8.2 Introduction to Data Processing in NMR/MRI

NMR and MRI data are composed of additive signals generated from the resonant frequency of individual nuclei present in a sample. Typically, there are many different resonance frequencies that we wish to distinguish. In NMR, these resonances will correspond to different chemical species, whilst in MRI they correspond to different spatial positions. In either case, we detect the time dependence of the signal and it is from this that we wish to infer certain information. However, the time dependent signal itself is an analogue signal that is very difficult to store and interpret. Instead, the time dependent signal is sampled digitally, which permits easy and robust data handling, and then Fourier transformed to obtain the distribution of frequencies present in the sample. In this section, we briefly outline the basic principles of these standard data handling approaches and also identify some of their limitations. These principles are covered in more detail in many of the excellent texts on NMR and MRI.[2–6]

8.2.1 Discrete Fourier Transform

The primary method of extracting useful information from an NMR or MRI signal is through the use of the discrete Fourier transform. Ignoring relaxation, the NMR signal, S, obtained from a system with m resonances can be described by:

$$S(t) \propto \sum_{j=1}^{m} \rho_j e^{-i2\pi \nu_j t} \tag{8.1}$$

where ρ_j is related to the amount of the jth resonance in the rf coil, ν_j is the resonant frequency of the jth resonance and t is time. The information of interest is the intensity and frequency of the distinct resonant frequencies, *i.e.* ρ_j and ν_j. The spectrum of frequencies is then extracted through inverse Fourier transformation of the time domain signal:

$$\rho(\omega) = \int_{-\infty}^{\infty} S e^{i2\pi\nu t} dt \tag{8.2}$$

However, the analytical Fourier transformation is difficult to apply to real signals as it requires integration over all time. Instead, the time domain signal is sampled digitally and the resulting discrete signal is then converted to the frequency domain using the discrete Fourier transform, which can be efficient computationally. The forward and inverse discrete Fourier transform are written:

$$S_n = \frac{1}{\sqrt{N}} \sum_{k=0}^{N-1} \rho_k e^{-i2\pi kn/N} \tag{8.3}$$

$$\rho_k = \frac{1}{\sqrt{N}} \sum_{n=0}^{N-1} S_n e^{i2\pi kn/N} \tag{8.4}$$

where ρ and S are now discrete functions and ρ_k is the intensity associated with the kth frequency component of the spectrum, S_n is the nth time point and N time points are acquired in total. The coefficients $\sqrt{N}$ in eqns (8.3) and (8.4) are sometimes combined and appear only in either the forward or inverse transform; the factor of $\sqrt{N}$ is used here to ensure symmetry of the forward and inverse transforms.

Since the discrete Fourier transform is a simple linear combination of the measured data points, S_n, it is straightforward to represent eqn (8.3) in matrix form. We therefore define $\boldsymbol{S}$ as a vector comprising the N time domain data points S_n and $\boldsymbol{\rho}$ as the vector of the intensities of the N frequencies. Eqn (8.3) is then written:

$$\boldsymbol{S} = \boldsymbol{A}\boldsymbol{\rho} \tag{8.5}$$

where $\boldsymbol{A}$ is an N by N matrix with entries given by $\frac{1}{\sqrt{N}} e^{-i2\pi kn/N}$, where n and k refer to the row and column indices of the elements of $\boldsymbol{A}$. From the periodicity of the coefficients of $\boldsymbol{A}$, it is easy to see that the inverse Fourier transform is simply given by the conjugate transpose of the forward transform. In the matrix representation, it is also clear that the Fourier transform is a linear operation, a point that will be useful in later analysis.

A key feature of the discrete Fourier transform is that the signal is assumed to be "band-limited". In other words, frequencies that are contained in the original time series of data can only be reconstructed unambiguously if they are contained within a limited range, or bandwidth. The bandwidth of the detected signals is determined by the inverse of the time between the data points in the time domain, as defined by the sampling theories of Nyquist[7] or Shannon.[8] The Nyquist sampling theorem is illustrated

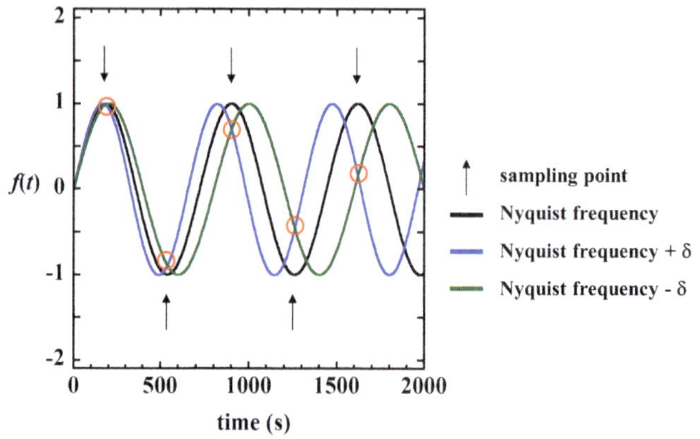

Figure 8.1 Illustration of the Nyquist sampling theorem. Curves are shown for three continuous signals, one at the Nyquist frequency, one δ greater than the Nyquist frequency and one δ less than the Nyquist frequency. These three signals are sampled at the five discrete times indicated by the arrows on the figure. The signal intensity of the component at the Nyquist frequency is clearly distinguishable from the other two frequencies. However, the signal intensities for the signals δ above and below the Nyquist frequency are identical at the time at which the signal is sampled. These two frequencies are therefore indistinguishable from each other.

in Figure 8.1, which shows signals at, just above and just below the Nyquist frequency. These signals are sampled at the five discrete times indicated by the arrows in Figure 8.1. At these points in time, the signals above and below the Nyquist frequency by an amount δ have the same intensity and it is impossible to distinguish between these two signals. It is therefore easy to see that any signal that occurs outside the Nyquist bandwidth will appear "folded back" into the spectral window. The resolution of the spectrum is determined by the number of points acquired, N, or equivalently the length of time for which the signal is acquired.

An additional consideration is that the number of calculations required to perform the discrete Fourier transform is $O(N^2)$, where N is the number of points in the digital data (*i.e.* N operations for each of N points). This becomes prohibitively time consuming when dealing with large values of N. Consequently, the Fast Fourier Transform (FFT) algorithm is often employed.[9] The FFT algorithm divides the calculation of the original N data points, into a series of calculations of progressively smaller numbers of data points. This process provides a computational short-cut to performing the Fourier transform, where the number of calculations required is reduced to $O(N \ln N)$. For realistic signals, which contain thousands or millions of data points, the FFT represents a significant saving of computational time, even with modern computers. Originally, the FFT necessitated that N be a power of 2 (*e.g.* 2, 4, 8, 16, 32, 64...), however with modern algorithms this constraint can be relaxed.[10] The discrete Fourier transform is quick and easy to calculate and accurately divides a signal into its component frequencies. However, the discrete Fourier transform has some limitations, particularly when considering noisy or decaying signals. We will return to these points when we discuss more advanced signal processing techniques in Sections 8.3 and 8.4.

8.2.2 Filtering

Filtering is critical to NMR and MRI to reduce noise in the spectral window and to minimise the contribution of noise from data points that are acquired after the signal has relaxed. In this section, we first introduce the concept of filtering and then explain how it is used to improve the SNR of a spectrum.

Filtering uses the convolution theorem to isolate the components of the spectrum that are of interest. The convolution theorem states that the product of two functions in the time domain is equal to the convolution of these two functions in the frequency domain. Similarly, the product of two functions in the frequency domain is equal to the convolution of these functions in the time domain. Mathematically, the convolution of two functions $f(t)$ and $g(t)$ can be expressed as:

$$f(t) \otimes g(t) = \int_{-\infty}^{\infty} f(\tau) g(t - \tau) \mathrm{d}\tau. \tag{8.6}$$

If these functions have Fourier transforms $F(\omega)$ and $G(\omega)$, respectively, then the convolution theorem can be written:

$$\mathcal{F}\{f(t) \otimes g(t)\} = \mathcal{F}\left\{\int_{-\infty}^{\infty} f(\tau) g(t-\tau) \mathrm{d}\tau\right\} = F(\omega) G(\omega), \tag{8.7}$$

where $\mathcal{F}$ denotes the Fourier transform. The convolution theorem forms the basis of filtering of signals in NMR (and other fields).

Filtering provides a method for reducing the noise in the spectral window. As stated in Section 8.2.1, if a signal appears at a frequency outside the spectral bandwidth, that signal will fold back into the spectrum. This is true even if the signal arises from random noise. Since in a conventional acquisition the noise is assumed to be white, that is, it has a constant amplitude with frequency, if the measured signal is not filtered then noise arising from the entire electromagnetic spectrum would fold into the spectrum. Thankfully, this does not happen as by digitising our signal we inherently introduce some level of filtering, which helps to eliminate much of the noise. The simplest filter function would simply average the time domain signal for the length of time between acquiring data points. Such a filter would introduce a *sinc* weighting to the signal intensity in the frequency domain, which will dramatically reduce noise arising from frequencies outside the spectral bandwidth. The *sinc* function arises as it is the Fourier transform of a "hat" function (*i.e.* the *sinc* and *hat* functions are conjugate Fourier pairs). We would like our filter to behave like a hat function in the frequency domain—that is, to have a constant intensity over a narrow and defined range of frequency and be zero outside this range. Therefore, our filter function would ideally be a *sinc* function in the time domain. However, the *sinc* function is infinite so, in practice, it must be truncated. Such filters can be achieved using an analogue filter on the analogue (continuous) time domain signal directly. However, modern spectrometers can digitize the signal at sufficiently high rates that it is often beneficial to apply the filter digitally as then it can be readily optimised for any desired frequency bandwidth.

In considering the discrete Fourier transform in Section 8.2.1, we ignored signal relaxation. In practice, the signal will decay with time owing to both T_2 and T_2^* relaxation, which leads to a broadening of the line in frequency space. Signal decay also means that data points do not contribute equally to the desired signal in the final spectrum. If the signal is acquired for too long, the additional measurements contribute primarily noise to the spectrum without contributing significant additional signal. Therefore, acquiring a signal for too long will lead to a noisy spectrum. Filtering provides a method of overcoming this limitation, though at the cost of an increase in the line width of the spectrum. A typical filter is an exponential decay function, as this minimises the contribution of the noisy data points, the use of which is illustrated in Figure 8.2. In practice, a "matched filter" is often used, where the exponential decay is chosen such that it matches the decay due to T_2^* relaxation. A matched filter can be shown to provide the optimum SNR and does not introduce too much broadening.

Much of the discussion to this point has focussed on applications and analysis of NMR signals. However, the same techniques are used when processing MRI data. However, we now introduce a technique that is predominantly used in MRI and therefore we introduce MRI formally in the next section.

8.2.3 Non-Cartesian Imaging

MRI and NMR data have traditionally been acquired on a Cartesian grid and reconstructed using the FFT. However, there are potentially significant advantages to acquiring data on a non-Cartesian grid, including increased robustness to flow artefacts, improved SNR, more efficient use of the coils used to generate gradients in the magnetic field, more efficient sampling, and reduced data acquisition time.[11] It is theoretically possible to reconstruct data acquired on a non-Cartesian grid using the discrete Fourier transform, however this requires N^2 operations and is therefore prohibitively time consuming for any N of interest. Over the last decade or so there have been significant advances in the algorithms available to reconstruct data acquired on a non-Cartesian grid, such that it is now possible to obtain a reconstruction in $O(N\log N)$ operations, *i.e.* of the same order as with conventional FFT reconstructions.[12–15] With the development of these fast reconstruction algorithms, non-Cartesian imaging is now common on high-field MRI systems[11] and is increasingly being used in high resolution multidimensional NMR.[16] To the best of the authors' knowledge, non-Cartesian NMR and MRI have

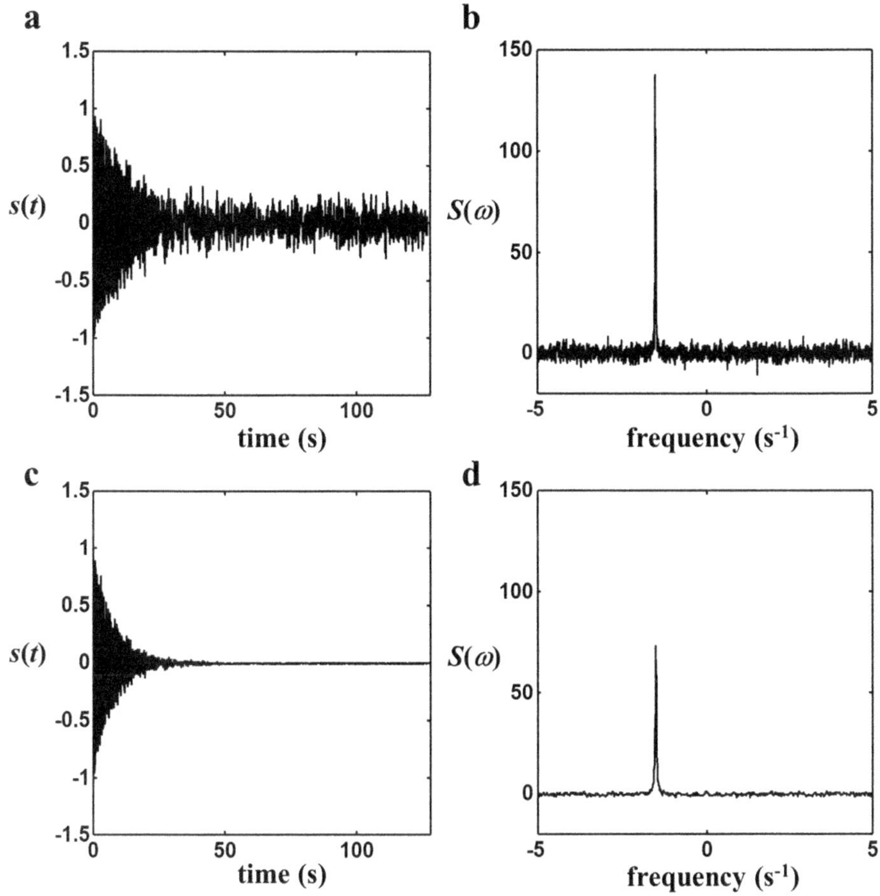

Figure 8.2 Illustration of the use of the convolution theorem as a filter for an NMR signal containing a single resonance at −1.5 Hz. (a) The original noisy signal is shown as a function of time. (b) Fourier transform of the signal in (a). The spectrum has a high intensity at a frequency of 1.5 Hz as desired, but there is significant noise in the spectrum. (c) The original signal has been multiplied by an exponential filter with a time constant equal to the time constant of the original signal. (d) Fourier transform of the signal in (c). The magnitude of the Fourier transform of the signal is approximately half of that in (b), as the filter has broadened the signal in the frequency domain. However, the noise has decreased in magnitude by a factor of ~4.5, giving a gain in SNR of ~2.2.

not yet been demonstrated using low-field hardware. Non-Cartesian NMR is unlikely to have a major impact on low-field NMR, as it is only beneficial when acquiring three dimensional or higher spectra, which are likely to require a high SNR. However, there are strong motivations for using non-Cartesian MRI techniques on low-field instruments in the future; therefore, we briefly review some aspects of reconstructing non-Cartesian MRI data here and simply note that these are equally applicable to multidimensional NMR.

The MRI signal in the presence of a linear variation in the magnetic field strength is given by:

$$S(t) = \int_V \rho(\boldsymbol{r}) e^{-i2\pi \boldsymbol{r}\cdot \boldsymbol{k}(t)} \mathrm{d}V, \tag{8.8}$$

where ρ is the spin density distribution in the volume V, $\boldsymbol{r}$ is the position vector $[x\,y\,z]$, and k defines the locations of the measured signal in k-space as given by:

$$\boldsymbol{k}(t) = \gamma \int_0^t \boldsymbol{G}\,\mathrm{d}\tau, \tag{8.9}$$

where γ is the gyromagnetic ratio of the nuclei under observation and $\boldsymbol{G}$ is the strength of the magnetic field gradient used for imaging. Cartesian sampling places strong demands on gradient strength and switching speed ("slew rate"), especially for ultrafast MRI techniques like echo planar imaging (EPI).[17] These demands motivated the first non-Cartesian MRI imaging techniques, such as spiral imaging whereby the rate of change of the gradient with time is significantly reduced. One of the simplest methods to sample k-space in a non-Cartesian manner is to vary the strength of the gradient in the x and y directions in an oscillatory manner, *e.g.* $G_x = t\cos(\omega t)$ and $G_y = t\sin(\omega t)$, then eqn (8.8) will produce a set of data distributed along a spiral, as indicated in Figure 8.3.

One of the principle advantages of spiral imaging is that it permits an efficient coverage of k-space. The highest intensity information in MRI images is typically located at close to $k = 0$, and spiral imaging will always sample this region of the signal more heavily, which can lead to improved

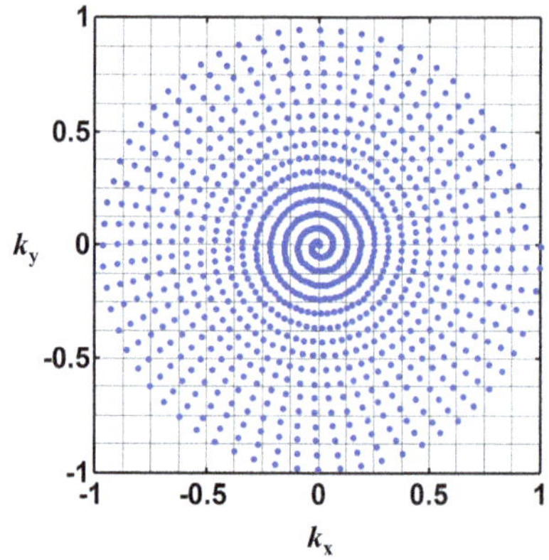

Figure 8.3 Illustration of a spiral sampling pattern acquired with a constant linear velocity. Data points are acquired at the points marked by the blue symbols, and are not located on the Cartesian grid. The density of samples near the origin is also shown to be significantly greater than in the outer portion of k-space. This change in sampling density can improve the SNR of the final image, but must be corrected for during the image reconstruction.

SNR from spiral acquisitions.[11] Further, spiral scans do not typically sample the corners of *k*-space which reduces the amount of data required by approximately 22% (the data cover a π/4 fraction of *k*-space). Images reconstructed without the information from the corners of *k*-space are almost indistinguishable from fully sampled images. Therefore, spiral sampling can provide a more efficient coverage of *k*-space and reduced total acquisition time.[11,18] However, there are some disadvantages to spiral MRI.

Limitations of slew rate and gradient strength mean that the gradient produced by a system will not follow the ideal trajectory shown in Figure 8.3 exactly. Further, this approach samples the same number of points along each turn, despite the distance covered through *k*-space increasing with distance from the origin. Therefore, the simple trajectory described above will not yield an optimal sampling of *k*-space. Various alternative methods have been suggested for generating the gradient waveforms to produce the most efficient sampling of *k*-space possible whilst considering limitations of the gradient strength and slew rate.[19–23] In practice, even with these optimised approaches, it is beneficial to measure the gradient strength, and hence *k*-space locations, using either a simple calibration experiment (see, for example ref. 24), or multiple receiver coils.[25]

It is also clear that the *k*-space data shown in Figure 8.3 are not evenly distributed on a Cartesian grid and this is why the FFT cannot be used to reconstruct the data directly. Instead, the data are used to interpolate the signal on a Cartesian grid and these "re-gridded" data are then used to reconstruct the image using the FFT. Mathematically, the ideal interpolation would be performed using a *sinc* function as the interpolation kernel.[12] However, the *sinc* function is infinite and therefore cannot be applied using real data. Alternative functions have since been considered including the truncated *sinc*[12] and Kaiser–Bessel[13] functions. Fessler and Sutton extended these ideas and developed a method for calculating the optimal kernel function by minimising the error between the signal reconstructed using the exact Fourier transform and that reconstructed using the re-gridding solution.[14] The approach of Fessler and Sutton can be used to generate almost arbitrarily small artefacts in the re-gridding solution and their software is available freely online.

An additional consideration when sampling *k*-space in a non-Cartesian manner is that the density of data points sampled will vary with location in *k*-space. Typically, spiral trajectories will result in an oversampling near the origin of *k*-space. Therefore, if the density of samples is not accounted for in the reconstruction, the resulting image will be heavily weighted by the low-frequency components (*i.e.* data obtained close to $k = 0$) and will appear heavily blurred. This blurring can be corrected through the use of a density compensation function, which weights the signal in the reconstruction according to the local sample density. Various algorithms have been proposed for calculating the density compensation function including calculating analytical functions,[26] the determinant of the Jacobian matrix between Cartesian and spiral coordinates,[27] and the Voronoi diagram, which calculates the area around each sampling point.[15] The choice of method will depend on the quality and complexity of the measured or input trajectory. The Voronoi approach is applicable to almost any sampling pattern and therefore is generally our method of choice; however, the calculation can be computationally intensive.

The major challenge associated with non-Cartesian MRI is that artefacts arising from imperfections in the gradients, B_0 field or chemical shift can be more pronounced than with Cartesian imaging. The principle reason for this is that with Cartesian imaging, these artefacts occur primarily in a single direction. Thus, with EPI it is common to see ghosting artefacts arising from imperfections in matching the location of data points when traversing *k*-space in the positive and negative read directions.[6] Although the artefacts encountered with Cartesian imaging may be severe, they can be identified readily. By contrast, with non-Cartesian imaging, the gradient in the magnetic field is varied simultaneously in two (or sometimes three) dimensions and hence artefacts accumulate in both of these directions. Therefore, artefacts cause the image to shift in all directions simultaneously, which leads to significant blurring of the image; the effects are similar regardless of whether the artefact arises from imperfections in the gradient, B_0 field or chemical shift. Imperfections in the gradients used will lead to a misalignment of the measurements in *k*-space, which can be overcome through calibration or measurement of the true *k*-space trajectory, as mentioned above. Chemical shift effects tend not to be very significant at low field, but these can be corrected by incorporating the chemical shift into the reconstruction.[3] The most significant artefacts for low field imaging are those arising from variations in B_0 and therefore we will concentrate on techniques to mitigate these here.

The simplest approach to correcting for inhomogeneity in the B_0 field is to map the variation in B_0 and then build this information into the reconstruction. In the presence of an inhomogeneous magnetic field, the signal model from eqn (8.7) is modified to:

$$S(t) = \int_V \rho(\boldsymbol{r}) e^{-i2\pi(\mathbf{r}\cdot\mathbf{k}(t)+t\gamma\phi(\mathbf{r}))}\, dV, \tag{8.10}$$

where ϕ describes the variation in the B_0 field with position $\boldsymbol{r}$. Therefore, if the B_0 variation is known, the signal can be recovered from the so-called conjugate phase reconstruction:

$$\rho(x,y) = \int_0^T S(t) e^{-i2\pi\boldsymbol{r}\cdot\boldsymbol{k}(t)+t\gamma\phi(\boldsymbol{r})}\, dt. \tag{8.11}$$

Thus, provided that ϕ is known, the image can be reconstructed even in the presence of an inhomogeneous B_0 field.

The B_0 field is mapped readily using a phantom that completely fills the rf coil and contains only a single resonance, such as water. Two spin-echo images of this phantom are then measured where the gradient echo and spin echo are shifted in time relative to each other between the two images. The signal in the two images (obtained after Fourier transformation of the measured data) is given by:

$$\mathrm{Im}_1(x,y) = \rho(x,y)e^{-i2\pi\gamma(\mathrm{TE}_1\phi(x,y))}$$
$$\mathrm{Im}_2(x,y) = \rho(x,y)e^{-i2\pi\gamma(\mathrm{TE}_2\phi(x,y))} \quad (8.12)$$

where ϕ is the field map, ρ is the spin density map and TE_i is the time between the centre of the spin echo and the gradient echo for the *i*th image. The field map can then be extracted readily from the difference in the phase of the image in each pixel.

The field map can be used to calculate the conjugate phase reconstruction directly. However, this calculation will be time consuming as each pixel must be reconstructed with its own off resonance frequency. In practice, field map correction is often simplified by segmenting the data either in time[28] or space,[29] or by approximating the field map variation with linear terms.[30] These simplifications to the field map correction significantly reduce the computational burden, however some residual blurring of the signal will remain.

In summary, non-Cartesian sampling permits more optimal sampling of *k*-space acquiring only those data necessary to reconstruct a high quality image. Furthermore, non-Cartesian sampling strategies tend to sample the low frequency region of *k*-space more heavily, which can improve the SNR. However, non-Cartesian sampling requires re-gridding, which is a significantly more complex reconstruction than simply using the DFT, and is susceptible to artefacts and blurring arising from B_0 inhomogeneity. The additional complexity associated with re-gridding has been largely overcome now with free access to numerous high quality and fast re-gridding algorithms. Furthermore, significant effort is currently being invested in minimising the effect of B_0 inhomogeneity, and it is anticipated that this will shortly become routine. Therefore, it is likely that the improved speed and SNR of non-Cartesian imaging will have significant benefit for low field imaging in the future.

8.3 Compressed Sensing

Compressed Sensing is a technique that will potentially lead to new, optimised data acquisition strategies for any given system, in terms of both spatial resolution and SNR in a given acquisition time. Compressed Sensing is the culmination of many years of research on the mathematics of sampling and signal compression. The concept of 'compression' in signal processing is familiar to most as we routinely store digital images in cameras and mobile phones in compressed form. Compression enables us to significantly reduce the data requirements for storage of a 'signal', without adversely affecting our ability to reproduce that 'signal'. In the context of NMR, a 'signal' could comprise, for example, an image or a spectrum. Compressed Sensing applies the ideas of signal compression to the measurement itself. One way of viewing this is that if it is possible to compress an image into a dataset comprising far fewer data points, then why acquire the 'redundant' data points in the first instance?

In order to accurately describe a spectrum or image, we must take discrete samples and, as noted in Section 8.2, sampling theory[7,8] established the principles of such sampling many years ago. Despite this, it has also long been known that in many cases it is possible to accurately reconstruct a signal with fewer samples than are required by sampling theory.[31–33] Compressed Sensing provides a mathematical framework to quantify how accurately a signal can be reconstructed when fewer samples are acquired than traditional sampling theory requires, as well as providing guidance as to how these measurements should be obtained in the first place.[34,35]

The key premise to Compressed Sensing is that the signal of interest contains only a small amount of "information". In this context, "information" characterizes how much data is required to reproduce a signal accurately. Mathematically, the information content of a signal can be described by the magnitude and location of the non-zero elements that comprise the signal in compressed form. This definition is counter-intuitive as it implies that a totally random signal, such as white noise, is described as having very high "information" content as it can only be reconstructed from knowledge of the exact value of each coefficient; by contrast a noise-free spectrum or image can often be represented by a small number of coefficients and therefore contains a relatively small amount of information. This characterisation of information is valuable because signals that are of interest typically contain a lot of structure—for example, pixels in an image will correlate strongly with nearby pixels. The idea of signal compression is to use the structure of the signal to reduce the number of data points required to describe the original signal. For example, in the 'jpeg' compression used on many digital cameras, the image is fully sampled and then transformed into a new basis using the discrete cosine transform. The large coefficients of the new basis are stored, while the small coefficients are discarded. The occurrence of localised structure in the image will typically mean that there will be many more small coefficients than large coefficients, and thus when the small coefficients are discarded, the amount of information representing the image is reduced or 'compressed.' Compressed sensing exploits the ideas behind compression to reduce the number of data points required to measure the signal in the first place. In other words, Compressed Sensing seeks to acquire the signal directly in compressed form.

8.3.1 Theory of Compressed Sensing

The mathematics of Compressed Sensing is well established. Here we introduce the key concepts; however, the reader is referred to the existing descriptions of the concept for further details.[34–39]

Formally, we wish to obtain a signal $\boldsymbol{x}$ (written as a column vector with n entries or values) using a set of noisy measurements $\boldsymbol{y}$ (written as a column vector with m entries or values) and a linear measurement system described by a matrix $\boldsymbol{A}$. In other words, we want to solve the set of linear equations given by:

$$\boldsymbol{y} = \boldsymbol{A}\boldsymbol{x} + \boldsymbol{\sigma} \tag{8.13}$$

where $\boldsymbol{\sigma}$ is the vector of the noise in the measurements. In the context of NMR/MRI the measurement matrix would correspond to the Fourier transform and the noise would be Gaussian. Ignoring noise for now, conventional sampling theory dictates that $m = n$ measurements are required to recover $\boldsymbol{x}$. However, Compressed Sensing theory tells us that it is possible to recover $\boldsymbol{x}$, even when $m \ll n$. Furthermore, Compressed Sensing theory provides guidelines on (1) the characteristics of $\boldsymbol{A}$, (2) the number of measurements required, and (3) the types of algorithm to use to recover $\boldsymbol{x}$ stably.

Compressed Sensing theory requires that the measurements and the basis used to represent the compressed signal are "incoherent". Incoherence requires that each measurement contains information from many coefficients of $\boldsymbol{x}$, and that the relative contribution from each coefficient of $\boldsymbol{x}$ changes for each measurement. This is equivalent to saying that the measurement basis cannot be represented by a small number of coefficients in the compressed basis. Qualitatively, the incoherence condition means that artefacts arising from undersampling appear as a noise-like contribution in the reconstruction. From this perspective, the true signal coefficients can be recovered as they will add in a coherent manner, whereas the undersampling artefacts will add in an incoherent manner and hence can be removed. Since it can almost guarantee very high incoherence, randomised sampling has played an important role in the development of Compressed Sensing. However, recent advances are leading to more advanced, semi-random sampling strategies[40] and structured sampling matrices.[39]

Compressed Sensing theory also requires that the signal can be compressed, and it is the degree to which a signal can be compressed that determines the number of measurements required. A signal $\boldsymbol{x}$ is compressible if it can be well described using $s \ll n$ non-zero coefficients (a signal is said to be "sparse" if this approximation is exact). The "information" content of the signal is described by the location and magnitude of the s non-zero coefficients. Thus, if a signal contains only a small number of non-zero coefficients, it contains little information. Many signals of interest are directly compressible. For example, in NMR spectroscopy the useful information in the signal is the magnitude and location of the peaks in the spectrum. However, the number of peaks will often be orders of magnitude less than the total number of possible locations for these peaks (determined by the bandwidth and resolution of the spectrum), hence the spectrum is readily compressible. Given that a signal can be represented by only s non-zero coefficients and with suitable measurement matrix $\boldsymbol{A}$, then Compressed Sensing theory states that $\boldsymbol{x}$ can be recovered from only m measurements, where m is given by:

$$m = Cs\log(n/s) \ll n \tag{8.14}$$

and C is a (small) constant.[41] This powerful result means that recovering a signal at higher resolution (*i.e.* greater n) does not require a linear increase in the number of measurements as conventional sampling theory requires, but instead is determined primarily by the information content (number of non-zero elements s) of the signal.

Lastly, to recover the signal $\boldsymbol{x}$ from the underdetermined system of equations described by eqn (8.13), it is necessary to devise a means to extract the compressed signal from the measured data. In the context of Compressed Sensing, we seek to find the solution with the fewest non-zero elements that is consistent with the measured data. The number of non-zero elements is commonly called the l_0-"norm" and is denoted $\|\boldsymbol{x}\|_0$. Therefore, we seek the solution with minimum l_0-norm. Unfortunately, finding this solution directly is intractable as it has combinatorial computational complexity. One of the key findings of Compressed Sensing theory[34,35] is that this solution can be well approximated by the minimum l_1-norm, which is defined as:

$$\|\boldsymbol{x}\|_1 = \sum|x_i|. \tag{8.15}$$

This astonishing finding can be understood relatively simply by considering the solution described by an l_p-ball—that is, the set of values that have a constant l_p-norm, as illustrated in Figure 8.4 for a simple signal described by two coefficients, x_1 and x_2.[38]

The minimum l_0 solution will always correspond to the point at which only one of the two possible variables is non-zero, as shown in Figure 8.4(a). By contrast, the established approach of least squares minimisation, *i.e.* using the l_2-norm (defined by $\|\boldsymbol{x}\|_2 = (\sum_i|x_i|^2)^{0.5}$) on the regularising term, penalizes large coefficients, and thus will generally find a smooth solution with both variables non-zero. In Figure 8.4(b) the least squares solution shows the minimum l_2-"ball" intersects with the line of feasible solutions at a point corresponding to both x_1 and x_2 being non-zero. One the other hand, l_1-norm minimisation results in a sparse solution because the "pointiness" of the l_1-ball means that with high probability the minimum-l_1 solution will lie on one of the axes; the small coefficients are then suppressed while the important larger coefficients carrying information about the signal in the sparse domain are preserved. Thus, the minimum l_1-norm solution will, with very high probability, yield the same solution as the minimum l_0-norm. The equivalence of the l_0- and l_1-minima is powerful as the intractable l_0-minimum solution is readily found by solving the l_1-minimisation problem:

$$\begin{aligned} &\text{minimise} \quad \|\hat{\boldsymbol{x}}\|_1 \\ &\text{subject to} \quad \|\boldsymbol{A}\hat{\boldsymbol{x}} - \boldsymbol{y}\|_2 \le N^{0.5}\varepsilon \end{aligned} \tag{8.16}$$

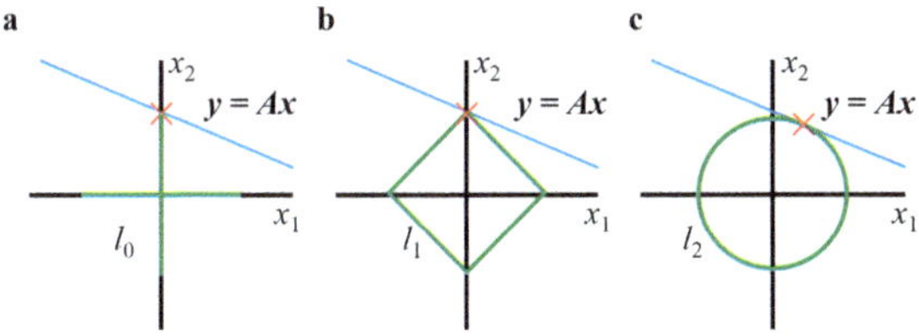

Figure 8.4 Illustration of the difference between l_0-, l_1- and l_2-norm minimisation of a signal $\boldsymbol{x}$ containing two coefficients x_1 and x_2. All of the solutions that satisfy the equation $\boldsymbol{y} = \boldsymbol{Ax}$ are illustrated by the blue line in the figure above. If the underlying solution is known to be sparse, then the minimum l_0-norm solution will yield the true solution. The l_0-ball in this example is given by the green lines in (a). A minimum l_0 solution is indicated by the red cross. However, this solution is intractable to find for problems with realistic numbers of dimensions. In many cases, the "pointiness" of the l_1-ball provides an equivalent solution, as illustrated in (b). Finding the minimum of the l_1-norm is a simple convex optimisation problem and therefore readily solved. By contrast, the more conventional l_2-minimum (least squares solution) will almost always yield a non-sparse solution, as illustrated in (c). This example is a slight over simplification as $\boldsymbol{x}$ is only a 2-dimensional vector. In this case, the minimum l_0 norm does not yield a unique solution. For realistic problems and where sufficient measurements are taken (*i.e.* $m \geq s\log(n/s)$), the l_0-norm yields a unique solution and the l_1-norm will yield an equivalent solution (adapted from ref. 38).

where $\hat{\boldsymbol{x}}$ is the reconstruction of the true signal $\boldsymbol{x}$ from the measured data $\boldsymbol{y}$, and ε characterizes the standard deviation of the noise in the data. Since the first proof that l_1-minimisation could be used to solve the l_0-norm minimization problem, various other algorithms have been used to address the recovery of signals in the compressed sensing framework, notably greedy algorithms and Bayesian techniques.[42] It can be shown that the solution to eqn (8.16) will yield an estimate of the true solution $\boldsymbol{x}$, even in the presence of noise in the measured data, with an accuracy bounded by:

$$||\boldsymbol{x} - \hat{\boldsymbol{x}}||_2 \leq C_1 s^{-1/2}||\boldsymbol{x} - \boldsymbol{x}_s||_1 + C_2\varepsilon \tag{8.17}$$

where $\boldsymbol{x}_s$ is the best s-sparse approximation of $\boldsymbol{x}$ and C_1 and C_2 are small positive constants.[43] Eqn (8.17) means that the solution to eqn (8.16) will yield an estimate of $\boldsymbol{x}$ that is almost as accurate as the best s-sparse approximation of $\boldsymbol{x}$. There are two major implications of this result: firstly, if $\boldsymbol{x}$ is s-sparse (*i.e.* $\boldsymbol{x} = \boldsymbol{x}_s$), then in the absence of noise the recovery will be exact; secondly, even in the presence of noise, any error in the reconstruction is not dramatically amplified.

Although the error in reconstruction of noisy data is bounded by eqn (8.17), the error will tend to be biased when using an l_1-norm reconstruction. We can illustrate this bias using the example of a noisy measurement of a one-sparse signal (*i.e.* one non-zero coefficient), as shown in Figure 8.5. Since there is only one non-zero coefficient, the location of this is likely to be

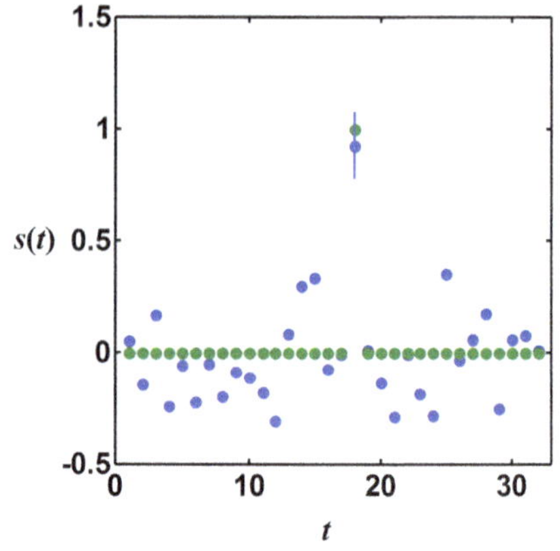

Figure 8.5 An intuitive explanation as to why Compressed Sensing tends to produce biased estimates of the true signal, resulting in a loss of contrast. The true function to be recovered is illustrated by the green circles, and contains only one non-zero element. The blue circles illustrate noisy measurements of the desired function. It is clear from these measurements that the true function contains only one value with a high intensity, as indicated by the blue line. However, based on the known noise of the measurements, the true value is likely to occur at any value along the blue line. Therefore, the minimum l_1-norm solution that is consistent with the measured data is the minimum value on the blue line, or about 0.85. Thus, a Compressed Sensing reconstruction will recover the true value, but with a bias. This bias can be removed using more advanced error measures[46] or through least squares fitting of the locations on the non-zero values identified by Compressed Sensing.[47]

recovered accurately. However, there will be a cluster of values for that coefficient that are consistent with the measurements to within the experimental noise level. The l_1-minimum solution will correspond to the minimum of these values for that coefficient. Thus, the result will be biased. In the more general case of multiple non-zero coefficients, this bias will tend to manifest as a loss of contrast.[44,45] The loss of contrast can be minimized using more advanced algorithms, such as Bregman iterations.[46] Alternatively, the solution to eqn (8.15) can be used to recover the support of $\boldsymbol{x}$, which should contain $k < m$ non-zero coefficients, therefore these non-zero coefficients can be fit to the measured data with a least squares model to remove the bias.[47]

Where Compressed Sensing can be implemented, the undersampling that becomes possible represents a substantial reduction in data acquisition time over the conventional "bandwidth" limited approach identified by Nyquist and Shannon.[7,8] For example, consider the extension of a two-dimensional NMR spectrum to a three-dimensional spectrum, where the number of measurements in the indirect dimensions are n_2 and n_3, respectively. In this case, the number of resonances to be identified remains constant, in other words, s is fixed, but these must now be identified in a three-dimensional array. From eqn (8.14) it is easy to see that only an additional ~$(1 + \log(n_3)/\log(n_2))$ measurements are required to resolve the s resonances into this third dimension, assuming s is small. For example, assuming 128 data points in each indirect dimension, only twice as many measurements are required to extend the measurement into the third dimension, which represents a 64 fold time saving over the conventional sampling approach. Many choices are available for the basis used to define the compressed representation, and so compressed sensing theory is very powerful. Examples of sparse bases include the discrete cosine transform, discrete wavelet transform, and spatial finite differences (note the l_1-norm of the spatial finite differences transform is commonly referred to as total variation). Recent work has extended these concepts even further to include signals that have a concise representation, such as signals that are structured-sparse, low-rank, or lie on a manifold.[48]

We illustrate the concept of Compressed Sensing in Figure 8.6 using a simulated NMR spectrum, adapted from the example of Lustig *et al.*[36] In this case we have a simple spectrum comprising three peaks. If the time domain data are fully sampled, then the spectrum is readily reconstructed using the Fourier transform, as shown in Figure 8.6(a). However, if the time domain data are undersampled, then the conventional Fourier transform would result in a noisy spectrum in which the peaks with the weakest signal intensity are lost in the "noise", as seen in Figure 8.6(b). However, the apparent noise in this spectrum is an artefact arising from the limited number of observations of the system. These artefacts are removed if we use a Compressed Sensing algorithm to reconstruct the spectrum, and the signal from the weakest peak is recovered, as shown in Figure 8.6(c). It is important to note that the same raw data were used to reconstruct the signal in both Figure 8.6(b) and (c), the only difference is the reconstruction procedure used. Thus, it is possible to reconstruct the image using only a subset of the measurements conventional sampling theory requires.

The most obvious advantage of employing a Compressed Sensing methodology is that by decreasing the number of data points that are required to obtain the desired information, datasets can be acquired much faster than is currently achievable. In the following section, we review some applications of Compressed Sensing to both high and low field NMR and MRI.

8.3.2 Applications of Compressed Sensing

There are relatively few examples of the application of Compressed Sensing on low field NMR instruments. Therefore, this review will present some of the most relevant examples from high field NMR/MRI as well as those applications on low field instruments.

Compressed sensing came to the fore in MRI following publication of the paper by Lustig *et al.* in 2007.[36] In this seminal work, Lustig *et al.* demonstrate that it is possible to reconstruct sparsely sampled MRI angiography and brain images from as little as 10% of the data points that the conventional Nyquist sampling criteria requires. Since this early demonstration of Compressed Sensing, the technique has been demonstrated in a variety of medical MRI applications including functional MRI,[49,50] cardiac imaging,[51,52]

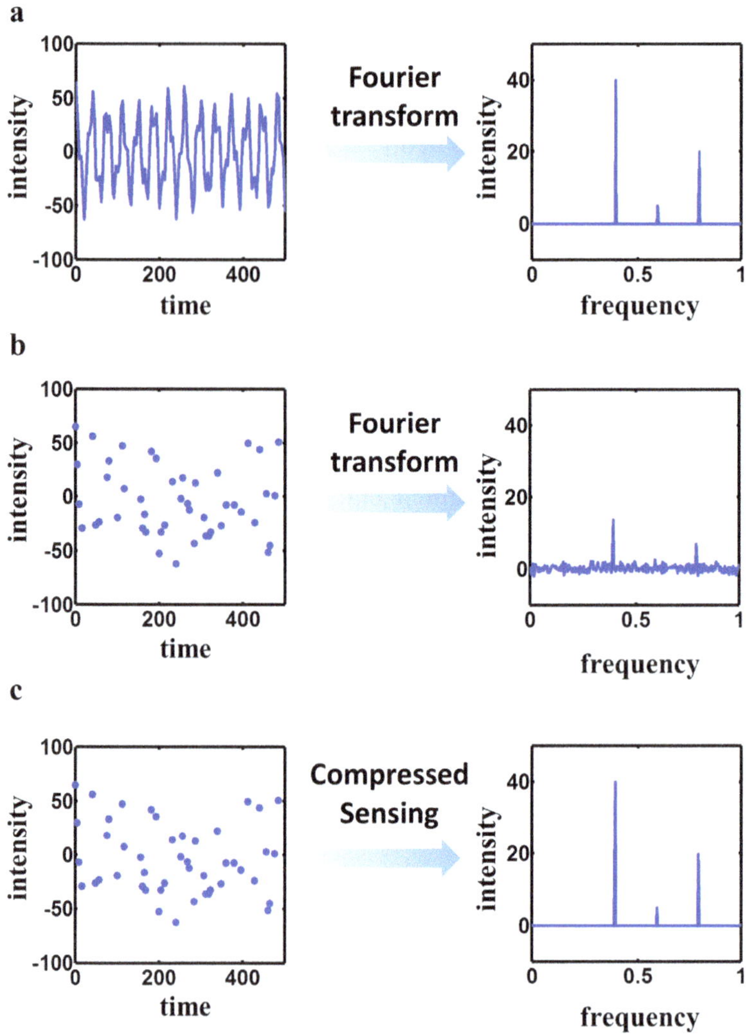

Figure 8.6 (a) A conventional NMR spectrum is obtained by sampling one complex time domain data point for every point in the resulting spectrum. (b) If the time domain data are not fully sampled, then a Fourier transform reconstruction will result in significant noise-like artefacts and a poor quality spectrum in which the weakest intensity components are lost. (c) If the same data as shown in (b) are reconstructed using Compressed Sensing, the original spectrum can be recovered. If the measurements do not contain noise, this recovery is exact.

and hyperpolarised ^{13}C imaging of metabolites.[53] Surprisingly, even with its relatively recent introduction, Compressed Sensing techniques are already appearing in clinical MRI applications.[54]

Following the success of Compressed Sensing in medical MRI, it was rapidly adopted in non-medical imaging applications. Early examples include the quantitative mapping of moisture in very low density food stuffs,[55] high resolution velocity imaging,[56,57] and multidimensional NMR spectroscopy for protein structure determination.[58,59] The fully sampled techniques that would otherwise have been used in these applications were inherently slow. Compressed sensing permits high quality, quantitative reconstructions of the desired signal from between 16% and 30% of the data required using Nyquist sampling theory, providing a comparable reduction in acquisition time. In one of these examples, Parasoglou *et al.*[55,60] selected the *k*-space points to be sampled according to the known signal distribution in the dry wafers. By incorporating prior knowledge of signal intensity in *k*-space, the authors were able to improve the quality of the reconstructed images significantly. The idea of exploiting the expected distribution of the intensity of the signal in *k*-space is currently an area of significant interest in mathematics.[61]

The benefit of Compressed Sensing can be even more significant in ultrafast MRI. For example, Tayler *et al.* used Compressed Sensing to reconstruct high-resolution velocity images in multiphase flow using a variable density spiral approach.[62] Figure 8.7 shows some examples of the types of velocity maps that can now be obtained of a freely rising gas bubble in water. In this case, Compressed Sensing permits a reduction in the data required by almost a factor of four, but this reduction in acquisition time also has the advantage of eliminating artefacts arising from susceptibility differences between the phases. If a conventional fully sampled data set were acquired, it would have necessitated the use of multiple excitations increasing the acquisition time by an additional factor of 2, which makes the technique too slow to capture the complex dynamics of this multiphase flow. There are many other examples of Compressed Sensing applications

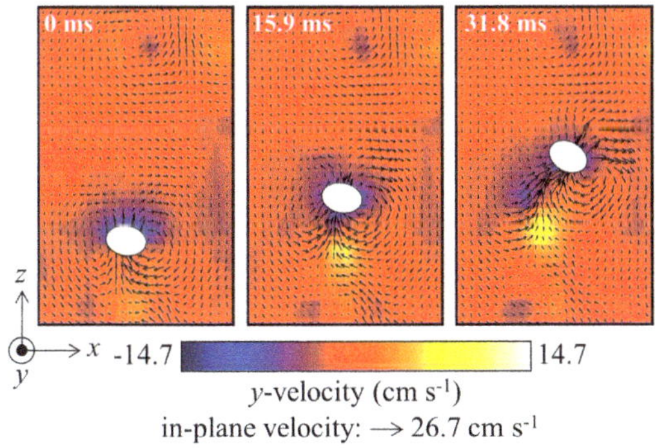

Figure 8.7 Three frames from a series of high-resolution velocity images of the water surrounding a freely rising gas bubble obtained using Compressed Sensing. Each frame was obtained through the combination of three separate images of the individual components of the velocity in the water. Each individual velocity component was acquired in 5.3 ms giving a total acquisition time for a three-component velocity image of 15.9 ms. Adapted from the results presented by.[62]

using high field NMR and MRI, however we will now switch the emphasis to low field applications.

The first demonstration of Compressed Sensing using low field MRI was in the application of T_2 mapping of porous solids,[63] and a similar application has also since been demonstrated by ourselves.[64] The first of these papers sought to extend earlier work on one-dimensional T_2 mapping reported in the same laboratory,[65,66] to two-dimensional acquisitions. In order to maintain a quantitative imaging technique on a system with a short T_2^*, a pure phase encoded imaging sequence was used, which is inherently very slow. Through the use of Compressed Sensing the experiment time was reduced by a factor of 5, with only an insignificant effect on the image quality. Interestingly, the authors note that the Gibb's ringing phenomena were still observed with high levels of undersampling. The Gibb's phenomena are an artefact of the limited extent of *k*-space sampling that can be achieved and therefore are undesirable. It is possible to eliminate such an affect using Compressed Sensing techniques, as has been demonstrated in multidimensional NMR acquisitions;[67] however, it can be difficult to achieve this in practice.

Low field MRI is increasingly appealing in medical MRI applications owing to the lower cost and the possibility of open magnet designs.[68–71] However, the use of low field MRI is frequently hampered by the long acquisition times that are required. In medical MRI, it is necessary to reduce the scan time to improve patient comfort and minimise distortions arising from movement of the patient. Conventional high field imaging is often able to overcome these challenges through the use of fast pulse sequences and parallel MRI. However, such approaches are less appealing on low field instruments as they generally require more homogeneous B_0 fields, more complex hardware, and a higher SNR. Lee *et al.* demonstrate that Compressed Sensing can be used as an alternative method for reducing scan time, without increasing the burden on the hardware.[72] They show that images of a phantom and a human head can be recovered from only 50% of the fully sampled data, with negligible loss of information.

Perhaps the most challenging low field MRI systems use unilateral magnet designs. Such systems are optimal for patient comfort and permit imaging of samples that would otherwise be impossible to study.[73] Unilateral MRI systems are characterised by strongly inhomogeneous B_0 fields, which necessitate the use of pure phase encoded imaging techniques.[74] The long imaging times associated with pure phase encoded imaging limit the applications to which unilateral MRI can be applied. Liberman *et al.* have recently demonstrated the application of Compressed Sensing on a unilateral low field instrument.[75] In their work, Compressed Sensing permits the scan time to be reduced by an order of magnitude without affecting the image quality, as illustrated in Figure 8.8. These results will potentially permit unilateral MRI devices to be used in a range of new applications in the future, including biomedical imaging. One point to note is that for the low-resolution images that can typically be acquired using a unilateral device, it is difficult to obtain a sparse representation of the images. Liberman *et al.* used a Daubechies-6

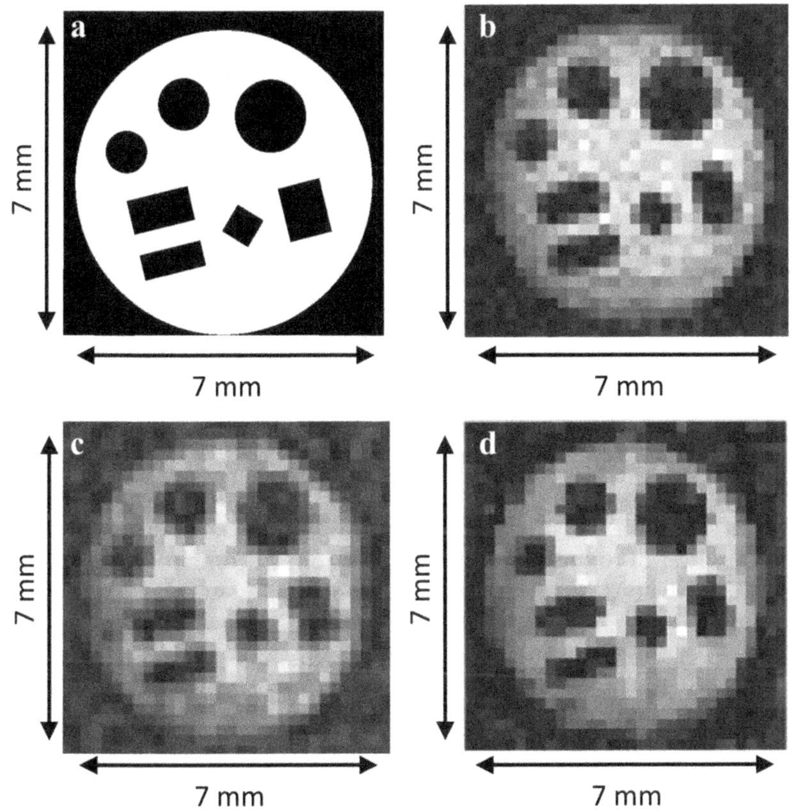

Figure 8.8 Demonstration of high-resolution images of a phantom obtained using a unilateral MRI system. Panel (a) shows a sketch of the phantom containing three circles of different sizes, and four rectangular objects of different size, aspect ratio and orientation. (b) Conventional fully sampled MRI image of the phantom obtained from the unilateral magnet with 12 averages. The total acquisition time for this image was 72 minutes. (c) Image of the phantom obtained with 2 averages using only the central 50% of *k*-space. The object is recovered, however contrast is poor and the resolution has decreased from that shown in (b). (d) Compressed Sensing reconstruction obtained after sampling 50% of *k*-space and acquiring 2 averages. The total acquisition time for these data was 6 minutes. The Compressed Sensing reconstruction shows good recovery of all major features and has retained similar contrast to that shown in (b), despite being acquired in an order of magnitude less time. Adapted from ref. 75 with permission from Elsevier.

wavelet transform as their sparse basis. The wavelet transform is particularly effective for natural images, such as photographs, that have a relatively high resolution (typically >256^2 pixels). In many medical imaging applications the wavelet transform has been found to be beneficial (*e.g.* see ref. 36 and 49). However, for low-resolution images, the spatial finite differences transform can yield a representation that is more sparse than a Daubechies wavelet transform. Therefore, although the results presented by Liberman *et al.* are impressive, it is possible that a modified Compressed Sensing approach could yield even higher quality images in this application, although for biomedical imaging applications the best approach would need further investigation.

8.3.3 Future Developments

The examples shown in the previous section illustrate that Compressed Sensing is beginning to be used widely with high field NMR and MRI, however there are relatively few examples of Compressed Sensing using low field hardware. The key idea behind Compressed Sensing is to transfer some of the complexity of imaging from the hardware to the software, and in this context it would appear well suited to low field applications. Image reconstruction with Compressed Sensing is significantly more complex than with conventional Nyquist sampling, however many software packages are becoming available to simplify this process and with indications that commercial high field scanners will shortly include Compressed Sensing capabilities,[76] it is only a matter of time before such calculations become routine.

The most obvious benefit of Compressed Sensing with low field MRI will be in reducing the acquisition time required to achieve a given spatial resolution, and thereby permitting the study of dynamic systems. However, when dynamic systems are to be studied it is clear that the acquisition time can be reduced even further by exploiting the temporal correlation of the signal.[51,52,77] The idea here is analogous to the difference in compression achieved with movies and still images. For example, when compressing a still image or photograph (*e.g.* through jpeg compression) it is typical to achieve a compression rate of 10 : 1 before introducing artefacts into the image; when compressing a movie (*e.g.* through mpeg compression), it is typical to achieve a compression rate of 50 : 1 or better. Similarly, if using Compressed Sensing on a dynamic system, it should be possible to sample significantly fewer data points per frame and then to exploit structure in both time and space in order to reconstruct the entire time series of images. At present, it is an open question how best to exploit the temporal information during reconstruction. However, this is an area of active research and it is likely that such an approach will make dynamic imaging on low field instruments possible in the near future.

In many low field MRI applications, the sensitivity of the measurement is a major challenge. Compressed Sensing has already been shown to be an effective method for improving the sensitivity of NMR[78] and MRI[55,56] in a given acquisition time. It is therefore likely that Compressed Sensing techniques will lead to significant improvements in the quality of data that can be obtained from low field instruments as well. It is still an open question as to whether it is better to undersample the data in *k*-space but acquire each data point with a higher SNR, or to fully sample *k*-space but with each data point having a lower SNR. In either case, it is possible to reconstruct the signal using the same l_1-minimisation technique that is used in Compressed Sensing. However, sparse sampling approaches have long been known to improve the SNR of multidimensional NMR data[79] and more recently similar results have been shown in the context of gas phase imaging.[56] These results indicate that sparse sampling and Compressed Sensing are promising methods for improving the sensitivity of MRI measurements, which will be important for low field imaging.

Lastly, one of the areas in which Compressed Sensing is likely to have a major impact in low field MRI is through more advanced field map correction. Doneva *et al.* recently demonstrated a field map estimation with simultaneous acquisition of water and fat images using Compressed Sensing.[80] If such an approach can be adapted to low field MRI, it could potentially increase the range and complexity of pulse sequences that can be used during image acquisition, which could dramatically increase the SNR achievable per unit time analogous to when techniques such as EPI were introduced to high field MRI.[81]

There is much scope for further development of Compressed Sensing techniques. However, this section has highlighted some of the many exciting applications of Compressed Sensing on both high- and low-field NMR and MRI. Emerging algorithms and theory will make implementation of Compressed Sensing approaches simple. Therefore, it is anticipated that Compressed Sensing will become a routine approach for multidimensional NMR and MRI on both high- and low-field instrumentation in the coming years.

8.4 Bayesian Analysis

Bayesian analysis is based on the work of Thomas Bayes,[82] who presented a method for analysing the probability of a random event occurring when presented with additional information about that event. To paraphrase an example from Sivia and Skilling,[83] if someone told you that the day is cloudy, it might be natural to assume there is some chance of rain. If you were to observe subsequently that those clouds were very dark, you would likely perceive that the chances of rain are greatly increased. Bayesian analysis presents a framework under which such deductive reasoning can be analysed quantitatively. Although the original work of Bayes dates back to the 18th century, Bayesian analysis was largely forgotten until the work of Jeffreys in 1939.[84] There are now typically 10^4 articles published annually that either use Bayesian analysis or seek to develop Bayesian mathematics. For interest, we also note that several researchers have derived Compressed Sensing reconstructions from a Bayesian perspective.[85–87] This section presents a basic introduction to Bayesian analysis techniques, for more in depth discussion the interested reader is referred to the many excellent texts on the subject.[83,88–90]

8.4.1 Theory of Bayesian Analysis

Bayesian analysis aims to infer the probability that a system is in any given state θ given some measurements d and information I. The distribution describing this probability is termed the posterior probability density and is written

$p(\theta|d,I)$, which is read the probability density of state θ given measurements d and prior information I. Note that here we use p to denote a probability density function; p is normally used to denote a probability and so strictly $p(\theta|d,I)$ should be obtained by integrating a probability density function over some defined limits. However, this slight abuse of notation is generally easier to use in practice. The posterior probability is determined from Bayes' theorem:

$$p(\theta|d,I) = \frac{p(d|\theta,I)\,p(\theta|I)}{p(d|I)} \tag{8.18}$$

where the terms on the right hand side of eqn (8.18) are the likelihood $p(d|\theta,I)$, the prior $p(\theta|I)$ and the evidence (also known as the marginal likelihood) $p(d|I)$. The posterior probability density fully characterises our knowledge of the system, therefore we can use it to estimate the most likely state of the system from the maximum of the distribution, the mean state of the system from the mean of the distribution, and most importantly the uncertainty bounds of our knowledge. Being able to unambiguously define our uncertainty is a major strength of Bayesian analysis.

Eqn (8.18) is a simple expression that forms the basis of Bayesian analysis, however it is deceptively simple. Indeed, it may prove impossible to apply Bayesian analysis to many problems of practical interest. In order to apply Bayesian theory to a problem, it is first necessary to specify a model for the problem, the sample space, hypothesis space, prior probabilities and sampling distribution.[89] Therefore, let us explore the expression in eqn (8.18) and the challenges associated with it in more detail.

Firstly, we draw the reader to the term I describing the information on which the analysis in eqn (8.18) is based. This term characterises the information that we are imposing during the formulation of the model or, in other words, our assumptions about the system. Such information is most typically related to the form of the model or the noise. The effect of this information term is striking and if it is not addressed correctly can lead to incorrect inference. Jaynes illustrates this using the example of Mrs Gloria Stewart demonstrating extrasensory perception (ESP) by correctly guessing a card many more times than could be expected by chance.[89] In this example, he states that the probability that she had guessed these values by chance was $\sim 10^{-139}$. Therefore, in light of such strong evidence the only logical conclusion would appear to be that ESP has been proven. However, despite this, Jaynes states that he is not inclined to change his perception that ESP is a hoax. The question arises, how can such steadfast denial of the evidence be justified in the light of Bayes' theorem? The key point that is missing in the calculation above is that the initial application of Bayes' theorem is flawed. In reality there are possibilities other than that she guessed the solution by chance or that she has ESP, for example systematic bias could be introduced by the tester, *etc.* By ignoring these possibilities in the analysis, we would incorrectly reach the conclusion that ESP has been proven. If instead we factor in these various other possibilities, it is very easy to demonstrate that the more likely situation is simply that bias has been introduced, either intentionally or unintentionally, during the testing process. Thus, if the information used to formulate the model is not correctly identified, it is very easy to reach an incorrect conclusion. Although crucial to the formulation of the problem, the notation I is frequently left out of mathematical descriptions for simplicity. We will follow this convention here but note that the reader must always remember that this term is implicitly present.

In order to apply eqn (8.18), it is necessary to find an expression for the term $p(d|\theta)$. This term is referred to as the likelihood. In essence, it captures our model of the process that we are considering. It is not possible to describe the likelihood for any general process, so we will illustrate this for the case of a simple NMR signal. Such a model was originally described by Bretthorst[90] and his work forms the basis of much of the subsequent work on Bayesian analysis in NMR. In this case, our signal model may be given by:

$$S(t) = Ae^{-i\omega t}e^{-t/T_2^*} + \epsilon \tag{8.19}$$

where $S(t)$ is our signal, A is the amplitude of the signal, ω is the resonance frequency and ϵ is a noise term. Already we see that we must make several assumptions about this signal if we are to analyse it in a Bayesian context. For example, here we have assumed that the signal oscillates with a single frequency and decays through an exponential process, but is otherwise of fixed amplitude. Perhaps less obviously, we have also assumed that we can detect the signal using quadrature detection and that our two channels are perfectly 90° out of phase with each other. We have also assumed that the noise is a simple additive process, typically with NMR data it is assumed that the noise is normally (Gaussian) distributed, where the normal distribution is given by:

$$p(x\,|\,\mu,\sigma) = \frac{1}{\sigma\sqrt{2\pi}}\,e^{-\frac{(x-\mu)^2}{2\sigma^2}} \tag{8.20}$$

for mean μ and standard deviation σ; since the normal distribution occurs so frequently, it is often abbreviated $\mathcal{N}(\mu,\sigma)$. Therefore, assuming the noise is independent, additive, Gaussian distributed noise, the likelihood can be written:

$$\begin{aligned} p\left(S|A,\omega,\sigma,T_2^*,t\right) &= p\left(\text{real}\left(S - Ae^{-i\omega t}e^{-t/T_2^*}\right)\right) \\ &= \frac{1}{\sigma\sqrt{2\pi}}e^{-\frac{\left(\text{real}\left(S-Ae^{-i\omega t}e^{-t/T_2^*}\right)\right)^2}{2\sigma^2}} \end{aligned} \tag{8.21}$$

where for this model, the state of the system θ is defined by A, ω, σ, and T_2^* and for simplicity we are only considering the real part of the measured signal. Thus, eqn (8.21) describes the probability of measuring the signal intensity

S given prior knowledge of the amplitude, frequency, standard deviation of noise and T_2^* relaxation time, as well as the information that the model can be described by eqn (8.18). In itself, describing the signal with eqn (8.21) is not directly useful as it requires us to already know everything that we would seek to find out about our system (*i.e.* the amplitude, frequency, *etc.*). This is where Bayes' theorem is useful as it enables us to use the model equation we have formulated to calculate the probability distribution for the parameters of interest (amplitudes, frequencies, *etc.*). However, before we can do this we must first consider the remaining terms in eqn (8.18).

The second term to consider from eqn (8.18) is the prior, $p(\theta|I)$. The prior describes our existing knowledge of the parameters that describe the state of our system. In the case of the NMR problem given above, our prior would need to capture what we know about the possible values for the amplitude, frequency, noise and T_2^* relaxation terms. Correct choice of these parameters should be determined based on our knowledge of the system. For example, if we are studying ^{1}H NMR on a 0.5 T magnet, then the frequency of the signal is likely to be around 21 MHz. Therefore, our prior should be centred on this frequency. The exact choice of the function for the prior is often not critical and a normal distribution is often chosen for simplicity. If we are not confident of the exact frequency, we would describe the prior with a broad distribution; on the other hand, if we knew the chemical we were studying, *e.g.* cyclohexane, then we could identify the frequency more precisely. It is also possible to use an improper prior such as a uniform distribution or Jeffreys' prior, which is uniform probability in the logarithm of the variable. Improper priors such as these are advantageous when little is known about the value of a parameter as they do not introduce bias to the analysis. In general, the more accurately the prior is defined, especially with regard to functional form, the more accurate the final inference will be, however it may come at the cost of making the problem intractable.

The final term to consider in eqn (8.18) is the denominator, which is often termed the evidence. The evidence is often ignored in Bayesian analysis problems because it simply acts as a normalisation constant. Thus, if one is certain that the form of the model used is appropriate, and the task is simply to estimate the value of the parameters of that model, then $p(d)$ is constant for all values of θ. Therefore, the posterior can be calculated with an arbitrary value of $p(d)$ and once the calculation is complete, the probability density can be normalised such that the integrated probability is 1. However, in certain situations, $p(d)$ must be calculated explicitly. The most common example of this is *model selection* where one seeks to identify the form of the model. For example, in the case of the NMR data above, alternative models might include the case when the data are described by two (or more) frequencies. In this case, if the number of frequencies is not known *a priori*, these alternative models must be compared and $p(d)$ must be calculated explicitly for each case.

It is already apparent that the approach outlined can be a powerful method for analysing experimental data, as it provides a framework under which all of our existing knowledge can be combined together to produce a robust, quantitative analysis. However, the real strength of Bayesian analysis requires us to consider one further piece of mathematical analysis: marginalisation.

Marginalisation is the process whereby the effect of certain variables can be removed from subsequent analysis and is described neatly by Sivia and Skilling.[83] As an example of where marginalisation is useful, we can consider the case of the NMR spectroscopy example above. In this case, we might be interested in the frequency of the spectrum as this will tell us what species are present, however we might not be interested in the amplitude of the signal and therefore we would like to remove this factor from the analysis. Parameters that affect the signal but are not of interest in subsequent analysis are termed "nuisance parameters" in Bayesian analysis and marginalisation provides a means to eliminate these from future calculations. Marginalisation is achieved mathematically by integrating over the probability distribution of that parameter. For example, if a system is governed by two parameters X and Y, but we are only interested in the response due to parameter X, we could marginalise the contribution from Y by calculating:

$$p(X) = \int_{-\infty}^{\infty} p(X,Y)\mathrm{d}Y \tag{8.22}$$

Marginalisation is powerful as it enables us to eliminate those quantities that necessarily affect the signal but are of no inherent interest.

Calculation of integrals, such as that shown in eqn (8.22), is one of the major challenges in Bayesian analysis. For certain problems, analytical solutions to eqn (8.22) are readily obtainable, for example if the calculation of the probability distribution involves only normally distributed variables. However, for most problems of practical interest, the calculation of the integral in eqn (8.22) is intractable analytically. At first sight, this would seem to dramatically reduce the significance of Bayesian analysis and constrain it to only those few systems where an analytical solution is possible. However, it turns out that the law of large numbers provides a method for computing such integrals numerically.

The law of large numbers states that the average of many independent random variables with common mean and finite variances tends to stabilise at their common mean.[91] Mathematically, this idea means that the integral given by eqn (8.22) can be approximated by:

$$p(X|I) = \lim_{m\to\infty} \frac{1}{m}\left\{p\left(X,Y^{(1)}\right) + p\left(X,Y^{(2)}\right) + \ldots + p\left(X,Y^{(m)}\right)\right\} \tag{8.23}$$

where each $p(X,Y^{(i)})$ represents a uniformly random draw from the support of Y. Using eqn (8.23), it is possible to numerically integrate any function through a simple summation. This approach of taking random samples to estimate parameters or integrals is known as a Monte Carlo strategy and much effort is devoted to efficient implementations of these approaches.[91] The power behind this approach is that the error in the estimation scales with $m^{-1/2}$, regardless of the dimensionality of the problem. This scaling makes Monte Carlo approaches strongly favourable over deterministic sampling, which typically require $O(m^D)$ samples for a D dimensional problem. However, there are two principle challenges with Monte Carlo integration: (i) although the standard deviation of the error may decrease with $m^{-1/2}$, in high dimensional space the error may still be formidably large; and (ii) it may not be possible to sample Y uniformly. Various approaches are available to help overcome these challenges and the reader is referred to one of the excellent texts or reviews on Bayesian analysis and Monte Carlo strategies for further information.[83,88,91,92]

In this section, we have outlined some of the basic principles of Bayesian analysis, including Bayes' theorem and marginalisation. We have also highlighted why this technique is so powerful. In the following section, we will illustrate the use of Bayesian analysis for applications in NMR and MRI.

8.4.2 Applications of Bayesian Analysis

The most fundamental problem in NMR/MRI is that of identifying a chemical spectrum, and as a result of this, of all problems in NMR/MRI that have been studied by Bayesian analysis spectrum identification has been studied most widely. The first demonstration of Bayesian techniques in NMR was developed by Bretthorst during his PhD under the supervision of Jaynes; this thesis was later revised and issued as a book,[90] and further developed in a series of publications.[93–101] In these early works, the emphasis was on establishing analytical solutions to the Bayesian analysis of NMR signals. As such, it was necessary to make a few simplifying assumptions, *e.g.*, that resonances were well separated. However, these assumptions were later relaxed by exploiting the orthogonality of the sine and cosine functions that give rise to the time domain signal in NMR[96] or through the introduction of numerical methods to approximate the calculations.[97,98] Bretthorst was able to show that for stationary (*i.e.* non-decaying) frequencies that are well separated, the Fourier transform is an ideal Bayesian estimator of the frequency components.[99] However, in cases where the signal decays or frequencies overlap, the Fourier transform is no longer optimal and Bayesian techniques can be advantageous. For example, Bretthorst and others have demonstrated the superiority of Bayesian analysis when data have low SNR,[100–103] to automate quantification of spectra,[97,98,104,105] especially for the concentration of metabolites where the data have a very high dynamic range,[98,104] and to improve the spectral resolution of complex, multidimensional NMR spectra.[106–110] The application of Bayesian analysis to solid state spectra has also been demonstrated, where the goal was to extract the chemical shift anisotropy and asymmetry parameters, in addition to the isotropic chemical shift.[111] Whilst most of these studies have been performed on the time domain data directly, a few examples suggest there may be benefits in applying the technique to the frequency domain signal.[104,112,113]

We illustrate the benefits of Bayesian analysis in NMR spectroscopy in the quantification of the composition of a mixture of butanone and cyclohexane using ^{13}C NMR in Figure 8.9. This system has two challenges. Firstly, since

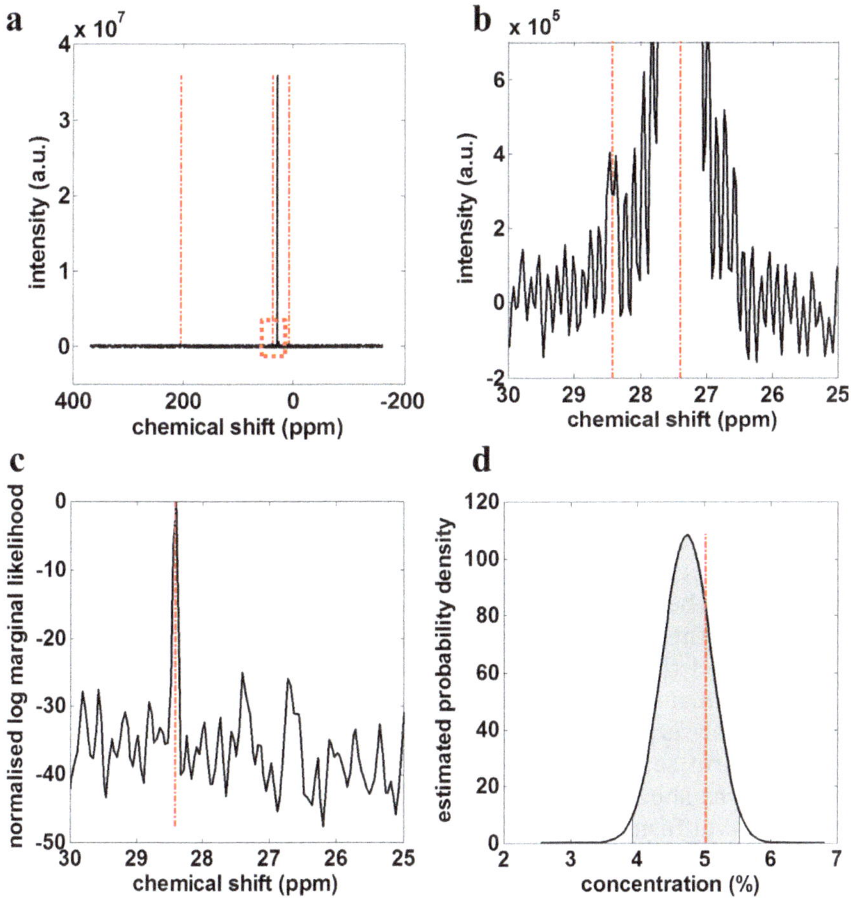

Figure 8.9 Comparison between Bayesian inference and Fourier transform analysis of a mixture containing 5% 2-butanone and 95% cyclohexane. Panel (a) shows the whole Fourier transform spectrum of the mixture. The SNR of the butanone peaks is about 6. The spectrum in (b) shows a close up of the part of the spectrum indicated by the red box in (a). One of the butanone peaks (28.43 ppm) is on the shoulder of the peak from cyclohexane (27:3 ppm), and thus its intensity will be overestimated in Fourier transform analysis. The acquisition time of the free induction decay was limited by the duration of the ^{1}H decoupling, which gives rise to significant truncation artefacts making quantitative analysis of the butanone even more challenging. (c) Log marginal likelihood as a function of the 28.43 ppm resonance frequency. It is well distinguished from adjacent frequencies and shows no indication of the truncation artefacts exhibited in (b). The posterior probability distribution of the concentration of butanone is shown in (d), as estimated from Bayesian inference. The shaded grey region in (d) indicates the 95% credible region, or in other words indicates the range of concentrations within which we can be 95% certain the true concentration lies. In this case, Bayesian inference estimates the concentration to be 4.73 ± 0.74 mol%, in good agreement with the true concentration of 5 mol%. The dashed lines in each plot indicate the true values of the resonant frequencies and concentration.

the acquisition is on natural abundance ^{13}C NMR the SNR is very low, which makes quantitative analysis challenging.[114] Secondly, the decoupling of the ^{1}H resonances can only be performed for a limited time with the experimental setup available, which leads to significant truncation artefacts and further complicates quantitative analysis. Bayesian analysis provides a method to overcome both of these limitations providing accurate, quantitative measurements. A further advantage of Bayesian analysis is that since we have the full posterior distribution we can also obtain quantitative estimates of the uncertainty in our measurements.

Bayesian analysis has also been used to improve the quality of data obtained from spectroscopic imaging.[115] In this example, the structural information provided by a non-spectrally encoded scan is used as a prior, which is shown to improve the quantitative recovery of metabolite signals when combined with spectroscopic imaging.

Whilst all of these examples have been applied to high field NMR measurements, the increased robustness to noise and overlapping peaks demonstrated suggests that Bayesian NMR will be of significant benefit when quantifying spectral information from low field instruments, and this is anticipated to be an area of growth in the future.

Soon after the application of Bayesian techniques to NMR spectroscopy, it was realised that the same model could be readily adapted to quantify flow and diffusion in phase encoded MR experiments.[116,117] In this case, Bayesian analysis permits sparse sampling of the phase encoding data using a non-uniform sampling of q-space (reciprocal displacement space). This approach was shown to provide a more precise quantification of the fluid velocity and diffusion (or dispersion) in simple laminar and pulsatile flows. The same technique was subsequently used to analyse dispersion in packed beds[118] and flow in a heat exchanger.[119] Bayesian techniques have also been used for flow mapping in medical MRI in combination with both phase encoded flow imaging[120] and tagged flow imaging,[121] where Bayesian techniques demonstrate superior performance with respect to under-sampling, noise and complexity of the flow.

The application of Bayesian analysis to flow and NMR spectroscopy have both yielded significant improvements in sensitivity and robustness, however these approaches do not fully exploit the benefits of marginalisation. One application of Bayesian analysis in NMR that does exploit marginalisation was first demonstrated in the characterisation of bubbly flows using NMR.[122,123] The objective of this work was to quantify the size distribution of bubbles in a gas–liquid two-phase flow. If the bubbles are assumed to be spherical, the problem can be described as identifying the size of a sphere irrespective of the position of the sphere. The k-space signal associated with a set of N bubbles, defined by the set of radii $\{r\}$ and locations $\{x_c\}$, can therefore be described by:

$$F(k|\{r\},\{x_c\}) = \sum_{j=1}^{N} H(r_j, k) e^{-2\pi i k x_{c,j}}, \tag{8.24}$$

where $H(r_j,k)$ is the k-space signal for the jth bubble centred on the axes as a function of the bubble size r_j and location in k-space k and $x_{c,j}$ is the location of the centre of the jth bubble. Since the bubble locations are of no interest, they are a nuisance variable and we can use marginalisation to remove the contribution of this variable from our signal model. Assuming the set of $x_{c,j}$ is independent and identically distributed, then after marginalisation, provided that N is sufficiently large, the magnitude of the signal will be described by a Rayleigh distribution and the likelihood function will take the form:

$$P\left(\left|\left(F(k)\right)\right| \mid \sigma\right) = \frac{\left|F(k)\right|}{\sigma^2} \exp\left(-\frac{\left|F(k)\right|^2}{2\sigma^2}\right), \tag{8.25}$$

where $\sigma^2 = N/2E(|H(r_j,k)|^2)$. The function $E(|H(r_j,k)|^2)$ can be derived analytically in some cases, but in general is calculated numerically after assuming the form of the distribution for r_j; in the examples presented to date,[122–125] r_j was assumed to be normally or log-normally distributed, though any simple functional expression could be used. This approach therefore only requires the estimation of the parameters of the size distribution to characterise the bubbly flow. The benefit of this approach is clear in recent applications using low field NMR.[124,125] For example, Figure 8.10 demonstrates the measurement

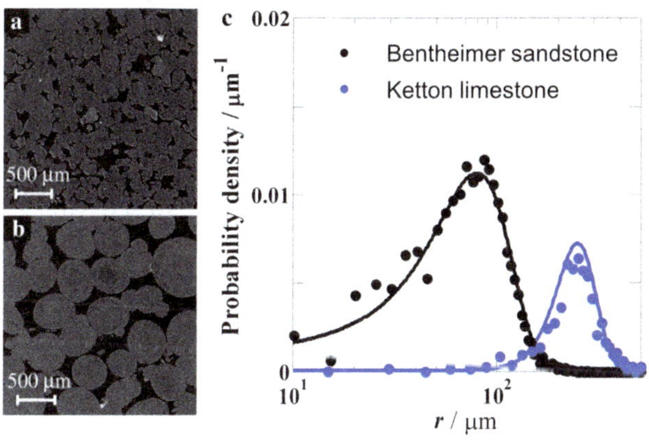

Figure 8.10 X-ray micro-tomography was used to obtain detailed images of the three dimensional structure of porous rock samples. Two-dimensional slices through these data sets are shown for: (a) a Bentheimer sandstone and (b) a Ketton limestone. The sandstone has a much finer grain size distribution than the limestone. These X-ray images have high contrast so it was straightforward to extract the size distribution of the grains from these images, which are shown by the data points in (c). MR measurements of the sample were also acquired for $k_x = 1 \times 10^3$ m^{-1} to 12×10^3 m^{-1}. The grain size distribution was extracted from these measurements using the likelihood function given in eqn (8.25), and shown by the solid lines in (c). The full Bayesian analysis also provides uncertainty bounds on the grain size distribution. Figure 8.10 is adapted from ref. 125.

of the grain size distribution in a porous rock with a spatial resolution of approximately 10 μm. These measurements were performed on a low field, permanent magnet instrument in an acquisition time of approximately 1 hour. By comparison, if a full three-dimensional image of the same rock sample had been acquired with the same resolution and SNR, it would have taken in excess of 20 years, and it is likely that in order to produce a viable image a higher SNR, and hence longer acquisition time, would be required.

These results demonstrate the dramatic benefit Bayesian analysis techniques can provide in analysing MR data. Bayesian analysis provides a robust inference framework that permits us to capture all of our existing knowledge of a process and its relationship to the measurements of interest. Further, through the use of marginalisation, it is possible to isolate the contribution to only those parameters of interest which can provide dramatic benefits in terms of acquisition time and sensitivity. Whilst relatively few low field MR measurements have been demonstrated using Bayesian analysis, it is clear that this is an area where Bayesian techniques can have significant benefit.

8.5 Conclusions and Outlook

In this chapter we have presented a summary of some of the conventional techniques used to process NMR and MRI data. There are significant limitations associated with these techniques, including difficulty in quantifying the signals, especially in the presence of noise, difficulty in resolving spectral components, and large data sampling requirements leading to long acquisition times. All of these factors have driven researchers to explore more advanced signal processing opportunities. A significant advantage of developing more advanced signal processing techniques is that they can be applied almost universally to any experimental protocol. Thus, whilst a technique might be developed for ^{1}H NMR, it is often relatively straightforward to extend such a technique to, *e.g.*, ^{13}C NMR and in some cases even to MRI.

The two new signal processing techniques that we have considered are Compressed Sensing and Bayesian analysis. Both of these techniques have been available or are derived from techniques that have been available for many years now. However, it is only with recent advances in theory and computational power that they are beginning to be widely adopted. We see both of these approaches as complementary.

In the case of Compressed Sensing, the diversity of applications in which it has already been shown to be beneficial demonstrates that it has strong appeal. We must caution that Compressed Sensing reconstructions in the presence of noise are still slightly inferior to those that would be obtained using full sampling and the Discrete Fourier Transform. Thus, if there is time to acquire a complete data set, this is still the recommended approach. However, reconstructions are now of sufficiently high standard that the cost of experimental measurement time alone will often be justification enough to use a Compressed Sensing strategy. Further, the recent introduction of the so-called Generalised Sampling Theorem indicates that in the future it may always be beneficial to use a Compressed Sensing reconstruction.

Bayesian analysis provides a framework in which all of our knowledge about a signal can be combined to extract the information of interest. As such, the benefits of using a Bayesian approach can be astonishing. In NMR spectroscopy, Bayesian analysis is now well established. There are a few analysis packages that are freely available for users to download, although at present these software packages are not especially user friendly and can be difficult to adapt to individual users requirements. In other applications, Bayesian techniques are less well established, but the benefits have been shown to be impressive, especially on low field instruments. The major disadvantage of the Bayesian approach is that in general it takes some effort to develop a suitable analysis procedure for any given problem. For this reason, it is likely that Bayesian analysis will not become an everyday tool, except in a few specific cases. Fortunately, one such example is likely to be NMR spectroscopy, which is one of the most widely used analytical techniques today.

References

1. B. Blumich, F. Casanova and S. Appelt, *Chem. Phys. Lett.*, 2009, **477**, 231–240.
2. M. H. Levitt, *Spin dynamics: Basics of Nuclear Magnetic Resonance*, John Wiley and Sons Ltd, Chichester, 2008.
3. D. G. Nishimura, *Principles of Magnetic Resonance Imaging*, 2010, http://www.Lulu.com.
4. P. T. Callaghan, *Principles of nuclear magnetic resonance microscopy*, Clarendon Press, Oxford, 1991.
5. J. Keeler, *Understanding NMR Spectroscopy*, John Wiley and Sons Ltd, Chichester, 2010.
6. E. M. Haacke, R. W. Brown, M. R. Thompson and R. Venkatesan, *Magnetic Resonance Imaging: Physical Principles and Sequence Design*, 1st edn, Wiley, New York, 1999.
7. H. Nyquist, *Trans. Am. Inst. Electr. Eng.*, 1928, 617–644.
8. C. E. Shannon, *Proc. IRE*, 1949, **37**, 10–21.
9. J. W. Cooley and J. W. Tukey, *Math. Comput.*, 1965, **19**, 297–301.
10. P. Duhamel and M. Vetterli, *Signal Process.*, 1990, **19**, 259–299.
11. B. M. A. Delattre, R. M. Heidemann, L. A. Crowe, J.-P. Vallée and J.-N. Hyacinthe, *Magn. Reson. Imaging*, 2010, **28**, 862–881.
12. J. D. O'Sullivan, *IEEE Trans. Med. Imaging*, 1985, **MI-4**, 200–207.
13. J. I. Jackson, C. H. Meyer, D. G. Nishimura and A. Macovski, *IEEE Trans. Med. Imaging*, 1991, **10**, 473–478.
14. J. A. Fessler and B. P. Sutton, *IEEE Trans. Signal Process.*, 2003, **51**, 560–574.
15. V. Rasche, R. Proksa, R. Sinkus, P. Börnert and H. Eggers, *IEEE Trans. Med. Imaging*, 1999, **18**, 385–392.

16. B. Coggins, R. Venters and P. Zhou, *Prog. Nucl. Magn. Reson. Spectrosc.*, 2010, **57**, 381–419.
17. P. Mansfield, *J. Phys. C: Solid State Phys.*, 1977, **10**, 55–58.
18. P. Szomolanyi, D. Goodyear, B. Balcom and D. Matheson, *J. Magn. Reson.*, 2001, **19**, 423–428.
19. C. B. Ahn, J. H. Kim and Z. H. Cho, *IEEE Trans. Med. Imaging*, 1986, **MI-5**, 2–7.
20. G. H. Glover, *Magn. Reson. Med.*, 1999, **415**, 412–415.
21. C. M. Tsai and D. G. Nishimura, *Magn. Reson. Med.*, 2000, **43**, 452–458.
22. P. T. Gurney, B. A. Hargreaves and D. G. Nishimura, *Magn. Reson. Med.*, 2006, **55**, 575–582.
23. B. A. Hargreaves, D. G. Nishimura and S. M. Conolly, *Magn. Reson. Med.*, 2004, **51**, 81–92.
24. J. H. Duyn, Y. Yang, J. A. Frank and J. W. van der Veen, *J. Magn. Reson.*, 1998, **132**, 150–153.
25. B. J. Wilm, C. Barmet, M. Pavan and K. P. Pruessmann, *Magn. Reson. Med.*, 2011, **65**, 1690–1701.
26. H. Meyer, B. S. Hu, D. G. Nishimura and A. Macovski, *Magn. Reson. Med.*, 1992, **13**, 202–213.
27. R. D. Hoge, R. K. Kwan and G. B. Pike, *Magn. Reson. Med.*, 1997, **38**, 117–128.
28. D. C. Noll, C. H. Meyer, J. M. Pauly, D. G. Nishimura and A. Macovski, *IEEE Trans. Med. Imaging*, 1991, **10**, 629–637.
29. L. C. Man, J. M. Pauly and A. Macovski, *Magn. Reson. Med.*, 1997, **37**, 785–792.
30. P. Irarrazabal, C. H. Meyer, D. G. Nishimura and A. Macovski, *Magn. Reson. Med.*, 1996, **35**, 278–282.
31. J. A. Hogbom, *Astron. Astrophys., Suppl. Ser.*, 1974, **15**, 417–426.
32. H. L. Taylor, S. C. Banks and J. F. McCoy, *Geophysics*, 1979, **44**, 39–52.
33. S. Gull and G. Daniell, *Nature*, 1978, **272**, 686–690.
34. E. J. Candes, J. K. Romberg and T. Tao, *Commun. Pure Appl. Math.*, 2006, **59**, 1207–1223.
35. D. L. Donoho, *IEEE Trans. Inf. Theory*, 2006, **52**, 1289–1306.
36. M. Lustig, D. L. Donoho and J. M. Pauly, *Magn. Reson. Med.*, 2007, **58**, 1182–1195.
37. M. F. Duarte, M. A. Davenport, D. Takhar, J. N. Laska, T. Sun, K. F. Kelly and R. G. Baraniuk, *IEEE Signal Process. Mag.*, 2008, **25**, 83–91.
38. R. G. Baraniuk, *IEEE Signal Process. Mag.*, 2007, 118–124.
39. M. F. Duarte and Y. C. Eldar, *IEEE Trans. Signal Process.*, 2011, **59**, 4053–4085.
40. B. Adcock, A. C. Hansen, C. Poon and B. Roman, 2013, arXiv:1302.0561, 1–44.
41. R. Baraniuk, M. Davenport, R. DeVore and M. Wakin, *Constr. Approx.*, 2008, **28**, 253–263.
42. J. A. Tropp and S. J. Wright, *Proc. IEEE*, 2010, **98**, 948–958.

43. E. J. Candès, *C. R. Math.*, 2008, **346**, 589–592.
44. S. Osher, M. Burger, D. Goldfarb, J. Xu and W. Yin, *Multiscale Model. Simul.*, 2005, **4**, 460–489.
45. M. Benning and M. Burger, 2012, arXiv:1211.2057v1.
46. T. Goldstein and S. Osher, *SIAM J. Imaging Sci.*, 2009, **2**, 323.
47. E. Candes and Y. Plan, *Proc. IEEE*, 2010, **98**, 925–936.
48. R. G. Baraniuk, V. Cevher and M. B. Wakin, *Proc. IEEE*, 2010, **98**, 959–971.
49. D. J. Holland, C. Liu, X. Song, E. L. Mazerolle, M. T. Stevens, A. J. Sederman, L. F. Gladden, R. C. N. D'Arcy, C. V. Bowen and S. D. Beyea, *Magn. Reson. Med.*, 2013, **70**, 1634–1643.
50. T. Hugger, B. Zahneisen, P. LeVan, K. J. Lee, H.-L. Lee, M. Zaitsev and J. Hennig, *PloS one*, 2011, **6**, e28822.
51. U. Gamper, P. Boesiger and S. Kozerke, *Magn. Reson. Med.*, 2008, **59**, 365–373.
52. R. Otazo, D. Kim, L. Axel and D. K. Sodickson, *Magn. Reson. Med.*, 2010, **64**, 767–776.
53. S. Hu, M. Lustig, A. Balakrishnan, P. E. Z. Larson, R. Bok, J. Kurhanewicz, S. J. Nelson, A. Goga, J. M. Pauly and D. B. Vigneron, *Magn. Reson. Med.*, 2010, **63**, 312–321.
54. S. S. Vasanawala, M. T. Alley, B. A. Hargreaves, R. A. Barth, J. M. Pauly and M. Lustig, *Radiology*, 2010, **256**, 607–616.
55. P. Parasoglou, D. Malioutov, A. J. Sederman, J. Rasburn, H. Powell, L. F. Gladden, A. Blake and M. L. Johns, *J. Magn. Reson.*, 2009, **201**, 72–80.
56. D. J. Holland, D. M. Malioutov, A. Blake, A. J. Sederman and L. F. Gladden, *J. Magn. Reson.*, 2010, **203**, 236–246.
57. V. S. Bajaj, J. Paulsen, E. Harel and A. Pines, *Science*, 2010, **330**, 1078–1081.
58. D. J. Holland, M. J. Bostock, L. F. Gladden and D. Nietlispach, *Angew. Chem., Int. Ed.*, 2011, **50**, 6548–6551.
59. K. Kazimierczuk and V. Y. Orekhov, *Angew. Chem., Int. Ed.*, 2011, **50**, 5556–5559.
60. P. Parasoglou, A. J. Sederman, J. Rasburn, H. Powell and M. L. Johns, *J. Magn. Reson.*, 2008, **194**, 99–107.
61. B. Adcock and A. C. Hansen, *J. Fourier Anal. Appl.*, 2012, **18**, 685–716.
62. A. B. Tayler, D. J. Holland, A. J. Sederman and L. F. Gladden, *Phys. Rev. Lett.*, 2012, **108**, 264505.
63. O. V. Petrov and B. J. Balcom, *J. Magn. Reson.*, 2011, **212**, 102–108.
64. J. Mitchell, T. C. Chandrasekera, D. J. Holland, L. F. Gladden and E. J. Fordham, *Phys. Rep.*, 2013, **526**, 165–225.
65. D. Xiao and B. J. Balcom, *J. Magn. Reson.*, 2013, **231**, 126–132.
66. D. Xiao and B. J. Balcom, *J. Magn. Reson.*, 2012, **220**, 70–78.
67. A. S. Stern, D. L. Donoho and J. C. Hoch, *J. Magn. Reson.*, 2007, **188**, 295–300.
68. H. T. Sanal, F. Cardoso, L. Chen and C. Chung, *Sports Med. Arthroscopy Rev.*, 2009, **17**, 31–39.

69. T. G. Sanders, M. B. Zlatkin, N. B. Paruchuri and R. W. Higgins, *Am. J. Roentgenol.*, 2007, **188**, 1094–1098.
70. F. Lakosi, G. Antal, C. Vandulek, A. Kovacs, R. Garamvolgyi, O. Petnehazy, G. Bajzik, J. Hadjiev, I. Repa and P. Bogner, *Pathol. Oncol. Res.*, 2009, **15**, 315–322.
71. F. Lakosi, G. Antal, C. Vandulek, A. Kovacs, G. L. Toller, I. Rakasz, G. Bajzik, J. Hadjiev, P. Bogner and I. Repa, *Pathol. Oncol. Res.*, 2011, **17**, 315–324.
72. D.-H. Lee, C.-P. Hong and M.-W. Lee, *J. Korean Phys. Soc.*, 2013, **62**, 328–332.
73. B. Blümich, J. Perlo and F. Casanova, *Prog. Nucl. Magn. Reson. Spectrosc.*, 2008, **52**, 197–269.
74. F. Casanova and B. Blümich, *J. Magn. Reson.*, 2003, **163**, 38–45.
75. A. Liberman, E. Bergman, Y. Sarda and U. Nevo, *J. Magn. Reson.*, 2013, **231**, 72–78.
76. C. M. Thiele and W. Bermel, *J. Magn. Reson.*, 2012, **216**, 134–143.
77. M. Usman, C. Prieto, T. Schaeffter and P. G. Batchelor, *Magn. Reson. Med.*, 2011, **1176**, 1163–1176.
78. M. J. Bostock, D. J. Holland and D. Nietlispach, *J. Biomol. NMR*, 2012, **54**, 15–32.
79. J. C. J. Barna and E. D. Laue, *J. Magn. Reson. (1969)*, 1987, **75**, 384–389.
80. M. Doneva, P. Börnert, H. Eggers, A. Mertins, J. Pauly and M. Lustig, *Magn. Reson. Med.*, 2010, **64**, 1749–1759.
81. M. S. Cohen and R. M. Weisskoff, *Magn. Reson. Imaging*, 1991, **9**, 1–37.
82. M. Bayes and M. Price, *Philos. Trans. R. Soc. London*, 1763, **53**, 370–418.
83. D. Sivia and J. Skilling, *Data Analysis: A Bayesian Tutorial*, 2nd edn, Oxford University Press, Oxford, 2006.
84. H. Jeffreys, *Theory of Probability*, Clarendon Press, Oxford, 1939.
85. M. Tipping, *J. Mach. Learn. Res.*, 2001, **1**, 211–244.
86. J. T. Parker and L. C. Potter, in *IEEE Radar Conference*, Ieee, Washington DC, 2010, pp. 1471–1475.
87. J. Xu, Y. Pi and Z. Cao, *IET Radar, Sonar Navig.*, 2012, **6**, 2–8.
88. J. M. Bernardo and A. F. M. Smith, *Bayesian Theory*, John Wiley and Sons Ltd, Chichester, 2004.
89. E. T. Jaynes, *Probability Theory – The Logic of Science*, Cambridge University Press, Cambridge, 2003.
90. G. L. Bretthorst, *Bayesian spectrum analysis and parameter estimation*, Springer-Verlag, New York, 1988.
91. J. S. Liu, *Monte Carlo Strategies in Scientific Computing*, Springer, New York, 2008.
92. U. von Toussaint, *Rev. Mod. Phys.*, 2011, **83**, 943–999.
93. G. L. Bretthorst, *Concepts Magn. Reson.*, 2005, **27**, 73–83.
94. G. L. Bretthorst, W. C. Hutton, J. R. Garbow and J. J. H. Ackerman, *Concepts Magn. Reson., Part A*, 2005, **27A**, 55–63.
95. G. L. Bretthorst, W. C. Hutton, J. R. Garbow and J. J. H. Ackerman, *Concepts Magn. Reson., Part A*, 2005, **27A**, 64–72.

96. G. L. Bretthorst, C. C. Hung, D. A. Davignon and J. J. H. Ackerman, *J. Magn. Reson.*, 1988, **79**, 369–376.
97. M. Merritt, G. L. Bretthorst, S. C. Burgess, A. D. Sherry and C. R. Malloy, *Magn. Reson. Med.*, 2003, **50**, 659–663.
98. W. C. Hutton, G. L. Bretthorst, J. R. Garbow and J. J. H. Ackerman, *Magn. Reson. Med.*, 2009, **62**, 1026–1035.
99. J. J. Kotyk, N. G. Hoffman, W. C. Hutton, G. L. Bretthorst and J. J. H. Ackerman, *J. Magn. Reson.*, 1992, **98**, 483–500.
100. G. Bretthorst, *J. Magn. Reson.*, 1990, **88**, 571–595.
101. R. F. Evilia, R. Effiong and S. L. Whittenburg, *Spectrosc. Lett.*, 1993, **26**, 1559–1570.
102. A. Rouh, A. Louis-Joseph and J. Y. Lallemand, *J. Biomol. NMR*, 1994, **4**, 505–518.
103. D. V. Rubtsov and J. L. Griffin, *J. Magn. Reson.*, 2007, **188**, 367–379.
104. H.-W. Koh, S. Maddula, J. Lambert, R. Hergenröder and L. Hildebrand, *J. Magn. Reson.*, 2009, **201**, 146–156.
105. C. Zheng, S. Zhang, S. Ragg, D. Raftery and O. Vitek, *Bioinformatics*, 2011, **27**, 1637–1644.
106. R. A. Chylla and J. L. Markley, *J. Biomol. NMR*, 1995, **5**, 245–258.
107. R. A. Chylla and J. L. Markley, *J. Biomol. NMR*, 1993, **3**, 515–533.
108. R. A. Chylla, B. F. Volkman and J. L. Markley, *J. Biomol. NMR*, 1998, **12**, 277–297.
109. J. W. Yoon, S. Godsill, E. Kupce and R. Freeman, *Magn. Reson. Chem.*, 2006, **44**, 197–209.
110. D. V. Rubtsov, C. Waterman, R. A. Currie, C. Waterfield, J. D. Salazar, J. Wright and J. L. Griffin, *Anal. Chem.*, 2010, **82**, 4479–4485.
111. J. R. Sachleben, *J. Magn. Reson.*, 2006, **183**, 123–133.
112. E. Aboutanios, Y. Kopsinis and D. Rubtsov, *Comput. Electr. Eng.*, 2012, **38**, 52–67.
113. W. Astle, M. D. Iorio, S. Richardson, D. Stephens, T. Ebbels, M. De Iorio, S. Richardson, D. Stephens and T. Ebbels, *J. Am. Stat. Assoc.*, 2012, **107**, 1259–1271.
114. F. Malz and H. Jancke, *J. Pharm. Biomed. Anal.*, 2005, **38**, 813–823.
115. J. Kornak, K. Young, B. J. Soher and A. A. Maudsley, *IEEE Trans. Med. Imaging*, 2010, **29**, 1333–1350.
116. D. Xing, S. J. Gibbs, J. A. Derbyshire, E. J. Fordham, T. A. Carpenter and L. D. Hall, *J. Magn. Reson.*, 1995, **106**, 1–9.
117. R. G. Wise, B. Newling, A. R. C. Gates, D. Xing, T. A. Carpenter and L. D. Hall, *Magn. Reson. Imaging*, 1996, **14**, 173–185.
118. M. H. G. Amin, S. J. Gibbs, R. J. Chorley, K. S. Richards, T. A. Carpenter and L. D. Hall, *Proc. R. Soc. A*, 1997, **453**, 489–513.
119. L. Sun and L. Hall, *Int. Commun. Heat Mass Transfer*, 2001, **28**, 461–466.
120. C. Binter, V. Knobloch, R. Manka, A. Sigfridsson and S. Kozerke, *Magn. Reson. Med.*, 2013, **69**, 1337–1345.
121. I. Smal, N. Carranza-Herrezuelo, S. Klein, W. Niessen and E. Meijering, *Med. Image Comput. Comput.-Assisted Intervention*, 2011, **14**, 573–580.

122. D. J. Holland, A. Blake, A. B. Tayler, A. J. Sederman and L. F. Gladden, *J. Magn. Reson.*, 2011, **209**, 83–87.
123. D. J. Holland, A. Blake, A. B. Tayler, A. J. Sederman and L. F. Gladden, *Chem. Eng. Sci.*, 2012, **84**, 735–745.
124. J. G. Ross, D. J. Holland, A. Blake, A. J. Sederman and L. F. Gladden, *J. Magn. Reson.*, 2012, **222**, 44–52.
125. D. J. Holland, J. Mitchell, A. Blake and L. F. Gladden, *Phys. Rev. Lett.*, 2013, **110**, 018001.

CHAPTER 9

Emerging Applications: Surface NMR

ANATOLY LEGCHENKO*[a]

[a]Institut de recherche pour le développement (IRD), Laboratoire d'Etude des Transferts en Hydrologie et Environnement (LTHE), F-38000 Grenoble, France

*E-mail: anatoli.legtchenko@ird.fr

9.1 Introduction

Surface NMR (SNMR) is a geophysical technique developed for groundwater investigations from the ground surface down to approximately 100 m. In a one-dimensional (1-D) implementation the subsurface is assumed to be horizontally stratified and the method is known as the Magnetic Resonance Sounding method (MRS). The method was developed in Russia in the early 1980s by a team of Russian scientists under the guidance of A.G. Semenov.[1] Their research program started from the Varian patent that proposed to use the Nuclear Magnetic Resonance (NMR) phenomenon for noninvasive detection of proton-containing liquids (hydrocarbons or water) in the subsurface.[2] A two-dimensional (2-D) implementation named the Surface Magnetic Resonance Tomography or 2D-SNMR was developed around 2005,[3] which was further extended to three-dimensional (3-D) applications named 3D-SNMR.[4] Today, SNMR imaging for groundwater is an emerging geophysical method and is still the subject of research and development of an international community composed of geophysicists and hydrogeologists. A new generation of

New Developments in NMR No. 5
Mobile NMR and MRI: Developments and Applications
Edited by Michael L. Johns, Einar O. Fridjonsson, Sarah J. Vogt, and Agnes Haber

Published by the Royal Society of Chemistry, www.rsc.org

commercially available SNMR instruments has been developed in France by IRIS Instruments (NUMIS^{poly} system, www.iris-instruments.com/), in USA by Vista Clara Inc. (GMR instrument, www.vista-clara.com/) and in Germany by Radic Research (MRS-MIDI-2, www.radic-research.de/). Instrumental developments have been also reported from China.[5,6]

Different laboratory-based NMR instruments and Nuclear Magnetic Resonance logging tools (NML) are widely used in petro-physics. The SNMR method is less known but it is based on the same theory and has much in common with standard NMR. However, the differences are also significant and consequently laboratory and NML experience cannot be applied directly. Indeed:

(1) SNMR uses a much lower static magnetic field (Earth's magnetic field) than laboratory or NML instruments. Many processes that influence the NMR response are magnetic field dependent and hence the results of laboratory and SNMR experiments may often have only qualitative correspondence.
(2) The volume investigated with SNMR is much larger than the volume investigated in the laboratory or in the borehole. Laboratory and NML instruments are able to resolve small volumes of a few cubic millimeters to a few tens of cubic centimeters. The minimum volume that can be characterized with SNMR is more than a few hundred cubic meters.
(3) Operating at low frequency (between 800 and 2600 Hz), SNMR is not able to quantify the total porosity because of difficulties measuring short timed signals.
(4) SNMR can be applied to any type of geology including unconsolidated materials. Serious difficulties of acquiring undisturbed samples and scale dependence of hydrogeological processes render comparisons of field results obtained with SNMR and results obtained in laboratories with rock samples difficult.

9.2 Experimental Setup

SNMR instruments operate in the Earth's magnetic field and the low sensitivity of the method is compensated by the use of a large investigated volume. For performing SNMR measurements, a wire loop of a few tens of meters in diameter is laid out on the ground. The loop is then energized by a pulse of alternating current $i(t) = I_0 \cos(\omega t)$ with frequency $\omega \approx \omega_0$ close to the Larmor frequency. Measurements of the magnetic resonance signal can be done after the pulse is terminated. For that, the same loop or one or more receiving loops can be used. The volume investigated with SNMR depends on the measuring setup and may vary from a few hundreds to a few tens of thousands cubic meters. The similarity between a well-known proton magnetometer and the SNMR is shown in Figure 9.1a. For measuring the magnitude of the Earth's magnetic field, a container with a few dm^3 of kerosene or some other hydrogen-rich liquid (such as water) is put inside the acquisition coil

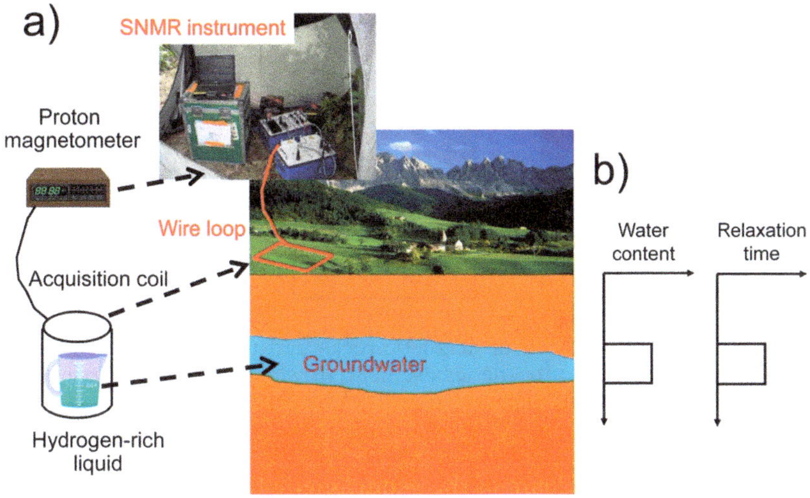

Figure 9.1 (a) Transformation of a proton magnetometer into a SNMR system. (b) Schematic presentation of SNMR results.

of the magnetometer and the frequency of the magnetic resonance signal is measured. The magnetic field can be calculated using the linear relationship between the Larmor frequency $\omega_0 = 2\pi f_0$ and the magnetic field $\omega_0 = \gamma B_0$, with γ being the gyromagnetic ratio. For performing SNMR measurements, the coil is enlarged and becomes a wire loop. Groundwater acts as the hydrogen-containing liquid (many cubic meters in volume). The Larmor frequency is known and the SNMR system can thus be tuned to the Larmor frequency, which allows observation of the magnetic resonance signal from groundwater. An absence of NMR signal at this frequency also means an absence of groundwater in a detectable quantity.

Excluding some special cases, hydrocarbons do not exist in the shallow part of the subsurface and thus only protons of groundwater generate the magnetic resonance signal. Inversion of acquired SNMR data provides a spatial distribution of the water content and of the relaxation time in the investigated water saturated formations (Figure 9.1b).

9.2.1 Earth's Magnetic Field

For performing SNMR measurements the Earth's magnetic field is employed as the static field B_0. However, the Earth's magnetic field, which is sufficiently homogeneous at the scale of SNMR set-up, is not constant around the world. Depending on the geographical location of the area investigated with SNMR, the geomagnetic field may change its intensity (from approximately 20 000 to 60 000 nT and the corresponding proton Larmor frequency from 800 to 2600 Hz) and inclination (from −90° to +90°). These variations have a significant impact on SNMR results.[7] Primarily, the amplitude of the NMR signal

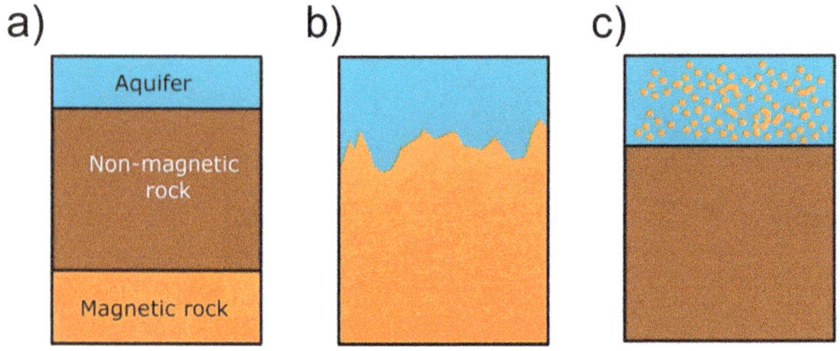

Figure 9.2 Possible location of magnetic rocks perturbing the Earth's magnetic field: (a) deep magnetic basement; (b) shallow magnetic basement; (c) magnetic material distributed in the aquifer.

is proportional to the square of the geomagnetic field $e_0 \sim B_0^2$. For example, in northern Siberia and Canada where B_0 is about 60 000 nT the same water saturation formation would generate a magnetic resonance signal four fold higher than in South America where B_0 is about 30 000 nT.

The Earth's magnetic field is often considered as constant at the loop scale. However in practice, the Earth's magnetic field can vary in time and it can be locally perturbed by rocks. The aquifer formation may be composed of magnetic rock like basalt, but it also may be composed of non-magnetic material overlying magnetic basement. In both cases, NMR measurements will be perturbed. Figure 9.2 shows possible locations of the magnetic rock relative to the investigated aquifer. When magnetic rocks are deep, this formation can be considered as a flat magnet and the magnetic basement creates a homogeneous field gradient (Figure 9.2a). However, when the basement is close to the aquifer its shape may be non-regular and perturbations of the geomagnetic field in the aquifer will also be complex (Figure 9.2b). In some cases, small magnetic particles may be mixed up with the aquifer material thus perturbing the geomagnetic field at the grain size scale, creating local field gradients that are difficult to quantify (Figure 9.2c).

Depending on the heterogeneity of the geomagnetic field, SNMR can be used either in FID (Free Induction Decay) or SE (Spin Echo) modes. In both cases, SNMR allows estimation of the water content w and T_2^* relaxation time. T_1 or T_2 relaxation times can be accessed in the FID mode or the SE mode respectively. Often both FID and SE are performed.[8,9] If the perturbation in the magnetic field is large then the SNMR method cannot be used (Figure 9.3 summarizes these different sub-surface conditions).

9.2.2 Instrumentation

All existing SNMR instruments consist of the following principal units (Figure 9.4): a current generator, tuning capacitors, DC–DC convertors, storage capacitors, amplifiers, analog–digital (A–D) converters and a microprocessor.

Earth's magnetic field

Homogeneous	Perturbed	Inhomogeneous	Highly inhomogeneous
FID T_1, T_2*	FID, SE T_1, T_2, T_2*	SE T_2, T_2*	SNMR cannot be applied
Limestone, chalk, sandstone, quartz	Small Inclusions of the magnetite or basalt, granite, deep magnetic basement	Important inclusions of the magnetite or basalt, shallow magnetic basement	Basalt

Figure 9.3 Selection of the measuring technique depending on the inhomogeneity of the Earth's magnetic field. The bottom row indicates example of rocks that may cause the perturbation.

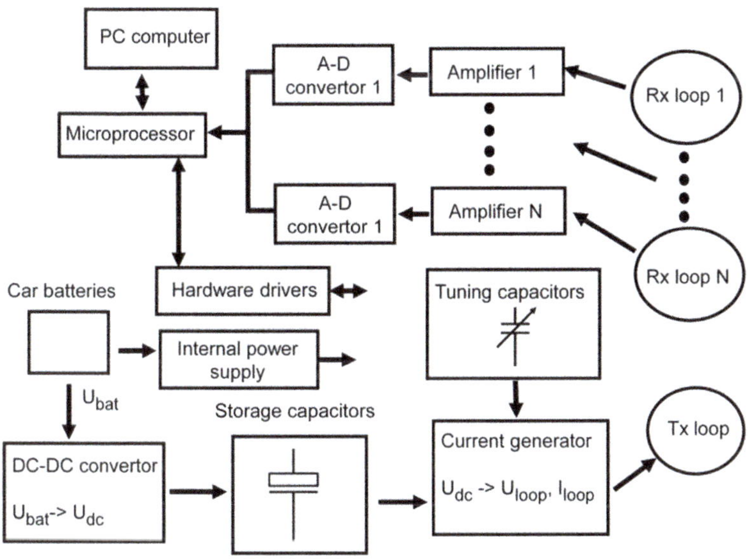

Figure 9.4 Schematic presentation of the principal units of a SNMR instrument.

The system is connected to a PC that is used for data processing and control of the instrument. Car batteries are used for the power supply and the instrument is connected to a transmitting/receiving (*Tx*/*Rx*) loop. One transmitting loop (*Tx*) and one or more separate receiving loops (*Rx*) with additional amplifiers and A–D convertors can also be used.[10]

To conduct experiments, the loop is tuned to resonance at the Larmor frequency. For that, the tuning capacitors are used. The storage capacitors are charged by the DC–DC converter to the voltage controlled by the computer. The maximum voltage of the storage capacitors depends on the design of the current generator and ranges between ±110 and ±430 V. In instruments designed for shallow explorations (less than 10 m), the current generator

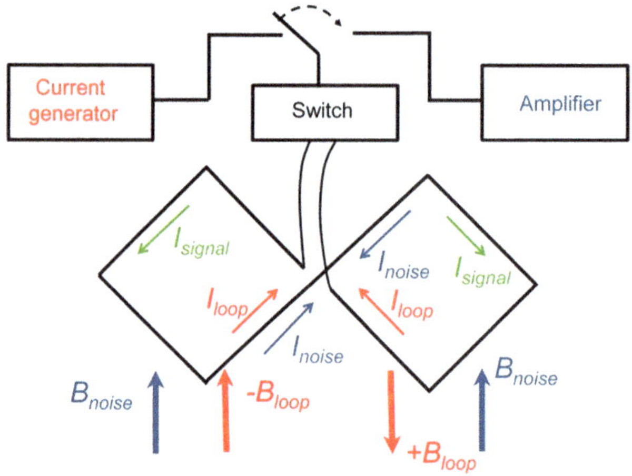

Figure 9.5 Schematic presentation of the figure-of-eight loop. The red color shows the transmitting stage and the blue and green colors show the receiving stage.

may be directly connected to the batteries without additional DC–DC units. Manufacturers report the instrumental noise to be less than 1 nV/√Hz, the maximum current in the loop ranging between 50 and 600 A and the maximum loop voltage to be between 500 and 4000 V. Experimental instruments are reported to have up to 7500 V of loop voltage and maximum current of 600 A. The noise and the signal are both variable and depend on different factors. Noise measured with the SNMR system is composed of instrumental noise generated by the instrument itself and ambient noise generated by a variety of natural and man-made sources. The ambient noise is always much higher than the instrumental noise. In practice, the ambient noise measured with a $75 \times 75\ m^2$ square loop and a 100 Hz band-pass of the amplifier was observed to be rarely less than 100 nV (10 nV/√Hz). In many cases, the ambient noise can be much higher.

The SNMR field setup is composed of one or more surface loops. One loop is the transmitting loop (*Tx*) and one or more loops are the receiving loops (*Rx*). The coincident loop configuration allows using the same loop for both transmitting and receiving (*Tx*/*Rx*). The volume affected by each *Tx* loop depends on the loop size and shape. The typical shapes for a loop are circular, square or a figure-of-eight.

Under low-noise conditions a square or circular loop is usually used. The figure-of-eight loop is composed of two equal circles or two equal squares (Figure 9.5). It has a smaller depth of investigation but it allows improvements to the signal to noise ratio.[11] The squares (or circles) of the figure-of-eight loop are connected in the opposite directions and the magnetic fields generated by the squares ($\mathbf{B}_{loop}$) are also oriented in the opposite directions. Consequently, the SNMR signals induced in each square (I_{signal}) have the inverse polarity, but, as the loops are twisted, the signals are added. Note that the noise induced in each square has the same polarity (I_{noise}) and thus the

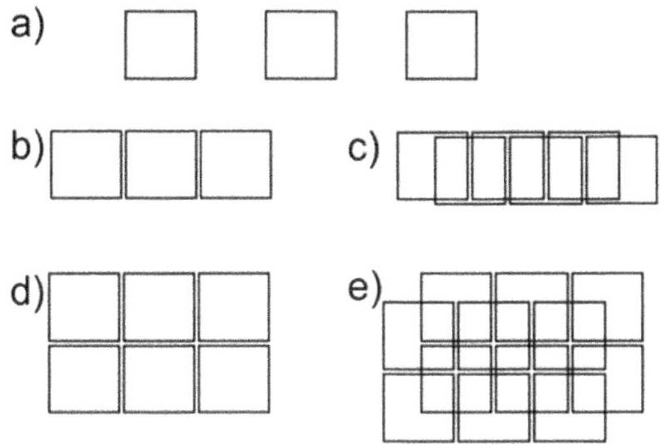

Figure 9.6 SNMR field setup: (a) 1-D survey; (b) 2-D minimal coverage; (c) 2-D optimal coverage; (d) 3-D minimal coverage; (e) 3-D optimal coverage.

noise is canceled. Using the figure-of-eight loop one may get up to ten fold improvement in the signal to noise ratio. For canceling the noise with the figure-of-eight loop the principal axe of the loop has to be set parallel to the noise source (Figure 9.5). Otherwise noise voltages induced in each square may be not equal and consequently the noise offset will be less efficient.

The loop sizes may vary between $d_l = 20$ m and $d_l = 150$ m (with d_l being the side length of a square loop or the diameter of a circular loop. Loops larger than 80 m usually have one turn and smaller loops may have between 2 and 5 turns of the wire. Each transmitting loop is made of an electrical cable with a cross-section of more than 6 mm^2. Such a thick wire allows generation of large current pulses (up to 600 A). Special attention should be paid to the electrical insulation of the wire because of a high pulse voltage (up to 4 kV).

If SNMR loops are separated at a distance of more than d_l (Figure 9.6a) then the measurements with each loop are independent and these measurements should be considered as a 1-D survey. When the loops are set side by side (Figure 9.6b) the measurements can be interpreted as a 2-D profile. The resolution of a 2-D survey may be improved by using half overlapped loops (Figure 9.6c). Note that setting loops closer than half of the overlapped loops will increase the labor and time required for measuring without a corresponding improvement in the resolution.

A 3-D field setup is shown in Figure 9.6d. For good resolution, the investigated area should be located within the area occupied by the SNMR loops. The half overlapped loops (Figure 9.6e) are a good compromise between resolution and time and labor consumption.

SNMR equipment is composed of four principal units with a weight from 20 to 25 kg each. For the fieldwork two or four car batteries and a loop composed of 300 to 400 m long wire are also required. In a noisy environment additional reference loops can be used. The total weight of the field equipment is 200 to 300 kg but the instrument is mobile and can be installed in a car or easily transported to the investigated sites (as is the case in Figure 9.7). Depending on the capacity, four car batteries are sufficient for 6 to 12 hours of measurements.

Figure 9.7 SNMR system in the field.

9.3 Theoretical Development and Modeling

9.3.1 Magnetic Field of the Surface Loop

During SNMR experiments, a magnetic field oscillating with the Larmor frequency $\mathbf{B}_1$ is generated by the surface loop. Note that the rock electrical resistivity may vary from less than 1 ohm-m to more than 100 000 ohm-m thus influencing the loop magnetic field within the investigated volume. For computing an electromagnetic field of a loop we approximate the transmitter loop by a vertical magnetic dipole of moment $M = I_0\pi a^2$. Its magnetic induction field is symmetrical relative to the dipole axis, and for a receiver located at the point $p(r,z)$ would be expressed as:[12,13]

$$\mathbf{B}(\mathrm{p}) = \mu[i\omega\mu\sigma(p)\mathbf{F}(p) - \mathrm{grad}(\mathrm{div}\mathbf{F}(p))] \tag{9.1}$$

where the electrical properties of the medium are represented by the magnetic permeability μ and the electrical conductivity σ. **F** is a Hertz potential which has only a vertical non-zero component, expressed as a Hankel transform

$$F_z(r, z) = \frac{M}{4\pi}\int_0^{\infty} K_d(\lambda, z)\lambda J_0(\lambda r)\mathrm{d}\lambda \tag{9.2}$$

where λ is the horizontal wave number, and K_d is a kernel function:

$$K_d(\lambda,z) = a_j(\lambda)e^{+m_j^z} + b_j(\lambda)e^{-m_j^z} \tag{9.3}$$

with $m_j = \sqrt{\lambda^2 + i\omega\mu\sigma_j}$, σ_j the conductivity of a layer j. The coefficients $a_j(\lambda)$ and $b_j(\lambda)$ are recursively determined and r,z are the radial and vertical distances from the dipole center, respectively.

Then the non-zero components of the magnetic induction field can be written as:

$$B_r = \frac{\mu M}{4\pi}\int_0^\infty \frac{\partial K_d(\lambda,z)}{\partial z}\lambda^2 J_1(\lambda r)\,\mathrm{d}\lambda$$
$$B_z = -\frac{\mu M}{4\pi}\int_0^\infty K_d(\lambda,z)\lambda^3 J_0(\lambda r)\,\mathrm{d}\lambda \tag{9.4}$$

where $J_0(\lambda r)$ and $J_1(\lambda r)$ are the Bessel functions.

Considering a loop with finite radius a, the kernel can be written as:[13,14]

$$K_l(\lambda,z) = \frac{J_1(\lambda a)}{\lambda(a/2)}K_d(\lambda,z) \tag{9.5}$$

and the components of the loop magnetic field become:

$$B_r = \frac{\mu I_0 a}{2}\int_0^\infty \frac{\partial K_d(\lambda,z)}{\partial z}\lambda J_1(\lambda a)J_1(\lambda r)\,\mathrm{d}\lambda$$
$$B_z = -\frac{\mu I_0 a}{2}\int_0^\infty K_d(\lambda,z)\lambda^2 J_1(\lambda a)J_0(\lambda r)\,\mathrm{d}\lambda \tag{9.6}$$

The computation of the Hankel transform, which is necessary for computing the loop magnetic field, requires specific numerical algorithms. The most widely used one is the lagged convolution filtering technique.[15,16] Comparison of different numerical routines, with refined filters for the oscillatory kernels as proposed by Anderson, showed that for all practical needs the standard Anderson's filter will be quite adequate.[17] However, any other numerical routine can also be used for computing the magnetic fields.

9.3.2 Imaging Equation

For computing magnetic resonance signal generated by the aquifer formation, let us consider a volume $\mathrm{d}V(x,y,z) = \mathrm{d}V(\mathbf{r})$ located below the measuring loop (Figure 9.8).

For simplicity we assume a coincident transmitting and receiving loop (*Tx*/*Rx* loop). The reference frame x,y,z of the Bloch model will be linked to

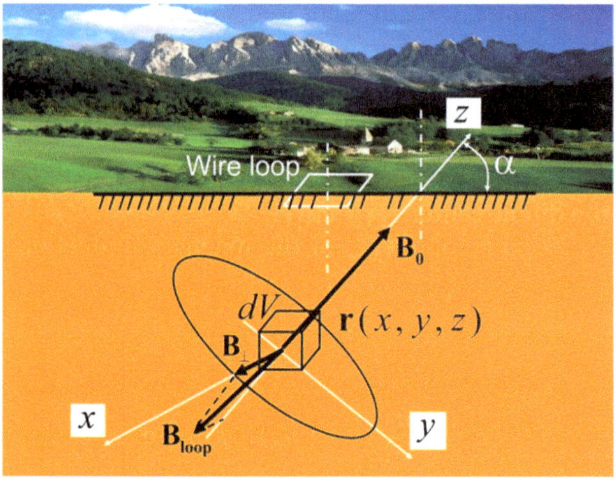

Figure 9.8 Investigated volume dV below the surface loop and the magnetic fields in the rotating frame of the Bloch model.

the Earth's magnetic fields $\mathbf{B}_0$ inclined at the angle α (z axis), and the projection of the loop magnetic field $\mathbf{B}_{\text{loop}}$ onto the plane transversal to $\mathbf{B}_0$ that is referred to as $\mathbf{B}_\perp = \mathbf{B}_1$ (x axis). The volume d$V(\mathbf{r})$ is assumed sufficiently small for the magnetic fields $\mathbf{B}_0$, $\mathbf{B}_\perp$ to be uniform and the local macroscopic spin magnetization $\mathbf{M}$ to be constant within this volume d$V(\mathbf{r})$. After the current in the loop (I_0) is terminated, the signal induced in the receiver loop is proportional to the sum of the flux of all the volumes d$V(\mathbf{r})$ and under near resonance conditions we get:[18]

$$e_0 = \frac{\omega_0}{I_0}\int_V B_\perp(\mathbf{r})M_\perp(\mathbf{r})w(\mathbf{r})\mathrm{d}V \tag{9.7}$$

where $0 \leq w(\mathbf{r}) \leq 1$ is the water content, and $\mathbf{r} = r(x,y,z)$ is the coordinate vector.

In a non-conductive subsurface the loop magnetic field $\mathbf{B}_\perp$ is linearly polarized, thus the free induction decay (FID) signal, assuming exact resonance, can be expressed as:

$$M_\perp = M_0 \sin(0.5\gamma B_\perp \tau_\mathrm{p}) \tag{9.8}$$

where τ_p is the duration of the current pulse in the loop.

When the Earth's magnetic field is perturbed by magnetic rocks, SNMR measurements can be performed in the spin echo (SE) mode. For computing the SE amplitude we assume the mathematical and geometrical interpretations presented in Hahn's original paper.[19] Under near resonance conditions, for pulses of equal intensity ($B_\perp$) and neglecting relaxation and molecular diffusion, $M_\perp$ can be expressed as:[20]

$$M_\perp = -M_0 \sin(\theta_1) \times \sin^2(0.5\theta_2) \tag{9.9}$$

where $\theta_1 = \gamma B_\perp \tau_{p1}$ and $\theta_2 = \gamma B_\perp \tau_{p2}$ are the flip angles for corresponding pulses. If a 90°, 180° pulse sequence is used then $\theta_2 = 2\theta_1 = \theta$ and consequently the SE amplitude can be computed using:

$$M_\perp = -M_0 \sin^3(\theta) \tag{9.10}$$

By performing measurements with different values of the pulse moment $q = I_0\tau_p$ one can manage the flip angle and thus gain access to imaging of the subsurface.

An electrically conductive subsurface causes a phase shift $\varphi^{Tx}{}_0$ of the transmitted magnetic field $\mathbf{B}^{Tx}{}_\perp$ relative to the current in the loop. The voltage induced in the receiving loop by an oscillating magnetic moment $\mathbf{M}_\perp$ is also dephased by the subsurface at an angle $\varphi^{Rx}{}_0$. In the case of the coincident *Tx*/*Rx* loop configuration $\varphi^{Rx}{}_0 = \varphi^{Tx}{}_0$ and consequently the imaging equation becomes:

$$e_0(q) = \omega_0 I_0^{-1} \int_V e^{j2\varphi_0^{Tx}} B_\perp(\mathbf{r}) M_\perp(\mathbf{r}, q) w(\mathbf{r}) \mathrm{d}V \tag{9.11}$$

9.3.3 Forward Modeling

9.3.3.1 Sensitivity of SNMR Loop

When performing 1-D measurements of water saturated formations we assume that the subsurface is horizontally stratified and thus the SNMR signal is attributed to the total volume affected by the measuring loop without consideration of the contributions from different parts of the subsurface. However, SNMR can also be applied in a 1-D configuration (single loop) in areas where the geological formations are not perfectly homogeneous. For example, this may be the case when investigating hard rock areas and the measuring loop is laid over a narrow fractured zone. If SNMR results indicate a good aquifer then hydrogeologists may ask where to drill a well. The loop affected area is about 100 × 100 m^2 and one expects intuitively that the borehole should be located in the center of the loop. Let us see whether such a recommendation will always be the best solution.

An important parameter that defines the sensibility map of an area investigated with one loop is the inclination of the Earth's magnetic field. For computing the contribution of the different parts of the subsurface to the SNMR signal we assume a square loop with a side of 80 m, a Larmor frequency of 2000 Hz, and resistivity of the subsurface of 100 ohm-m. We performed numerical modeling for the inclination of the Earth's magnetic field of 0°, 45° and 90°. The modeling results are shown in Figures 9.9–9.11. The area affected by the SNMR loop is presented by the isosurface at the level set arbitrarily at 8 nV. A horizontal slice at the depth of 10 m and two vertical cross-sections show the amplitude of the signal generated by the different areas of the investigated volume.

The modeling results presented in Figures 9.9–9.11 show that the sensitivity of the loop is progressively diminishing when depth increases and that

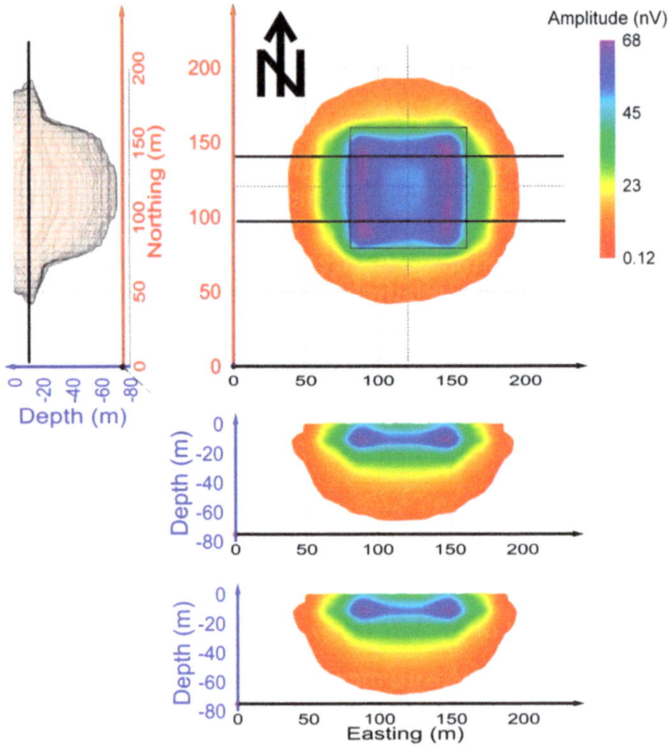

Figure 9.9 The sensitivity of an SNMR loop computed for one value of the pulse moment and considering the inclination of the Earth's magnetic field of 0°. The central graph shows a horizontal slice at the depth of 10 m below the loop. Position of the square loop is shown by the black square. The isosurface of the amplitude is plotted at the level of 8 nV (left hand plot). The volume within the 8 nV isosurface could be considered as the area investigated with one loop. The two bottom plots show the vertical distribution of the amplitude. The solid black lines in the central plot show the position of the two vertical cross-sections.

for the same depth the sensitivity varies laterally as a function of the inclination of the Earth's magnetic field. Depending on its location within the area affected by the loop the same volume of water may produce SNMR signal that varies four fold. When the Earth's magnetic field is close to horizontal, the central area has the maximal sensitivity. However, when the Earth's field is close to vertical the central area around the loop axis does not produce any signal. It is obvious that in such a case the recommendation to locate a well in the center of the SNMR loop may be unjustified. When the inclination of the geomagnetic field changes from horizontal to vertical, the most sensitive area is displaced toward the southern part of the loop in the northern hemisphere and towards the northern part of the loop in the southern hemisphere.

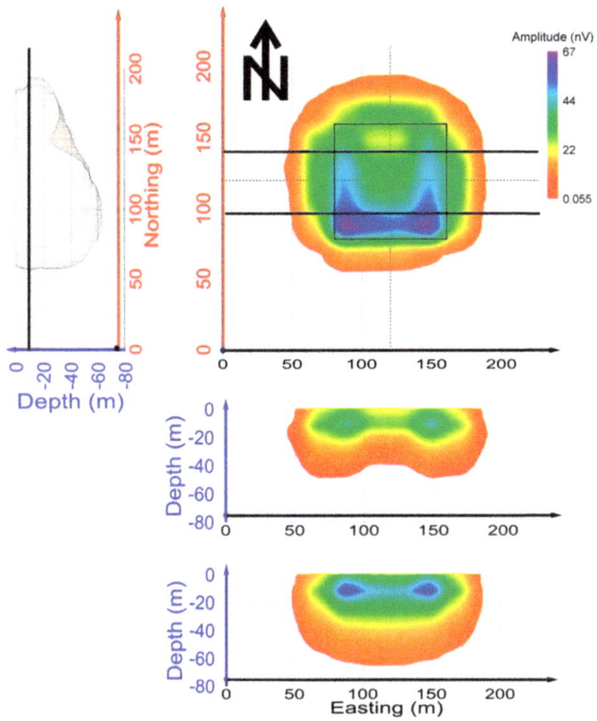

Figure 9.10 The sensitivity of an SNMR loop computed for one value of the pulse moment and considering the inclination of the Earth's magnetic field of 45°.

9.3.3.2 *Typical SNMR Responses*

To learn more about SNMR signals we will use a forward modeling routine based on eqn (9.11) and consider a 1-D sounding setup with the coincident *Tx*/*Rx* loop. We assume measurements in the FID mode (measuring the free induction decay signals) with the maximum pulse moment of 12 000 A-ms and dead time of the equipment of 40 ms.

Examples of the SNMR signal computed considering the half-space of bulk water are shown in Figure 9.12a. These results may be considered as an estimate of the maximum possible amplitude of the magnetic resonance signal as a function of the Earth's magnetic field: (a) a horizontal and weak geomagnetic field typical for equatorial areas, (b) an average magnitude geomagnetic field with the inclination of 45° corresponding to Europe and (c) a strong vertical geomagnetic field in polar areas. Note that depending on the geographical location the same volume of water produces different signals (remember that $e_0 \sim B_0^2$) and hence the efficiency of the method should be higher in areas with a large geomagnetic field. The SNMR response from a thin (2 m thick) water saturated formation (w = 20%, T_2^* = 150 ms) located at

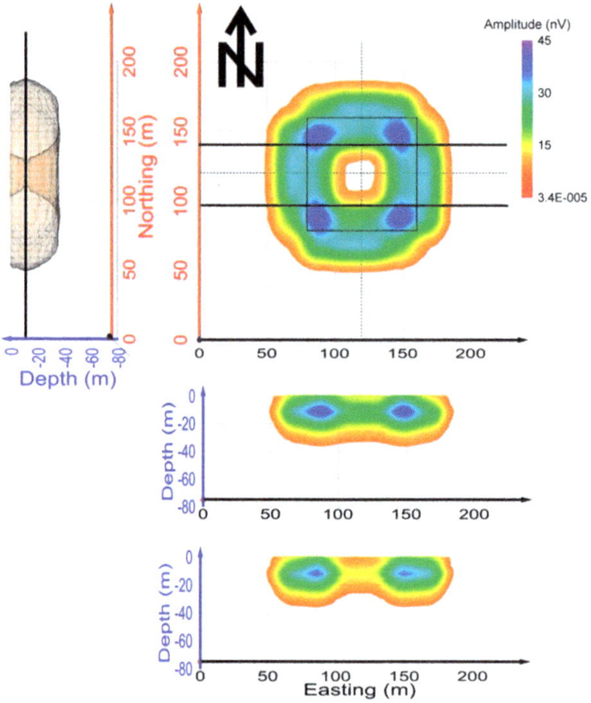

Figure 9.11 The sensitivity of SNMR loop computed for one value of the pulse moment and considering the inclination of the Earth's magnetic field of 90° (note that the vertical cross-sections appear asymmetric relative the loop axis because of asymmetric meshing when modeling).

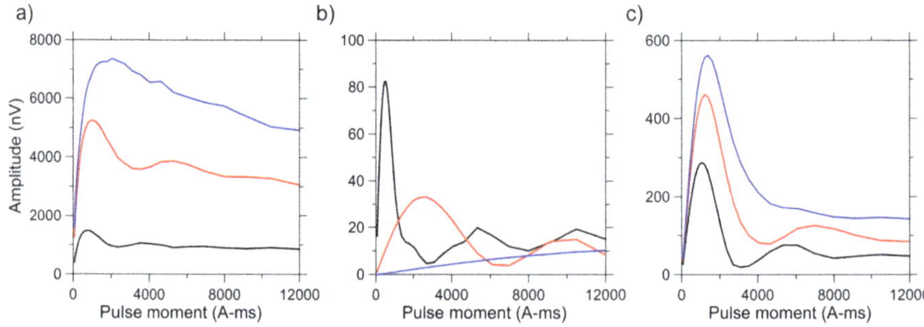

Figure 9.12 Amplitude of the SNMR signal computed *versus* pulse moment for different measuring conditions: (a) the half-space of bulk water (w = 100%, T_2^* = 1000 ms) and the Earth's magnetic field of the following magnitude and inclination respectively: (1) 23 474 nT (f_0 = 1000 Hz) and 0° (black line); (2) 46 948 nT (f_0 = 2000 Hz) and 45° (red line); (3) 61 033 nT (f_0 = 2600 Hz) and 90° (blue line); (b) a 2 m thick water saturated formation (w = 20%, T_2^* = 150 ms): (1) depth to water 5 m (black line); (2) depth to water 30 m (red line); (3) depth to water 80 m (blue line); (c) a water saturated formation of different thickness (w = 20%, T_2^* = 150 ms): (1) 10 m (black line); (2) 20 m (red line); (3) 30 m (blue line).

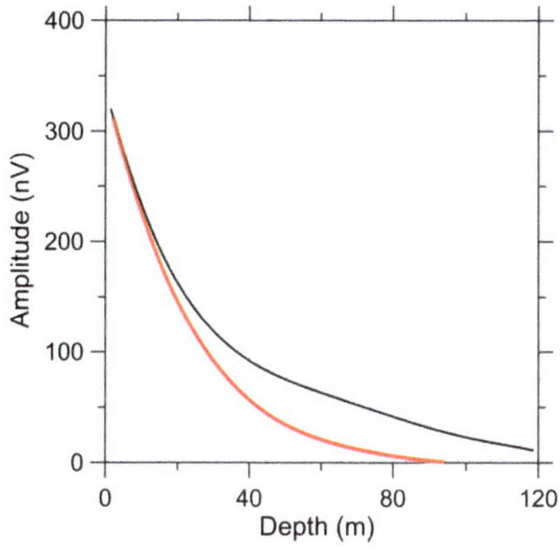

Figure 9.13 Amplitude of the SNMR signal *versus* depth computed considering a 1 m thick layer of bulk water (w = 100%, T_2^* = 1000 ms), a square loop of 80 m side, a Larmor frequency of 2000 Hz and an inclination of the Earth's magnetic field of 45°. The half-space resistivity was assumed to be 100 ohm-m (black line) and 5 ohm-m (red line).

different depths (5, 30 and 80 m) is shown in Figure 9.12b. One can see that shallow water produces a sharp signal and the SNMR response from a deep layer is almost a straight line (the initial part of the sine function). Note the relationship between the maximum of the signal amplitude and the pulse moment: a deeper layer has a maximum at a larger pulse moment.

Figure 9.12c shows how SNMR response depends on the thickness of the investigated water saturated formation. In this example a water saturated layer (w = 20%) is located at a depth of 10 m. The thickness of the layer varies from 10 to 30 m. One can see that a larger thickness of the formation causes the maximum amplitude of SNMR response to shift towards a larger pulse moment and, additionally, the $e(q)$ plot widens.

The magnetic field transmitted by the loop progressively diminishes with increasing depth. As a result, there are effectively smaller flip angles for deep layers than for shallow ones. The SNMR signal generated by groundwater and measured with a surface loop is also attenuated by the subsurface. These effects limit the depth of investigation with the method. Figure 9.13 shows the amplitude computed considering a 1 m thick layer of bulk water (w = 100%, T_2^* = 1000 ms) plotted *versus* depth of this layer. When the subsurface is more conductive the SNMR signal is more attenuated as is shown by the red line in Figure 9.13.

In the SE mode the SNMR signal has similar features as in the FID mode, but it has smaller amplitude. Figure 9.14 shows the initial amplitude of the FID and SE signals. The SE amplitude was computed without taking into account the transverse relaxation and assuming the second pulse twice as large as the first pulse. One can see that the SE signal from a shallow water saturated formation has the amplitude of about 75% of the FID amplitude.

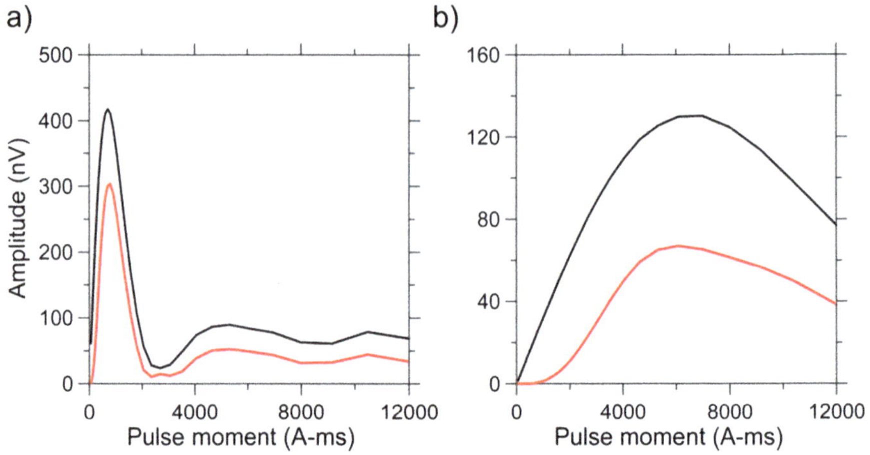

Figure 9.14 Amplitude of the SNMR signal *versus* pulse moment computed for two 10 m thick water saturated formations ($w = 20\%$) neglecting transverse relaxation effects and considering FID (black line) and SE (red line) signals. The Larmor frequency was set to 2000 Hz, the inclination of the Earth's magnetic field was 45° and the resistivity of the subsurface was 100 ohm-m. Depth to water was: (a) 5 m; (b) 50 m.

For a deep formation the SE amplitude is only about 50% of the FID amplitude. Note that when measuring SE signals, the pulse duration should be 2 to 3 fold shorter than when measuring FID signals (15 ms instead of 40 ms). Shorter pulses are necessary because of the large spectrum of the SE signal. Consequently, the maximum pulse moment available for SE measurements is half that of FID measurements. The depth of investigation in the SE mode is also smaller than in the FID mode.

Thus, the SE signal is smaller and it is attenuated by transverse relaxation and diffusion.[8] Consequently, the practical performance of the SNMR method in the SE mode is not as good as in the FID mode. However, in the presence of magnetic rocks the FID signal may be strongly affected or possibly undetectable and the use of the spin echo considerably improves the SNMR performance.[9]

9.4 Data Inversion

The SNMR inverse problem is nonlinear. A straightforward approach for linearizing the SNMR inverse problem consists of taking into account the electrical conductivity of the subsurface known from other geophysical measurements and to assume the near resonance conditions. This assumption is often justified when rocks are non-magnetic and SNMR system is tuned to resonance at the Larmor frequency. Small variations around the resonance (typically ±0.5 Hz) have only a small effect on the SNMR signal and can usually be neglected.

If $\varphi_0 = 0$ then the SNMR signal is real and inversion can be performed using only amplitudes; otherwise, inversion should be carried out considering a complex SNMR signal. Assuming that the stratification is horizontal, eqn (9.11) can be written as a Fredholm integral equation of the first kind:[21]

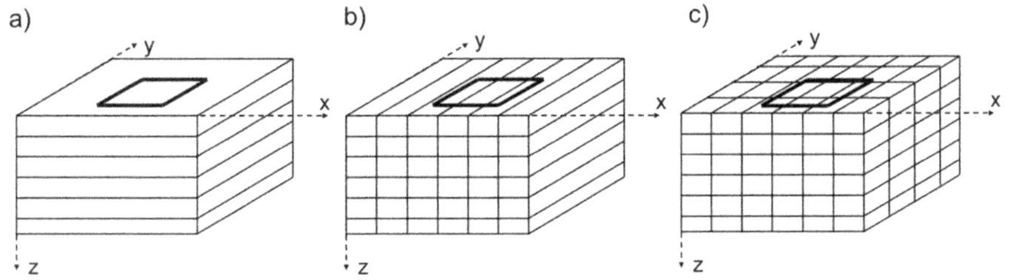

Figure 9.15 Visualization of SNMR results: (a) 1-D survey: the water content and the relaxation times are represented as infinite horizontal layers; (b) 2-D survey: the water content and the relaxation times are represented as parallelepipeds infinite in the direction transverse to the profile; (c) 3-D survey: the water content and the relaxation times are represented as cells smaller than the loop size.

$$e_0(q) = \int_0^{z_{max}} K_{1D}(q,z)w(z)\mathrm{d}z \tag{9.12}$$

where $K_{1D}(q,z)$ is the one dimensional kernel:

$$K_{1D}(q,z) = \omega_0 I_0^{-1} \int_{x,y} B_\perp(\mathbf{r}) e^{\mathrm{j}2\varphi_0^{Tx}(\mathbf{r})} M_\perp(q,\mathbf{r})\mathrm{d}x\,\mathrm{d}y \tag{9.13}$$

Numerical results show that distant protons produce a negligibly small signal and hence integration can be limited by the area of approximately $x^2 + y^2 < (3a)^2$ and $z_{max} \approx 1.5a$ can be considered as an estimate of the maximum possible depth of investigation, where a is the side for a square loop.

The investigated subsurface can be presented as shown in Figure 9.15. Depending on the survey design SNMR results can be presented as 1-D, 2-D or 3-D distributions of water content and/or relaxation times.

Let I_q be the number of pulse moments and I the number of data readings. When using only amplitudes $I = I_q$ and when using complex signals $I = 2I_q$. Let J be the number of model layers representing the subsurface (1-D inversion). In this case, the integral in eqn (9.12) can be approximated by a system of algebraic equations. In matrix notation the approximating equation can be written as:

$$\mathbf{Aw} = \mathbf{e_0} \tag{9.14}$$

where $\mathbf{A} = [a_{i,j}]$ is a rectangular matrix of $I \times J$ with the elements:

$$a_{i,j} = \begin{cases} \mathrm{Re}\left(\int_{z_j}^{z_{j+1}} K_{1D}(q_i,z)\mathrm{d}z \right), & \text{for} \quad i <= I_q \\ \mathrm{Im}\left(\int_{z_j}^{z_{j+1}} K_{1D}(q_{i-I_q},z)\mathrm{d}z \right), & \text{for} \quad i > I_q \end{cases} \tag{9.15}$$

with $\mathbf{e_0} = (e_{01},e_{02},...,e_{0i},...,e_{0I})^T$, $e_{0i} = e_0(q_i)$ being the set of experimental data and $\mathbf{w} = (w_1,w_2,...,w_j,...,w_J)^T$, $w_j = w(\Delta z_j)$ being the vertical distribution of water content.

For discretizing eqn (9.12) we define a set of pulse moments q_i for measuring SNMR signal $e_{0i} = e_0(q_i)$ with $i = 1,2,...,I$ (data set), and a set of cells Δz_j that approximate the subsurface with $j = 1,2,...,J$ (model of the subsurface). In the 1-D case, each cell is a horizontal layer of the thickness $\Delta z_j = z_{j+1} - z_j$ selected so that $z_j < z_{j+1}$ and $z_{\max} = \sum_{j=1}^{J} \Delta z_j$.

The 3-D SNMR measuring setup is composed of L transmitting loops (Tx) and K receiving loops (Rx). Let I_l be the number of pulse moments for the corresponding Tx loop. In this case, the number of data readings will be $I = 2 \times K \times L \times \sum_{l=1}^{L} I_l$, where the factor 2 appears when the complex signal is used. We divide the subsurface into J cells as shown in Figure 9.15c. So $J = J_x \times J_y \times J_z$, with J_x, J_y, J_z being the number of cells in each direction. Similar to the 1-D case, 3-D SNMR inversion requires resolution of a matrix equation $\mathbf{Aw} = \mathbf{e_0}$. Matrix $\mathbf{A} = [a_{i,j}]$ is a rectangular matrix of $I \times J$ with the elements:

$$a_{i,j} = \begin{cases} \mathrm{Re}\left(\omega_0 I_0^{-1} \int_{V_j} B_\perp(\mathbf{r}) e^{\mathrm{j}2\varphi_0^{Tx}(\mathbf{r})} M_\perp(q_i,\mathbf{r}) \mathrm{d}V \right), & \text{for} \quad i <= \sum_{l=1}^{L} I_l \\ \mathrm{Im}\left(\omega_0 I_0^{-1} \int_{V_j} B_\perp(\mathbf{r}) e^{\mathrm{j}2\varphi_0^{Tx}(\mathbf{r})} M_\perp\left(q_{i-\sum_{l=1}^{L} I_l},\mathbf{r} \right) \mathrm{d}V \right), & \text{for} \quad i > \sum_{l=1}^{L} I_l \end{cases} \tag{9.16}$$

Thus, the subsurface is represented by cells of volume V_j and each element $a_{i,j}$ contains the SNMR signal computed for the corresponding cell. Depending on the selected inversion scheme cells may contain only amplitudes or the real and imaginary parts of the complex signal.

Inversion of such SNMR data is ill-posed. Different methods for resolving ill-conditioned inverse problems can be found in the literature.[22] In this book we limit our presentation by the most popular Tikhonov regularization method.[23] The water content distribution $\mathbf{w}$ is the solution of the matrix equation $\mathbf{Aw} = \mathbf{e_0}$. To find an approximate solution, this method supposes minimization of Tikhonov's functional:

$$\mathrm{TF}(\alpha) = \|\mathbf{Aw} - \mathbf{e_0}\|_{L_2} + S^{\mathrm{damp}} = \min \tag{9.17}$$

where $S_{1D}^{\mathrm{damp}} = \alpha_z \times \left(\frac{\partial}{\partial z} \mathbf{w} \right)^2$, $S_{2D}^{\mathrm{damp}} = \alpha_x \times \left(\frac{\partial}{\partial x} \mathbf{w} \right)^2 + \alpha_z \times \left(\frac{\partial}{\partial z} \mathbf{w} \right)^2$ and $S_{3D}^{\mathrm{damp}} = \alpha_x \times \left(\frac{\partial}{\partial x} \mathbf{w} \right)^2 + \alpha_y \times \left(\frac{\partial}{\partial y} \mathbf{w} \right)^2 + \alpha_z \times \left(\frac{\partial}{\partial z} \mathbf{w} \right)^2$ are the estimates of the smoothness of the solution and $\alpha_x,\alpha_y,\alpha_z > 0$ are the smoothing factors in each direction.

9.5 Experimental Verification

9.5.1 Detection of Groundwater

Detection of water saturated geological formations is the most reliable information provided by SNMR. The SNMR signal is generated by groundwater molecules and a simple observation of the SNMR response can be interpreted as a signature of groundwater. However, SNMR is sensitive to the total volume of water below the measuring loop and, in practice, the capacity for detecting water is limited by the signal to noise ratio. The detectable volume of water can be estimated by a product of the SNMR water content and the thickness of the aquifer. For example, let us consider a 50×50 m^2 loop (two turns) and a 10 nV threshold. Modeling results show that SNMR is able to detect a 20 m thick weathered rock aquifer with a water content of 1% down to 20 m (top of the aquifer) and with a water content of 2% down to approximately 50 m. If the fractured part of the hard rock aquifer is characterized by an average effective porosity of 0.1% to 0.2%, then the total water volume is small and it would be barely detectable.[24] In the presence of magnetic rocks, the Earth's magnetic field is often heterogeneous and SNMR measurements may be perturbed. This problem can be partly resolved by using the spin echo measurements, but the SE signal is smaller and hence the performance of the method in the SE mode is not as good as in the FID mode.

A very convincing demonstration of the sensitivity of the method has been obtained during investigation of the Tête Rousse glacier in the French Alps.[25] Figure 9.16 shows the amplitude of the SNMR signal observed with a loop located over the water-filled cavern (solid line). One can see that SNMR signal was observed well above the noise level thus suggesting water in the glacier. After water was pumped out, SNMR measurements with the same loop and at the same location had signal amplitude of about 15 nV, which is close to the noise level.

9.5.2 Aquifer Geometry

The aquifer geometry can be derived from the distribution of the water content in the subsurface provided by the SNMR inversion. However, the resolution of the SNMR method is limited by the uncertainty in the inversion results. In general, the available resolution allows definition of the geometry with uncertainty of a few meters for shallow aquifers and uncertainty of more than 10 m for deep aquifers. In practice, two to three aquifers can be reliably identified using SNMR results.

An example of SNMR measurements carried out in France using a 56 m side figure-of-eight loop is presented in Figure 9.17. The reliable depth of investigation for this loop is about 60 meters. The borehole in this area detected

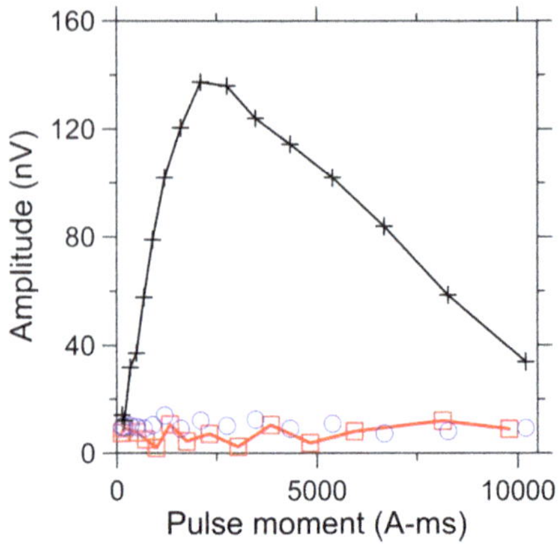

Figure 9.16 Experimental demonstration of the capability of the SNMR method to detect groundwater: the amplitude of the SNMR signal *versus* the pulse moments before (black) and after (red) drainage of the cavern in the Tête Rousse glacier. Noise measurements are shown by blue circles.

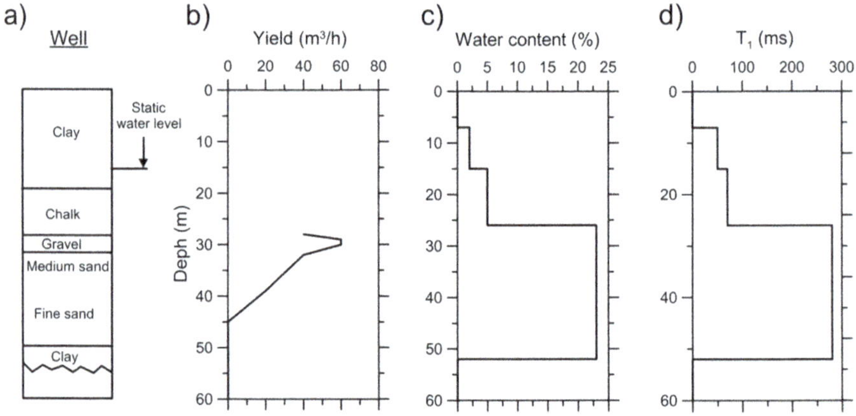

Figure 9.17 Example of 1-D SNMR inversion near borehole in France: (a) borehole log; (b) pumping rate *versus* depth; (c) SNMR water content *versus* depth (after[26]). (d) T_1 *versus* depth.

an aquifer between approximately 28 and 48 m (Figure 9.17a). The pumping rate *versus* depth measured during pumping shows the contribution of different parts of the subsurface to the total yield (Figure 9.17b). The water content (Figure 9.17c) and the relaxation time (Figure 9.17d) derived from SNMR measurements show that in general the aquifer was well detected, but the resolution was insufficient for resolving different parts of the aquifer.

Figure 9.18 shows signals for the example presented above. The measured amplitude and phase of the SNMR signal (crosses) and the theoretical signal computed after the inversion results that fits these data (red line) are plotted

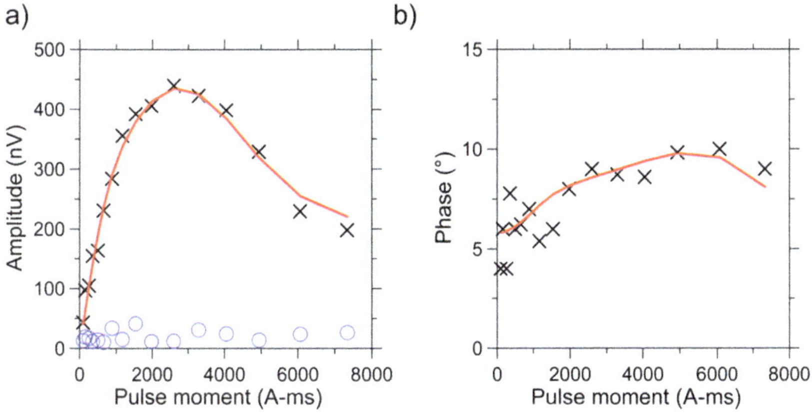

Figure 9.18 Measured SNMR signals and inversion fit: (a) amplitude; (b) phase. The black crosses show measured data and the red line shows inversion fit. The blue circles show the noise measurements.

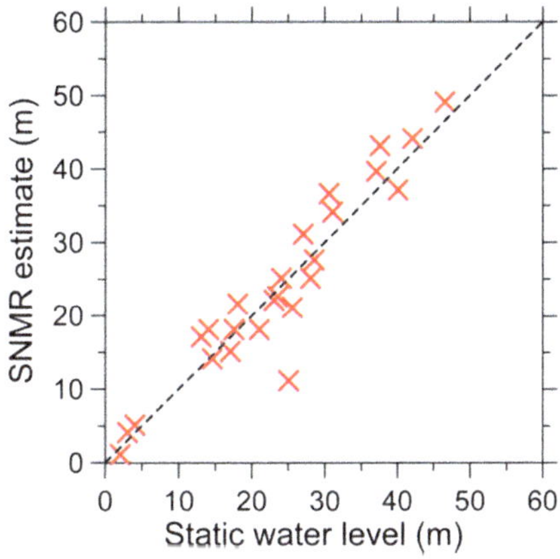

Figure 9.19 Comparison of the depth to water in unconfined aquifers derived from 1-D SNMR inversion with the static water level measured in boreholes (after[27]).

versus pulse moment. The signal to noise ratio was good and the inversion results can be considered as reliable.

Experimental evaluation of the SNMR capacity to detect the top of the aquifer has been reported in ref. 9. The summary is presented in Figure 9.19. The depth to the top of the aquifer derived from the SNMR inversion is depicted *versus* the static water level measured in boreholes. These experimental results demonstrate that SNMR is able to resolve the depth to groundwater. Observed limited resolution of the method was found in good agreement with the mathematical estimates of the uncertainty in the SNMR inversion results. Note that for confined aquifers and in the heterogeneous subsurface

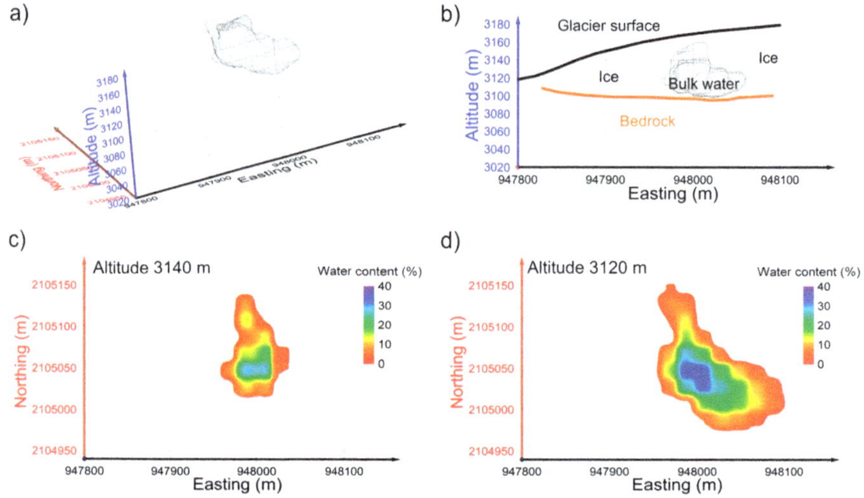

Figure 9.20 The results of 3-D inversion of SNMR data show location of the water-filled cavern in the Tête Rousse glacier. The cavern body is shown as: (a) isosurface of the cavern at the level of 13% water content; (b) N–S projection of the glacier body; (c) a horizontal slice of the water content at an altitude of 3140 m; (d) a horizontal slice of the water content at an altitude of 3120 m.

the static water level observed in boreholes may not correspond to the top of the aquifer estimated with the SNMR.

For the 1-D SNMR inversion, the subsurface can be considered as homogeneous if water saturated formations do not vary with a length scale less than three times the loop size. However, heterogeneous formations like karst or fractured rock aquifers may change their hydrodynamic properties at smaller distances. For investigating and modeling heterogeneous aquifers SNMR should be applied in the 2-D and 3-D modes. Reported application of 2-D SNMR measuring and inversion procedures show that the accuracy of the results was significantly improved.[3,28–31] The 3-D SNMR application is a new development and only one case study has been reported.[4,32]

For demonstration of the 3-D inversion we present results of the investigation of the water accumulation in the Tête Rousse glacier.[4,25,32] An example of measured signals is presented in Figure 9.16. The measuring setup was composed of nine half overlapped square loops with side length of 80 m each. The inversion was carried out using cells of $20 \times 20 \times 5$ m^3. Figure 9.20 shows the results of the 3-D inversion before the cavern was drained.

Note that the size of this cavern is smaller than the resolution of the inversion and the water content was averaged over a volume larger than the true cavern volume. Consequently, the water content was found to be only 40% instead of 100% expected in the real cavern. However, comparison of the SNMR prediction of the pumping rate *versus* depth and pumping results shows very reasonable correspondence in-between (Figure 9.21a).

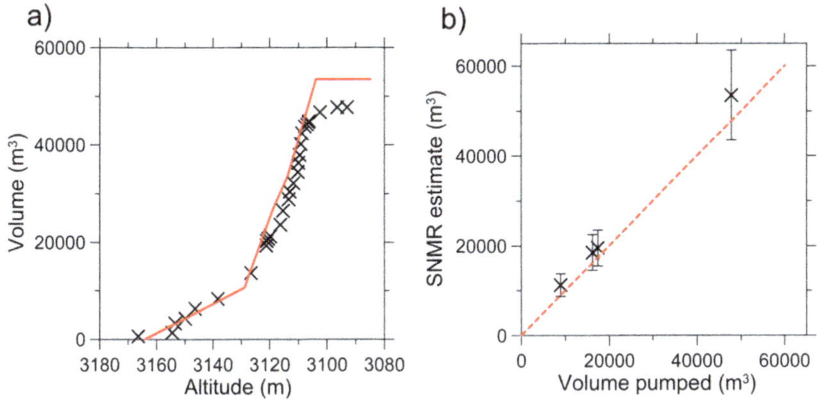

Figure 9.21 Verification of the SNMR results by pumping: (a) the volume of pumped water *versus* water level in the cavern (crosses) and the SNMR prediction of the pumping rate (red line); (b) comparison of the volume of water pumped from the cavern with the SNMR estimate of the volume of water in the cavern.

Water was pumped from the cavern four times and each time the SNMR estimate of the water volume was compared with the volume pumped out (Figure 9.21b). Taking into account the complexity of the cavern geometry and the uncertainty in SNMR results, we consider this correspondence as good. This example shows that despite the limited resolution 3-D SNMR results were confirmed to be reliable and very useful for local authorities for making decisions about drainage of this cavern. Note that the cavern was detected with the SNMR alone and the drilling program was guided by the 3-D SNMR results. Actually, no other surface geophysical method is able to provide similar information about water in the subsurface.

9.5.3 Aquifer Porosity and Hydraulic Conductivity

SNMR measurements allow for prediction of the hydraulic conductivity using the relaxation times of the magnetic resonance signal:[33]

$$T_1 \approx \frac{V}{\rho_1 S} \text{ and } T_2 \approx \frac{V}{\rho_2 S} \tag{9.18}$$

where V/S is the pore volume to surface ratio, and ρ_1, ρ_2 are the surface relaxivity rates for T_1 and T_2, respectively. Consequently, the hydraulic conductivity (K) can be estimated using the NMR provided water content w and either T_1 or T_2:

$$K = c \times T_1^b \times w^a \text{ and } K = c \times T_2^b \times w^a \tag{9.19}$$

where a,b,c are empirically determined values that may be different for different rocks.

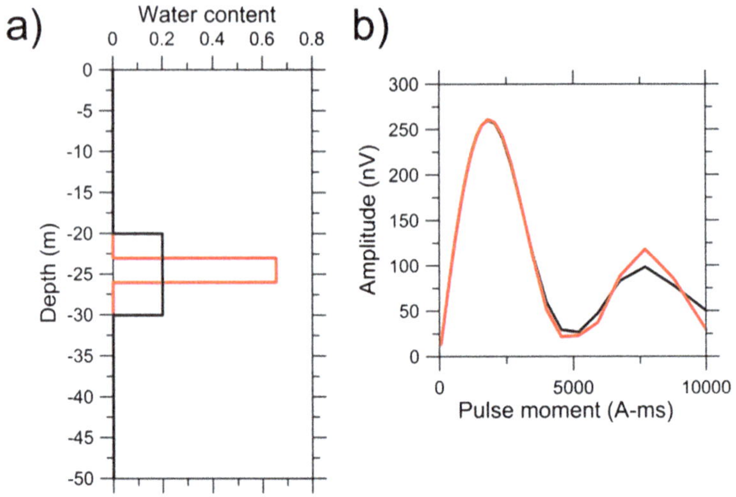

Figure 9.22 Numerical demonstration of the equivalence problem in the SNMR inversion: (a) two equivalent models (model 1 is shown with a black line and model 2 with a red one); (b) corresponding amplitude of the SNMR signal *versus* pulse moment.

For sandstones a good permeability prediction was obtained on rock samples using $a = 4$ and $b = 2$. However, better results for sand aquifers were obtained with the SNMR using $a = 1$ and $b = 2$[26] using the estimator proposed by Seevers.[34]

$$K = C_p \times T_1^2 \times w \tag{9.20}$$

where C_p is an empirical prefactor obtained after calibration of SNMR with pumping tests. T_2 is rarely used in SNMR and only the use of T_1 and T_2^* has been reported in the literature. T_2^* is easier to measure and it can be safely used by replacing T_1 in eqn (9.20) when investigating rocks with a low magnetic susceptibility ($<10^{-5}$ SIU).

In practice, for characterizing aquifers the transmissivity T is often used. The transmissivity can be estimated using the thickness of aquifer Δz also derived from SNMR inversion as:

$$T = K \times \Delta z = C_p \times T_1^2 \times w \times \Delta z \tag{9.21}$$

In NMR logging, the estimator $K \approx T_2^2 \times w^4$ is preferred as being more accurate,[35] but in SNMR the estimator $K \approx T_1^2 \times w$ has the advantage of being more stable. Indeed, SNMR inversion is ill-posed and has a non-unique solution. The equivalent solutions may be different, but all of them fit experimental data equally well. It is known that w and Δz separately are less stable than their product $w \times \Delta z$. Consequently, when using eqn (9.21) the transmissivity is estimated using the product $w \times \Delta z$, but when using $a = 4$ and $b = 2$ in eqn (9.19) the estimator will be $w^4 \times \Delta z$. Now, let us see how these two estimators influence the transmissivity prediction assuming the same value of T_1. For demonstration purposes we propose two different models (Figure 9.22a).

Table 9.1 Normalized transmissivity estimate for two equivalent models presented in Figure 9.22 and computed assuming $a = 1$ and $a = 4$ in eqn (9.19).

Model	Water content w	Thickness Δz (m)	Estimator $W^4 \times \Delta z$	Estimator $w \times \Delta z$
Model 1	0.2	10	0.029	1.0
Model 2	0.655	3	1.0	0.98

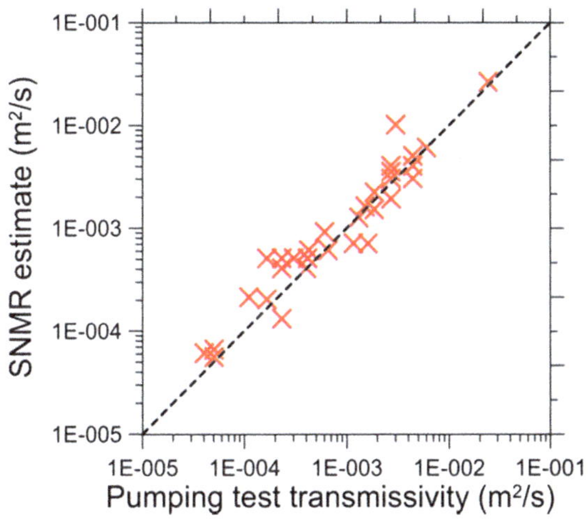

Figure 9.23 Experimental verification of the SNMR estimate of the transmissivity by the transmissivity obtained with pumping tests (after[27]).

The models are fully equivalent and Figure 9.22b shows the amplitude of the SNMR signal *versus* the pulse moment computed for these models. Corresponding estimators computed for these models (Table 9.1) demonstrate that $w \times \Delta z$ is a more stable parameter than $w^4 \times \Delta z$.

An experimental verification of the SNMR prediction of the transmissivity of aquifers is presented in Figure 9.23.

9.6 Conclusions

In this chapter we have presented an emerging application of NMR phenomena in geophysics. The surface NMR (SNMR or MRS) method is a non-invasive tool for 1-D, 2-D and 3-D groundwater investigations from the ground surface down to a depth of 100–150 m. SNMR operates in the Earth's magnetic field at a frequency between 800 and 2600 Hz and the investigated volume is much larger than with any other application of NMR (a few hundreds to a few thousands of m^3). SNMR enables reliably locating water-saturated formations (aquifers) and estimating their effective porosity and hydraulic conductivity. However, for these estimates calibration in different types of geology using boreholes and pumping tests is required.

The depth of investigation and resolution of the method are dependent on the measuring set-up. As for many other surface geophysical methods, SNMR resolution is limited by the equivalence problem (different models may produce equal signals). Water-saturated formation can be resolved with an uncertainty of a few meters close to the surface and the uncertainty increases progressively with depth.

SNMR is already an established technique but improvements in the accuracy of the results are still necessary for rendering this method more efficient. The major limitation is a poor signal to noise ratio, which requires a large amount of water in the subsurface. The problem of the high sensitivity of the method to electromagnetic interferences limits SNMR efficiency in hard rock and karst geological environments and in industrial areas with a high density of electrical installations.

References

1. A. Semenov, M. Schirov, A. Legchenko, A. Burshtein and A. Y. Pusep, Great Britain Patent B, 1989, 2198540.
2. R. Varian, *US Pat.*, 3019383, 1962.
3. M. Hertrich, M. Braun, T. Gunther, A. G. Green and U. Yaramanci, *IEEE Trans. Geosci. Remote Sens.*, 2007, **45**, 3752–3759.
4. A. Legchenko, M. Descloitres, C. Vincent, H. Guyard, S. Garambois, K. Chalikakis and M. Ezersky, *New J. Phys.*, 2011, **13**, 025022.
5. Z. Xiaohuz, L. Jun, W. Yingji and S. Sun Feng, *Chin. J. Sci. Instrum.*, 2006, 7, 006.
6. W. Yingji, L. Jun, R. Liangliang, G. Dongxu and J. Xiaochen, *Chin. J. Sci. Instrum.*, 2008, **8**, 012.
7. A. Legchenko, A. Beauce, A. Guillen, P. Valla and J. Bernard, *J. Environ. Eng. Geophys.*, 1997, **2**, 173–190.
8. A. Legchenko, J.-M. Vouillamoz and J. Roy, *Geophysics*, 2010, **75**, L91–L100.
9. J.-M. Vouillamoz, A. Legchenko and L. Nandagiri, *Near Surf. Geophys.*, 2011, **9**, 135–144.
10. A. Legchenko and G. Pierrat, *Near Surf. Geophys.*, 2014, **12**, 297–308.
11. D. Trushkin, O. Shushakov and A. Legchenko, *Geophys. Prospect.*, 1994, **42**, 855–862.
12. A. A. Kaufman and G. V. Keller, Volcanic Processes, *Model.*, 1983, 207.
13. J. Wait, *Geo-Electromagnetism*, Academic Press, San Diego, CA, 1982.
14. J. Ryu, H. F. Morrison and S. H. Ward, *Geophysics*, 1970, **35**, 862–896.
15. W. L. Anderson, *Geophysics*, 1979, **44**, 1287–1305.
16. W. L. Anderson, *Geophysics*, 1989, **54**, 263–266.
17. P. Valla and A. Legchenko, *J. Appl. Geophys.*, 2002, **50**, 217–229.
18. A. Legchenko and P. Valla, *J. Appl. Geophys.*, 2002, **50**, 3–19.
19. E. L. Hahn, *Phys. Rev.*, 1950, **80**, 580.
20. A. L. Bloom, *Phys. Rev.*, 1955, **98**, 1105.
21. A. V. Legchenko and O. A. Shushakov, *Geophysics*, 1998, **63**, 75–84.

22. A. Tarantola, *Inverse problem theory and methods for model parameter estimation*, siam, 2005.
23. A. Tikhonov and V. Y. Arsenin, *Methods for solving ill-posed problems*, John Wiley and Sons, Inc, 1977.
24. A. Legchenko, M. Descloitres, A. Bost, L. Ruiz, M. Reddy, J. F. Girard, M. Sekhar, M. Kumar and J. J. Braun, *Ground Water*, 2006, **44**, 547–554.
25. C. Vincent, M. Descloitres, S. Garambois, A. Legchenko, H. Guyard and A. Gilbert, *J. Glaciol.*, 2012, **58**, 866–878.
26. A. Legchenko, J. M. Baltassat, A. Bobachev, C. Martin, H. Robain and J. M. Vouillamoz, *Ground Water*, 2004, **42**, 363–373.
27. J.-M. Vouillamoz, J.-M. Baltassat, J.-F. Girard, J. Plata and A. Legchenko, *Bol. Geol. Min.*, 2007, **118**, 531–550.
28. J. Vouillamoz, A. Legchenko, Y. Albouy, M. Bakalowicz, J. Baltassat and W. Al-Fares, *Groundwater*, 2003, **41**, 578–586.
29. M. Boucher, J.-F. Girard, A. Legchenko, J.-M. Baltassat, N. Dörfliger and K. Chalikakis, *J. Hydrol.*, 2006, **330**, 413–421.
30. J.-F. Girard, M. Boucher, A. Legchenko and J.-M. Baltassat, *J. Appl. Geophys.*, 2007, **63**, 103–116.
31. J. Lehmann-Horn, J. Walbrecker, M. Hertrich, G. Langston, A. McClymont and A. Green, *Geophysics*, 2011, **76**, B165–B172.
32. A. Legchenko, C. Vincent, J. Baltassat, J. Girard, E. Thibert, O. Gagliardini, M. Descloitres, A. Gilbert, S. Garambois and A. Chevalier, *Cryosphere*, 2014, **8**, 155–166.
33. K.-J. Dunn, D. J. Bergman and G. A. LaTorraca, *Nuclear magnetic resonance: Petrophysical and logging applications*, Elsevier, 2002.
34. D. Seevers, SPWLA 7th Annual Logging Symposium, 1966.
35. W. Kenyon, *Log Analyst*, 1997, **38**, 21–43.

CHAPTER 10

Rock Core Analysis: Metallic Core Holders for Magnetic Resonance Imaging Under Reservoir Conditions

MATTHEW OUELLETTE[a,b], MING LI[b,c], GUANGZHI LIAO[b], ESAM M. A. HUSSEIN[a], LAURA ROMERO-ZERON[c], AND BRUCE J. BALCOM*[b]

[a]Department of Mechanical Engineering, University of New Brunswick, Fredericton, NB, Canada E3B 5A3; [b]MRI Research Centre, Department of Physics, University of New Brunswick, Fredericton, NB, Canada E3B 5A3; [c]Department of Chemical Engineering, University of New Brunswick, Fredericton, NB, Canada E3B 5A3
*E-mail: bjb@unb.ca

10.1 Introduction

Owing to the diminishing supply and increasing demand for petroleum, there is an increasing reliance on unconventional oil sources and marginal reservoirs, and a need to improve extraction efficiency from existing reservoirs.[1] Therefore, improved scientific understanding of rock–fluid interactions within reservoirs is desired.[1] This is facilitated by extracting rock core samples from reservoirs, in the form of cylinders on the scale of 100 mm in

New Developments in NMR No. 5
Mobile NMR and MRI: Developments and Applications
Edited by Michael L. Johns, Einar O. Fridjonsson, Sarah J. Vogt, and Agnes Haber

Published by the Royal Society of Chemistry, www.rsc.org

diameter and tens of meters in length. These core samples reflect the variation of a reservoir throughout the different stratified layers. Therefore, rock core plugs are cut at different layers along the rock core. These plugs are typically 38 or 25 mm (1.5 or 1") diameter, and up to 100 mm long. Core plugs are studied for various petrophysical properties, pertaining to rock–fluid interaction: porosity, pore size, permeability, relative permeability, saturation and wettability. Core plug measurement data is often employed in computer models to simulate the reservoir characteristics, to determine the financial viability of developing a reservoir, or to devise strategies to enhance oil recovery.[2]

It is necessary to test core plug samples at conditions representative of reservoirs. The pressure in a reservoir affects the porosity and permeability of the rock formations, and temperature variation influences fluid behavior, such as the wetting behavior of oil and water, or viscosity.[3–5] Therefore, a core holder is used to maintain a core plug at high pressure and temperature conditions during testing, resembling those of a reservoir. The pressure varies greatly, from just above atmospheric pressure in a shallow reservoir, to hundreds of MPa deep in a reservoir.[4] The temperature within reservoirs may range from 20 °C near the surface, to 100 °C deep in reservoirs.[3,6]

Hydrostatic type core holders are commonly employed to maintain core plugs at elevated temperature and pressure conditions. Radial pressure is applied to create a hydrostatic pressure condition, mimicking the depth of the core plug in a reservoir. Air, nitrogen or fluorinated oil are used to apply radial pressure by filling the internal volume of the vessel. This is also called the confining fluid. Once at reservoir conditions, various tests may be performed. Hydraulic pumps are commonly used to apply an axial pressure, to force fluid flow axially through the length of the core plug.

10.2 Imaging Techniques for Rock Cores

There are a number of techniques for imaging porous media, the most common of which is X-ray computed tomography (CT).[7] This technique relies on measuring the attenuation of X-ray photons at different directions. Therefore, CT provides mainly density indication, since X-ray attenuation is density dependent. Micro-CT systems can provide high-resolution images, but they are limited in the size of object they can image.[8,9] While micro-CT can distinguish between the solid material in a rock core (3000–4000 kg m^{-3} in density), and fluid in pores (about 1000 kg m^{-3} density), the requirement of multiple exposures around the object makes CT systems unsuitable for use under transient conditions. In X-ray radiography, a single exposure to radiation is required, but no detailed spatial information is provided along the direction of radiation travel.[10] Neutron radiography is also well suited for imaging in the presence of oil and water, owing to the high sensitivity of neutrons to the hydrogen present in these fluids.[11] However, neutron radiography usually requires a nuclear research reactor, which limits its widespread application. Acoustic methods are also used for rock core imaging, but they only provide information on structure and mechanical properties, and are unable to image through voids or interfaces.[12]

Magnetic resonance imaging (MRI) is a diagnostic imaging technique commonly used in clinical medicine. MRI has the advantage of being sensitive only to particular nuclei, mainly hydrogen, enabling very high contrast of hydrogenous fluids in porous media. Therefore, MRI has been employed in porous media studies to determine saturation, porosity, pore size distribution, pore geometry and permeability.[13–15] This work aims at employing a metallic core holder for imaging, by judicious selection of a metal that does not interfere significantly with the MRI process. In order to demonstrate how this obstacle was overcome, a brief introduction to MRI is given below.

10.3 Magnetic Resonance Imaging and Metal Compatibility

10.3.1 Principles of MRI

Magnetic resonance is a phenomenon in which nuclei absorb electromagnetic radiation at a resonant frequency when subjected to a magnetic field. Magnetic resonance occurs only in atomic nuclei with an odd number of protons or neutrons and that possess a net magnetic moment. Hydrogen is one of these nuclei. In a sample of MR observable nuclei, the nuclei are distributed among the possible spin states. When a static magnetic field B_0 is applied, the population difference between spin states creates a net magnetization vector, M_0. Nuclear spins may be excited between spin states by a radio-frequency (RF) field, produced by a coil applied perpendicular to B_0. This causes displacement of the net magnetization vector from equilibrium. The magnetization vector then returns to equilibrium with characteristic MR signal relaxation times T_1, T_2, and T_2^*. Magnetization perpendicular to the static field induces a voltage in the RF coil, which is detected as the MR signal.

In MRI, switched magnetic field gradients are employed to spatially resolve the MR signal in one, two or three dimensions.[16,17] An image is obtained by performing a Fourier transform of the MR signal, to produce the spatial distribution of the density of observable nuclei. Two imaging methods commonly employed by the MRI Research Centre at the University of New Brunswick are called: Single Point Ramped Imaging with T_1 Enhancement (SPRITE[18]), and Spin-Echo Single Point Imaging (SE-SPI[19]). Both are employed for examining hydrogen containing materials not easily imaged by conventional methods due to their short relaxation times, such as concrete, polymers, composites, food materials, and porous media.[18,19] SPRITE is employed to acquire 2D images in this work, and SE-SPI is used to acquire 1D profiles.

A metallic core holder must be compatible with the three magnetic fields employed in the MRI process: the strong static magnetic field B_0, the RF excitation field B_1, and the switched magnetic field gradients.

10.3.2 Compatibility with Static Field

The static magnetic field can produce a significant magnetic force that pulls the material into the magnet, if the material exposed to the field has a high magnetic permeability.[20] Therefore, paramagnetic or diamagnetic materials with a very low magnetic permeability must be considered when fabricating a pressure vessel for the core holder. Such materials include: aluminum, brass, copper beryllium, austentic stainless steel alloys, and titanium.

In MRI, the object under evaluation is placed within the magnet's bore. This restricts the size of the core holder. The magnet used in these studies, an Oxford Maran DRX HF 8.5 MHz magnet (Oxford Instruments Ltd., Oxfordshire, UK), accommodates up to a 100 mm diameter core holder. However, the core holder must also be supported in the magnet. In this work, a stand was employed to hold the imaged object upright and centered in the magnet. The requirement for this supporting structure limits the maximum outside diameter of the core holder, to approximately 75 mm (3 in).

10.3.3 Compatibility with RF Excitation Field

A metallic vessel acts as an electromagnetic shield for the RF probe, preventing RF signals from reaching samples within a metallic core holder. This necessitates placing the RF probe inside the vessel, as opposed to outside the vessel, as with non-metallic core holders. Placement of the RF probe inside the core holder increases MR sensitivity, due to the smaller size of the probe coil.[21,22] It also reduces background interference and noise, as reported in related work.[23]

The placement of the RF probe inside the core holder, along with the need for a fluid distribution system to study fluid imbibition and drainage, further constrain the size of the rock core. In this work, the rock core and fluid distribution system required 40 mm diametrically, while the RF probe assembly requires roughly 20 mm of thickness. Therefore, the minimum inside diameter of the core holder wall was approximately 60 mm.

10.3.4 Compatibility with Magnetic Field Gradients

Switched magnetic field gradients are essential to imaging and motion sensitization in MRI. Switching of the magnetic field gradient induces eddy currents in the core holder metal. These eddy currents, in turn, create their own magnetic field, which distorts the magnetic field experienced in the sample space, and the currents therefore degrade MR images.[24] Such distortions challenged previous MR studies employing metallic vessels, which were unable to acquire high quality images.[25–30] However, the intensity of eddy currents decay exponentially, with a time constant τ expressed as:[31]

$$\tau = \Gamma K \mu \sigma \tag{10.1}$$

where σ and μ, respectively, the electrical conductivity and magnetic permeability of the material, K is a constant and Γ is a geometric factor. A very

Table 10.1 MRI related properties of non-magnetic metals and a non-metal (PEEK).[32,33]

Material	Electrical conductivity σ ($\times 10^{-6}$ S m^{-1})	Thermal conductivity k (W m^{-1} K^{-1})	Yield strength Y (MPa)
Aluminum	37.7	150	200
Brass	15.6	109	100
BeCu	58.8	401	60
N60SS	1.02	16	483
Titanium	2.56	22	400
PEEK	—	**0.25**	**96**

rapid eddy current decay is desirable, to minimize magnetic field distortions. Therefore, as eqn (10.1) suggests, a low electrical conductivity material is needed.

10.3.5 Metal Selection

Many metals have low magnetic permeability; a few were identified in Section 10.3.2. One needs to then select a metal that has low electrical conductivity, in addition to having a high thermal conductivity, k, to facilitate control of the temperature of the core holder, and a high yield strength, Y, to withstand high pressures.

Table 10.1 shows the relevant properties of metallic materials with low magnetic susceptibility, along with a non-metal (polyether ether ketone: PEEK) for comparison. Copper beryllium and aluminum were employed in previous high pressure and high temperature MR studies, but not in imaging (MRI) studies due to their high electric conductivity.[23,25,26,29] As Table 10.1 indicates, Nitronic 60 stainless steel (N60SS) has an electrical conductivity lower than all other non-magnetic metals considered.[32,33] Nitronic 60 stainless steel also has a very high yield strength, but a low thermal conductivity, compared to other metals. Given, however that its thermal conductivity far exceeds that of non-metallic polymers, widely used in MRI studies, Nitronic 60 stainless steel is considered to be a well suited metal for construction of an MRI compatible core holder.

10.4 Design

The MRI compatibility considerations not only determined the type of metal to be used, N60SS, but also dictated the inside and outside diameter of the core holder, 64 and 76 mm, respectively. Now, one can proceed with the conventional aspects of designing the metallic pressure vessel of the core holder for testing under high temperature and pressure conditions. The resulting temperature distribution produces thermal stresses that must be combined with those resulting from pressure and magnetic forces. Therefore, the temperature distribution within the core holder was first investigated as discussed below.

10.4.1 Temperature Regulation

An internal temperature of 50 °C was chosen for the design of this core holder, which is within the desired range for reservoir applications.[3,6] This temperature was reached by regulating the temperature of the exterior of the core holder. The exterior temperature was controlled by pumping heated water through thermally conductive tubing in contact with the vessel.[23] The heat source maintains a section at the center of the exterior vessel wall, 100 mm long, at a constant temperature. This results in a constant temperature, $T(r_o)$, around the surface of the vessel, where r_o is the outside radius.

Temperature regulation of the vessel is a transient process, as the sample and vessel wall are initially at room temperature. Within the vessel, heat is transported *via* conduction, and can be numerically simulated. The core holder is considered as an annular cylinder, with isothermal conditions in the polar direction due to the radial symmetry caused by the application of the heat source evenly around the circumference. The heat conduction equation is:[34]

$$\frac{1}{\alpha}\frac{\partial T}{\partial t} = \frac{1}{r}\frac{\partial}{\partial r}\left(r\frac{\partial T}{\partial r}\right) + \frac{\partial}{\partial z}\left(\frac{\partial T}{\partial z}\right) \tag{10.2}$$

where T is the temperature at radius r and axial position z and α is the thermal diffusivity of the material, given as $\alpha = k/\rho c_p$ where k is the thermal conductivity of the material (W m^{-1} K^{-1}), ρ is its density (kg m^{-3}) and c_p is its specific heat capacity (J kg^{-1} K^{-1}). As eqn (10.2) indicates, a larger thermal diffusivity provides a larger rate of heat transfer, $\partial T/\partial t$, enabling the vessel to more quickly approach steady state. Also, it allows the temperature gradient across the radius, $\partial T/\partial r$, to be reduced.

At surfaces in contact with fluid (specifically the outside wall in contact with outside air), there is heat loss due to convection. The heat equation for convection at a surface is given by:

$$\frac{\mathrm{d}T}{\mathrm{d}t} = \frac{hA}{\rho c_\mathrm{p}}\left(T - T_\infty\right) \tag{10.3}$$

where A is the area for heat transfer, and T_∞ is the temperature of the fluid (air). Combining eqn (10.2) with known boundary/initial conditions and eqn (10.3) creates a two-dimensional problem, which was solved using a finite difference method.

A Nitronic 60 cylinder 300 mm long, with an inner radius of 31.8 mm, and thickness of 6.3 mm, was simulated. Table 10.2 shows relevant physical properties and parameters.[33] A mesh of 13 by 600 points was simulated for up to two hours of heating after application of the heat source, and is representative of the entire cylindrical wall.

Figure 10.1 shows results from the numerical simulation. It is seen that the vessel wall rapidly reaches equilibrium in the radial direction, with the inside (top) surface quickly reaching the outside (bottom) surface's temperature. It is also seen that there is an axial temperature gradient at steady state (after roughly an hour).

Table 10.2 Parameters used in stress calculations.

Geometric dimensions		
Inside radius	r_i	31.8 mm (1.25 in)
Outside radius	r_o	38.1 mm (1.5 in)
Length	L	300 mm (11.8 in)
Material properties		
Yield strength	Y	483 MPa
Modulus of elasticity	E	200 GPa
Poisson ratio	v	0.298
Magnetic susceptibility	χ_m	0.0030
Thermal expansion	α_T	17.3×10^{-6} °C^{-1}
Specific heat	c_p	470 J kg^{-1} K^{-1}
Density	ρ	8000 Kg m^{-3}
Loading conditions		
Temperature difference	ΔT	Computed – initial
Static magnetic field	B_0	0.204 T
Pressure	ΔP	Determined by SF

10.4.2 Stress Analysis

Stress analysis for the core holder requires determination of the stresses produced by the fluid pressure, heating of the vessel and the magnetic field. Cylindrical coordinates (r, θ, z) are used in the calculations, with the same geometric parameters as those given in Table 10.2, due to constraints of the MRI system. The goal of stress analysis is the find the safety factor, SF, of the core holder design. The geometry and temperature desired in the vessel is constrained, therefore, the safety factor is to be found as a function of the internal pressure in the core holder.

10.4.2.1 Pressure Stress

Stress due to the pressure is caused by a pressure difference, ΔP, between the inside and outside of the vessel. Given that the pressure outside the vessel is the atmospheric pressure (*i.e.* about 0.1 MPa), ΔP is equal to the gauge pressure inside the vessel. The pressure stresses $\sigma_{P,r}$ in the radial direction, $\sigma_{p,\theta}$ in the circumferential direction and $\sigma_{P,z}$ in the axial direction are given by:[35]

$$\sigma_{P,r} = \Delta P\left(\frac{r_i^2}{r_o^2 - r_i^2}\right)\left(-\frac{r_o^2}{r^2} + 1\right) \tag{10.4}$$

$$\sigma_{P,\theta} = \Delta P\left(\frac{r_i^2}{r_o^2 - r_i^2}\right)\left(\frac{r_o^2}{r^2} + 1\right) \tag{10.5}$$

$$\sigma_{P,z} = \Delta P\left(\frac{r_i^2}{r_o^2 - r_i^2}\right) \tag{10.6}$$

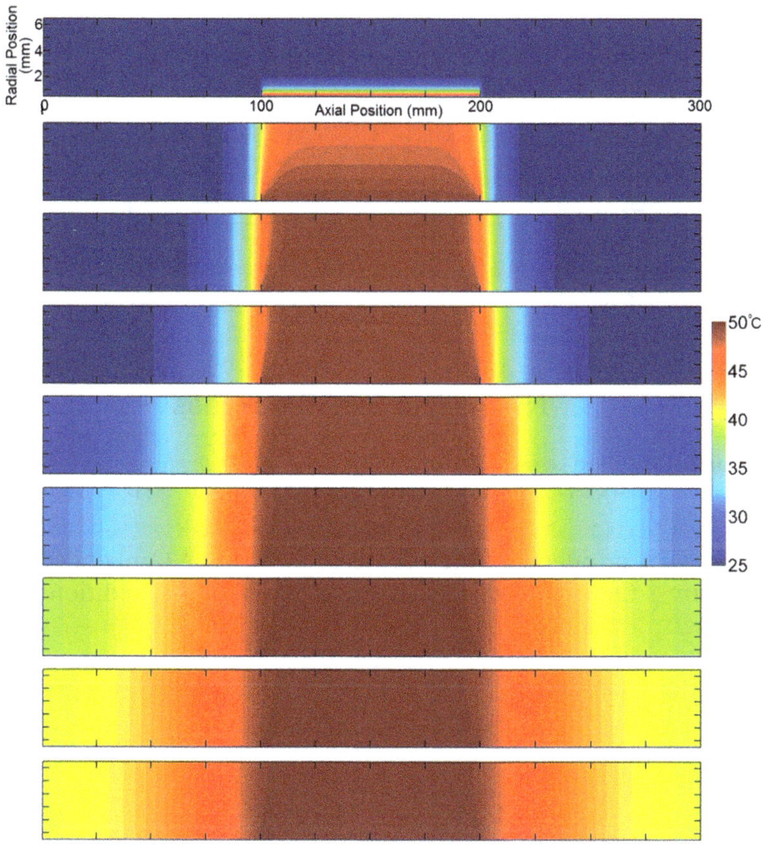

Figure 10.1 Calculated temperature distribution within the core holder wall, in the radial (vertical) and axial (horizontal) directions. Each sub-figure corresponds to a different time interval, from top to bottom: 0 s, 10 s, 30 s, 1 min, 5 min, 10 min, 30 min, 1 h and 2 h. The middle third of the wall is maintained at a constant temperature, to simulate the application of the heat source. There is a rapid conduction of heat in the radial direction, then a slow spread in the longitudinal direction.

where r_i and r_o are, respectively, the inside and outside radii of the vessel, and ΔP is the pressure difference. Employing eqn (10.4)–(10.6) and the parameters in Table 10.2, pressure stresses, normalized to ΔP, are calculated and plotted as a function of the normalized wall thickness, $(r - r_i)/(r_o - r_i)$, in Figure 10.2. Both the radial and circumferential pressure stresses are maximum at the inside wall, in compression and tension, respectively, and decrease towards the outside of the cylinder.

10.4.2.2 *Thermal Stress*

Thermal stresses arise from the thermal expansion of the cylinder, governed by the thermal expansion coefficient, α_T. The cylinder is unrestricted and free to expand in all three dimensions. Therefore, the cylinder expands

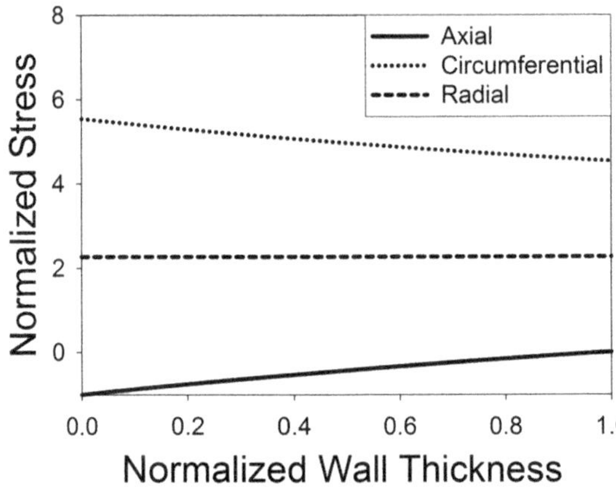

Figure 10.2 Pressure stresses, normalized to the internal gauge pressure of in the vessel, as a function of normalized wall thickness: $(r - r_i)/(r_o - r_i)$.

uniformly, with an identical thermal stress term $E\alpha_T\Delta T$ appended to each pressure stress term,[35] provided that stress equilibrium conditions are sufficiently met:

$$\frac{\partial \sigma_z}{\partial z} + \frac{\partial \sigma_{rz}}{\partial r} + \frac{\sigma_{rz}}{r} = 0 \tag{10.7}$$

$$\frac{\partial \sigma_r}{\partial r} + \frac{\sigma_r - \sigma_\theta}{r} + \frac{\partial \sigma_{rz}}{\partial z} = 0 \tag{10.8}$$

The thermal stresses were found at all points, with the value of ΔT being the temperature distributions shown in Figure 10.1, minus the initial temperature. The temperature distribution is nearly constant across the radius, and therefore the thermal stresses are also constant across the radius. Since the pressure stresses are maximum at the inside radius of the wall, the thermal stress at the inside radius were plotted as a function of longitudinal position, and are shown in Figure 10.3. As the figure shows, the thermal stresses are maximum at the center where the heat source is applied, and taper off towards the ends of the core holder. The equilibrium equations were monitored, and found to not exceed 10^{-9}.

10.4.2.3 Magnetic Stress

When placed in the strong magnetic field, B_0, the vessel experiences a force in the direction aligned with the field. This force, F_m, arises from the magnetic field gradient induced across the vessel when outside of the homogeneous region of the magnet, and is given by:[36]

$$F_m = \frac{\chi_m V}{\mu_0}\left(B_0 \frac{dB_0}{dz}\right) \tag{10.9}$$

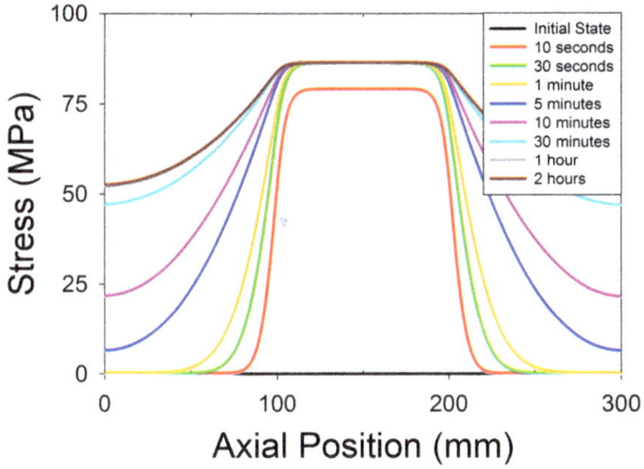

Figure 10.3 Thermal stresses at the inside radius ($r = r_i$), in the axial direction of the core holder, plotted at different time intervals after application of heat source.

where χ_m is the magnetic susceptibility of the vessel, V is its volume, μ_0 is the magnetic permeability in a vacuum and B_0 is the strength of the magnetic field. The value for $B_0 \frac{dB_0}{dz}$ is estimated as $B^2/1.8$ from ref. 36. Assuming the worst case scenario of a solid metal cylinder placed in the magnet (*i.e.* the largest volume), it is found that the magnetic force is 7.4 N, resulting in a corresponding stress of 6.5 kPa, which is four orders of magnitude lower than the yield strength of the material. Therefore, stress arising from the magnetic field is deemed insignificant and is omitted from calculations.

10.4.2.4 Endcap Considerations

The endcaps on the vessel are held in place with four bolts that prevent movement in the axial direction, which introduce stress concentration around the holes in the cylinder where the bolts intrude, and requires analysis of the bolts themselves. Since more than one screw is needed to secure the cap to the vessel, four equally spaced bolts were at each endcap. The bolts used were 5/8"-11 UNC threaded, with a corresponding cross-sectional area of 157 mm^2 (0.243 in^2). Each bolt is made from 316 stainless steel, a ductile material with a corresponding yield strength of 290 MPa. Failure of a bolt is caused by shearing from the axial force applied to it, determined by:

$$F_z = \Delta P A_{\text{endcap}} \tag{10.10}$$

where A_{endcap} is the cross-sectional area of the endcap, and ΔP is the difference between the internal and external pressures acting on the endcap. The resulting force acts as a transverse shearing force on the cross section of the bolt, given by:[37]

$$\sigma_{rz} = \sigma_{zr} = \tau_{\text{bolt}} = \frac{4F_{\text{zbolt}}}{3A_{\text{bolt}}} \tag{10.11}$$

where A_{bolt} is the total cross-sectional area of the screw. The stress τ_{screw} is found to be 4.25ΔP. Since the bolts are made from a ductile material and are experiencing shear stress, only ductile failure is considered.

The presence of each bolt also creates an area of stress concentration in the walls of the vessel: four equally spaced bolts were at each endcap. The stress concentrations are a localized effect, only present near the endcaps. There are two stress concentrations, one in the longitudinal direction that increases σ_z, and the other in the polar direction that increases σ_θ. The stress concentration is modeled as an infinite row of holes of radius r spaced L distance apart (*i.e.* modeling the cylinder as if it were unraveled). The value of L is the circumference divided by the number of holes, equal to 59.8 mm, and r corresponds to a hole for the screw size, 7.9 mm (5/8"). The stress concentration factors K_θ and K_z, in the circumferential and axial directions are given by:[38]

$$K_\theta = 3 - 1.061\left(\frac{2r}{L}\right) - 2.136\left(\frac{2r}{L}\right)^2 + 1.877\left(\frac{2r}{L}\right)^3 \tag{10.12}$$

$$K_z = 3 - 3.057\left(\frac{2r}{L}\right) + 0.214\left(\frac{2r}{L}\right)^2 + 0.843\left(\frac{2r}{L}\right)^3 \tag{10.13}$$

These relationships resulted in stress concentration factors $K_\theta = 2.6$ and $K_z = 2.2$. The stress concentrations are a localized effect, only present near the endcaps. Therefore, they are only applicable to stresses occurring at the ends of the core holder, and not the center of the core holder.

10.4.2.5 *Principal Stresses*

Since there are no shearing stresses along the radial, polar, and longitudinal axis of the considered pressure vessel, these axes are the principal axes. The combined principal stresses are then the summation of the pressure and thermal stress at each of these axes.

10.4.2.6 *Failure*

Nitronic 60 stainless steel exhibits both ductile and brittle behavior in its stress–strain relationship. There is a linear (ductile) stress–strain relationship until roughly half of the yield strength, then a dramatic non-linear (brittle) behavior prevails.[33] Therefore, both brittle and ductile failure modes were analyzed. Ductile failure was assessed using the maximum octahedral shear stress failure criterion, which indicates that yielding begins when the elastic energy of distortion reaches a critical value.[35] This criterion is also known as the maximum distortion strain energy criterion or the Von Mises method, which predicts that failure occurs when the octahedral shear stress,

τ_{oct}, exceeds the maximum allowable value, $\frac{\sqrt{2}}{3}Y$, where Y is the yield stress. Mathematically it is described by:[35]

$$SF = \frac{\tau_{oct}}{\frac{\sqrt{2}}{3}Y} \tag{10.14}$$

with

$$\tau_{oct} = \frac{1}{3}\sqrt{\left(\sigma_1 - \sigma_2\right)^2 + \left(\sigma_2 - \sigma_3\right)^2 + \left(\sigma_3 - \sigma_1\right)^2} \tag{10.15}$$

where σ_1, σ_2 and σ_3 are the principle stresses. The above failure criteria has to be applied at the point at which τ_{oct} is maximum, *i.e.* at the weakest point in the structure.

Failure by brittle deformation occurs when the strain energy per unit volume of any portion of the stressed member reaches the failure value of strain energy per unit volume. It is described as:[35]

$$SF = \frac{2EU_0}{Y^2} \tag{10.16}$$

where U_0 is the strain energy. This is equivalent to:

$$2EU_0 = \sigma_1^2 + \sigma_2^2 + \sigma_3^2 - 2\nu(\sigma_1\sigma_2 + \sigma_2\sigma_3 + \sigma_3\sigma_1) \tag{10.17}$$

Both the octahedral shear stress and strain energy were found to be maximum at $r = r_i$, where pressure stresses are maximum. Moreover, there are three possible locations for failure along the axial direction of the core holder. The first is at the center (hottest spot), where there are maximum thermal stresses due to the heat source application, and no stress concentration. The next is near the ends, due to the stress concentration created by bolt holes. The third location is within the bolts, which secure the endcap and prevent it from moving axially. Figure 10.4 shows safety factors at the three failure locations, as a function of the internal gauge pressure of the core holder. Below an internal pressure of 11.5 MPa, the center of the vessel is the weakest location, due to the dominance of thermal stresses, which are independent of the pressure stresses. Above 11.5 MPa, the stress concentrations in the ends overcome the thermal stresses and cause it to be the predicted failure location. In both cases, brittle failure is the predicted mode. With the standard pressure vessel and boilers safety factor of 3.5,[39] the maximum allowable pressure is 11.4 MPa (gauge).

10.5 Fabrication and Testing

The designed core holder was fabricated in-house, from a solid rod of Nitronic 60 stainless steel. Cutting was done at a low speed, to avoid phase transition from austenite to ferromagnetic martensite.[33] A cutting speed of

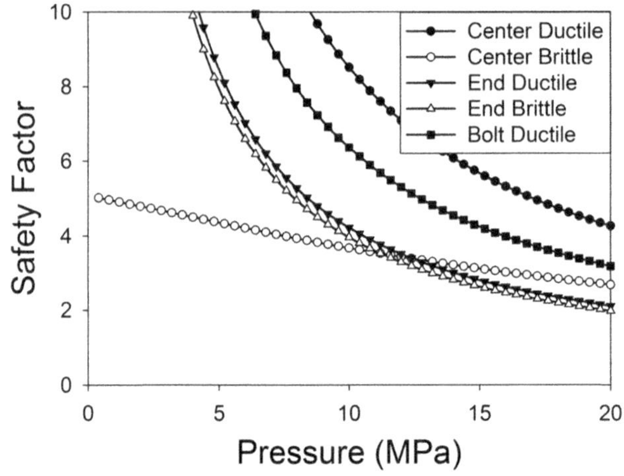

Figure 10.4 Safety factor plotted *versus* internal pressure ΔP of core holder, for the center of the vessel (hottest spot), the ends near the bolt holes (stress concentration) and within the bolts themselves.

40 rpm was employed, with a feed of 0.10 mm per revolution and a cut depth of 1.57 mm. The RF probe was fabricated from two concentric cylinders of glass fiber reinforced teflon, and M31CL epoxy. The RF probe resided in a cavity machined into the inner cylinder. The RF coil was made from copper and designed to create an RF field, B_1, perpendicular to the direction of the static magnetic field B_0.[22] The two cylinders were then etched and bonded together with epoxy. Figure 10.5 shows a cross-sectional diagram of the metallic core holder, and Figure 10.6 shows a photograph.

Tests were performed to verify that positioning a smaller RF probe inside of the metal vessel results in an increased signal-to-noise ratio (SNR), as compared with a non-metallic vessel using the RF probe outside of the vessel. This performance was tested by comparing the SNR of a basic free induction decay measurement for a water-saturated Berea sandstone sample in each probe. The RF probe outside of the non-metallic vessel had a diameter roughly twice that of the metallic core holder's, and employed a copper shield. It was found that the normalized SNR were 2.1 and 1, respectively. This increased SNR facilitates less signal averaging, to reduce measurement times.

To test the eddy current effect, the magnetic field gradient waveform was measured in the presence of the metallic core holder. This measurement determines the distortions from the eddy currents and how rapidly eddy currents decay. Eddy currents for the N60SS core holder were found to decay with a time constant τ that is 40 times shorter than that of aluminum.[23] This rapid eddy current decay indicated that the eddy current interference is minimal with the MRI process, and the metallic core holder is viable for imaging tests. To further verify this, imaging tests were performed on a water-saturated Berea sandstone, using a SPRITE method.[18] Figure 10.7 shows the resulting image, where no eddy current distortions are seen.

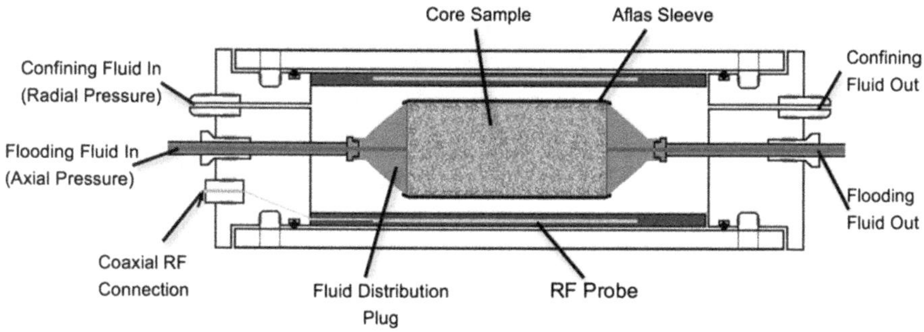

Figure 10.5 Cross-sectional diagram of a hydrostatic type core holder, used to maintain a rock core plug at elevated temperature and pressure conditions. The RF probe is placed inside of the core holder. A radial pressure is exerted by the confining fluid to create hydrostatic stress, while axial pressure causes flow through the core plug.

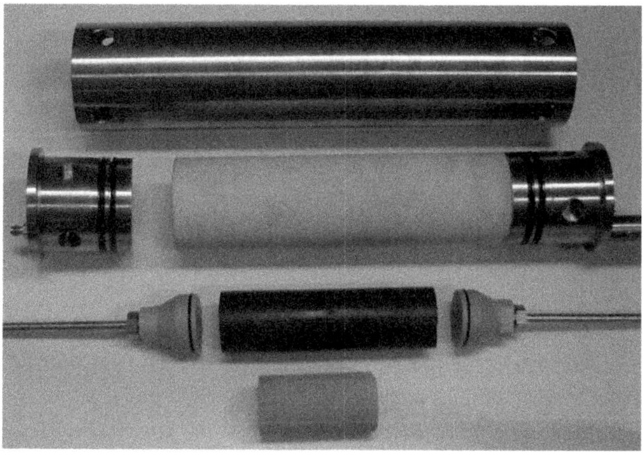

Figure 10.6 Photograph of MRI-compatible core holder fabricated from Nitronic 60 stainless steel. The core plug sample (bottom), is held adjacent to the two fluid distribution plugs by an Aflas sleeve (second from bottom), to make a connection to the inlet and outlet flow pipes. The encapsulated sample is positioned inside the RF probe (third from bottom). The entire assembly goes into the vessel (top). The vessel is sealed by o-rings on the endcaps, held in place by four bolts.

10.6 Imbibition Displacement Experiments

Once the core holder was deemed suitable for imaging tests, a constant flow rate dynamic displacement experiment was conducted, with brine employed as the wetting phase, and S20 standard oil and crude oil employed as the non-wetting phase. Berea sandstone core plugs, known to be homogeneous

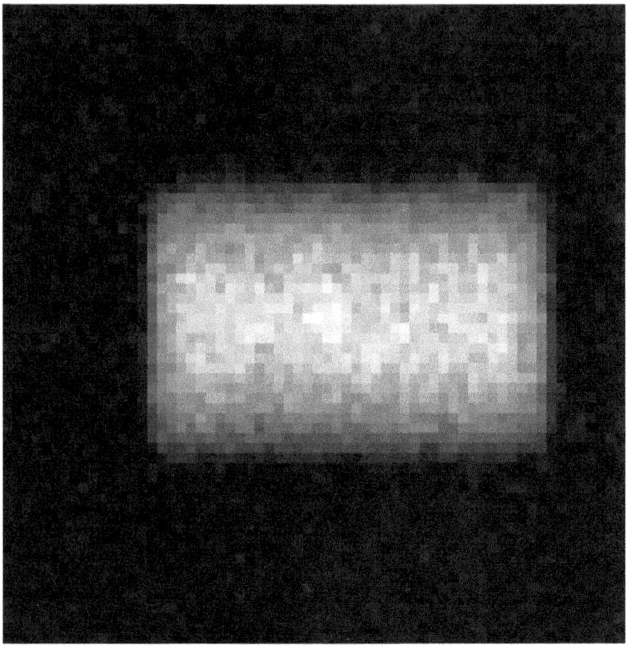

Figure 10.7 A 2D image of a Berea sandstone sample saturated with water to verify imaging capability due to minimized eddy currents. A SPRITE method is employed, with 128 scans, an encoding time of 150 μs, gradient step duration of 1.5 ms, and field of view of 80 mm × 80 mm. It took 24 min to acquire the image.

and initially water-wet, were employed in this study. The core plugs were 48.5 mm in length and 37.8 mm in diameter, with a porosity of 20.5% and an absolute permeability of 106 mD.

The core plugs were first cleaned, then dried for more than 24 hours in an oven at 100 °C to remove any residual fluid in the core plug. Drying ceased when the sample mass stabilized. After being fully saturated with brine (H_2O with 2% NaCl), measurements were performed to determine the maximum (100%) saturation profile. Saturation is defined as the fluid-filled pore-volume at a particular time relative to the initially fully saturated pore volume. Bulk measurements were also performed to obtain the MRI relaxation times, which dictate parameters used in the MRI process: $T_2^* = 821$ μs, $T_1 = 238$ ms, $T_2 = 167$ ms. To ensure that only signal from the oil phase was detectable during flooding experiments, D_2O was utilized to displace the H_2O water phase, since it does not produce an MRI signal at the resonance frequency of ^1H. The core plug was flooded with four pore volumes of D_2O brine, until there was no MRI signal from H_2O. To prepare the sample for imbibition displacement, a drainage experiment at a flow rate of 0.1 mL min^{-1} was performed with S20 standard oil (viscosity = 28.42 cP, density = 847.4 kg m^{-3}). The drainage process continued until no further wetting phase was produced.

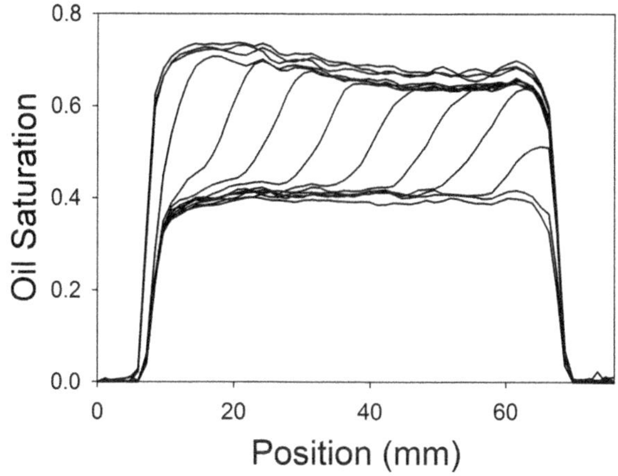

Figure 10.8 Imbibition displacement of an oil saturated Berea rock with heavy water, from left to right. A confining pressure of 2.1 MPa is used, the axial pressure varied to keep a constant flow rate of 0.1 mL min^{-1}. A SE-SPI method was used to obtain the images, each profile took 4 minutes to acquire, the time interval between each profile was 4 min. The gradient was applied with 0.3 ms ramp up and ramp down times, and 0.8 ms plateau, the field of view was 76 mm. Twenty time domain points were acquired for each echo, 256 echoes were collected, 2 scans.

Imbibition displacement was then performed at a confining pressure of 2.1 MPa, with injection of the wetting phase. Initially, only the non-wetting phase was produced at the outflow face. As injection continued, wetting fluid was produced when flow breakthrough occurred. Then, production of the non-wetting phase decreased but the wetting phase increased. Dynamic saturation profiles were measured at four-minute intervals by an SE-SPI imaging method, and are reported in Figure 10.8. The fluid production for both phases was monitored, residual oil saturation was reached when no further non-wetting phase was produced. From the saturation profiles in Figure 10.8, the unmovable water saturation can be determined as 28% at the beginning, and the residual oil saturation was found to be 38% at the end point.

The imbibition displacement experiment was repeated, in this case a back pressure regulator was used to maintain the outlet pressure at 0.35 MPa (50 psi). Figure 10.9 shows the results from this experiment. The shape of the profiles change due to higher pressures within the core plug. From the saturation profiles in Figure 10.9, the unmovable water saturation was determined to be 15% at the beginning, and the residual oil saturation was calculated as 31% at the end point. As expected, the saturation range of both fluids was extended with the application of a higher axial pressure, in comparison to the lower pressure of Figure 10.8.

Another Berea sandstone sample was tested, to acquire a series of 2D images. The Berea sample was 50 mm long, and 38 mm in diameter, with a permeability of 80 mD, and porosity of 21%. The core plug was first saturated with

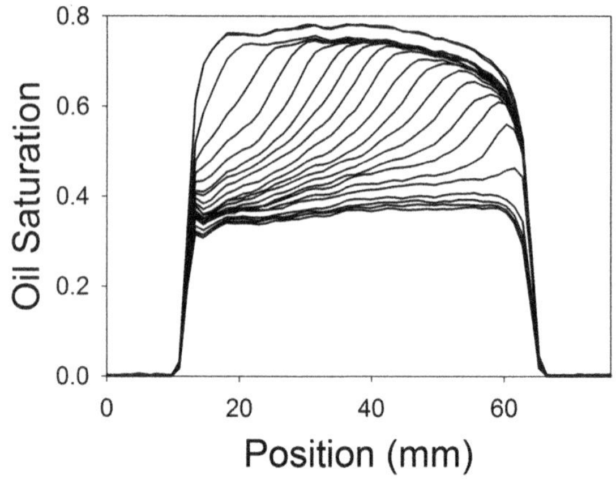

Figure 10.9 Imbibition displacement of an oil saturated Berea rock with heavy water, from left to right. A confining pressure of 2.1 MPa was used, the axial pressure varied to keep a constant flow rate of 0.05 mL min^{-1}. A back pressure regulator was used to maintain the outlet pressure at 0.35 MPa. A SE-SPI method was used to obtain the images, each profile took 4 minutes to acquire, the time interval between each profile was 4 min. The gradient is applied with 0.3 ms ramp up and ramp down times, and 0.8 ms plateau, the field of view is 76 mm. Twenty time domain points were acquired for each echo, 256 echoes were collected, 2 scans.

a high salinity D_2O brine (KCl 7.237 g L^{-1}, NaCl 4.267 g L^{-1}, $MgCl_2$ 0.320 g L^{-1} and $CaCl_2$ 2.180 g L^{-1}), by soaking the core plug in the brine for three days to reach equilibrium. The brine was then displaced with crude oil (aged at 85 °C and atmospheric pressure for 15 days). The oil saturated core plug was then flooded with a low salinity D_2O brine (the high salinity brine diluted 100 times) at a constant flow rate of 0.2 mL min^{-1}, and a confining pressure of 2.1 MPa. Images were acquired using a SPRITE imaging method, shown in Figure 10.10. The sample was held vertically in the magnet and flooded from bottom to top in the lab reference frame; the same flow direction is shown in Figure 10.10. Image intensity is proportional to saturation of crude oil in the core plug. The initial oil saturation was 82% and the residual oil saturation is 43%, after flooding for 510 minutes with D_2O brine.

10.7 Conclusions

A metallic MRI-compatible core holder was designed to safely withstand pressures up to 11.4 MPa and temperatures up to 50 °C, given the constraints of the MRI system. The core holder was manufactured from Nitronic 60 stainless steel because it was found to be a suitable MRI-compatible metal, minimizing eddy currents, which prevented imaging in previous studies with metallic vessels. It also has a high yield strength and higher thermal conductivity, beneficial for accommodating high pressure and temperature systems.

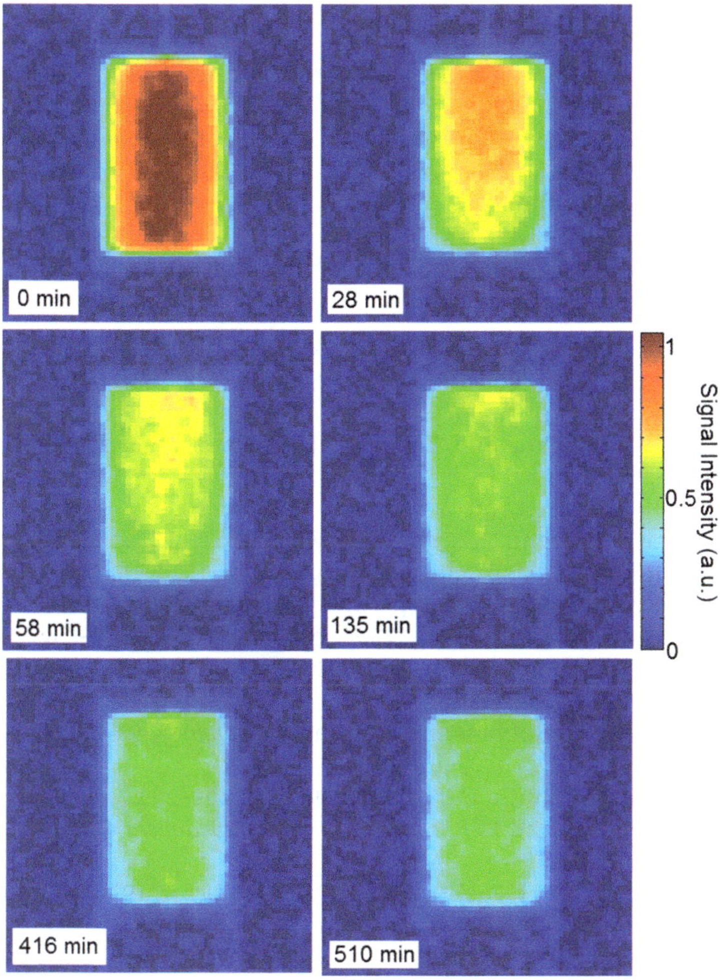

Figure 10.10 Series of two-dimensional images of a crude oil saturated Berea sandstone, flooded with low salinity D_2O brine. Intensity is proportional to crude oil saturation. The flow direction is from bottom to top of each sub-figure. A confining pressure of 2.1 MPa was used, and a flow rate of 0.2 mL min^{-1}. A SPRITE method is used to obtain the images. The field of view was 80 mm × 80 mm, the gradients were applied with a 1.5 ms duration. An encoding time of 150 μs was used, with 256 scans averaged.

Experiments were performed on various Berea sandstone samples, using a confining pressure of 2.1 MPa, flooding with water and oil. With the metallic core holder, one and two dimensional saturation distribution profiles and images during flooding were acquired, using SE-SPI and SPRITE imaging methods. These profiles and images spatially and temporally depicted fluid imbibition and drainage processes, from which residual oil and water saturations can be determined.

Acknowledgements

B. J. B. thanks the Canada Chairs program for a Research Chair in MRI of Materials (2009–2016), NSERC of Canada for a Discovery grant, the Atlantic Innovation Fund, Conoco Phillips, Saudi Aramco and Green Imaging Technologies. We thank R. MacGregor, B. Titus and J. Merrill of UNB for their assistance and suggestions.

References

1. F. Birol, *World energy outlook 2012*, International Energy Agency, 2012.
2. T. Guangming, D. Ogbe and D. Hatzignatiou, *SPE Form. Eval.*, 1995, **10**, 122–128.
3. J. Schembre, G. Tang and A. Kovscek, *SPE Reservoir Eval. Eng.*, 2006, **9**, 239–250.
4. J. Edmond, *Int. J. Rock Mech. Min. Sci.*, 1972, **9**, 161.
5. D. H. Gray, I. Fatt and G. Bergamini, *SPE J.*, 1963, **3**, 95–100.
6. V. Alvarado and E. Manrique, *Energies*, 2010, **3**, 1529–1575.
7. D. Wildenschild and A. P. Sheppard, *Adv. Water Resour.*, 2013, **51**, 217–246.
8. Z. Yang and P. Xiaofeng, *J. Eng. Thermophys.*, 2005, **26**, 850–852.
9. M. Tuller, C. M. Vaz, P. O. Lasso, R. Kulkarni and T. A. Ferre, *AGU Fall Meet. Abstr.*, 2010, D1013.
10. H. G. Chotas, J. T. Dobbins and C. E. Ravin, *Radiology*, 1999, **210**, 595–599.
11. F. de Beer, *S. Afr. J. Geol.*, 2006, **109**, 541–550.
12. C. Sayers and R. Smith, *Ultrasonics*, 1982, **20**, 201–205.
13. Q. Chen, F. Rack and B. J. Balcom, *New Tech. Sediment Core Anal.*, 2006, **267**, 193–207.
14. P. A. Osnent and K. J. Packer, *Philos. Trans. R. Soc., A*, 1990, **333**, 441–452.
15. K. Romanenko and B. J. Balcom, *AIChE J.*, 2012, **58**, 3916–3926.
16. M. Bernstein, K. King and X. Zhou, *Handbook of MRI Pulse Sequences*, Elsevier Academic Press, Burlington, 2004.
17. B. Blumich, *NMR Imaging of Materials*, Oxford University Press, Oxford, 2000.
18. M. Halse, J. Rioux, S. Romanzetti, J. Kaffanke, B. MacMillan, I. Mastikhin, N. Shah, E. Aubanel and B. J. Balcom, *J. Magn. Reson.*, 2004, **169**, 102–117.
19. L. Li, H. Han and B. J. Balcom, *J. Magn. Reson.*, 2009, **198**, 252–260.
20. J. Schenck, *J. Magn. Reson. Imaging*, 2000, **12**, 2–19.
21. D. Hoult, *Concepts Magn. Reson.*, 2000, **12**, 173–187.
22. J. Mispelter, M. Lupu and A. Briquet, *NMR probeheads for biophysical and biomedical experiments: theoretical principles and practical guidelines*, Imperial College Press, London, 2006.
23. H. Han, M. Ouellette, B. MacMillan, F. Goora, R. MacGregor, D. Green and B. J. Balcom, *J. Magn. Reson.*, 2011, **213**, 90–97.
24. F. Goora, H. Han, B. Colpitts and B. J. Balcom, *IEEE Trans. Magn.*, 2012, **48**, 2440–2448.
25. B. Benedek and M. Purcell, *Chem. Phys.*, 1954, **22**, 2003–2012.

26. J. Jonas, *Science*, 1982, **216**, 1179–1184.
27. J. Grunwaldt, R. Wandeler and A. Baiker, *Catal. Rev.*, 2003, **45**, 1–96.
28. R. Angel, N. Ross, F. Seifert and T. Fliervoet, *Nature*, 1996, **384**, 441–444.
29. I. Horvath and J. Miuar, *Chem. Rev.*, 1991, **91**, 1339–1351.
30. S. Lee, K. Mibe, Y. Fei, G. Cody and B. Mysen, *Phys. Rev. Lett.*, 2005, **94**, 165507.
31. R. Stoll, *The analysis of eddy currents*, Clarendon Press, Oxford, 1974.
32. J. Davis, *ASM Handbook Committee Metals Handbook*, ASM International, Ohio, 2008.
33. ARMCO, *Nitronic 60 stainless steel bar and wire UNS-S21800 product data bulletin, nitronic 60 stainless steel data sheet*, 1990.
34. H. Grober, S. Erk and U. Grigull, *Fundamentals of Heat Transfer*, McGraw-Hill, New York, 1961.
35. A. P. Boresi and R. J. Schmidt, *Advanced Mechanics of Materials*, John Wiley and Sons, Wyoming, 2004.
36. J. Schenck, *Med. Phys.*, 1996, **23**, 815–850.
37. W. Nash and M. Potter, *Strength of Materials*, McGraw-Hill, New York, 2011.
38. R. Pearson, *Stress Concentration Factors*, John Wiley and Sons, 1974.
39. K. Rao, *ASME Boiler and Pressure Vessel Code (BPVC), Section VIII Division 1*, American Society of Mechanical Engineers, 2012.

CHAPTER 11

Outlook: Quo Vadis, NMR?

EVA PACIOK[a] AND BERNHARD BLÜMICH*[a]

[a]ITMC, RWTH Aachen University, Worringerweg 1,
52074 Aachen, NRW, Germany
*E-mail: bluemich@itmc.rwth-aachen.de

11.1 Magnets

The component that dominates NMR instruments today is the magnet. Most of today's conventional magnets are made from type II superconductive wires of extreme complexity and fragility.[1,2] The difficulty with superconductive wire production lies in transforming the brittle ceramic high-temperature superconducting (HTS) metal oxides, such as $(Bi,Pb)_2Sr_2Ca_2Cu_3O_x$ (BSCCO) and $(Y,Eu,Er)Ba_2Cu_3O_{7-x}$ (YBCO, REBCO), or low-temperature alloy superconductors (LTS), such as NbTi and Nb_3Sn, into kilometre-long and thin, continuous and flexible films, whilst achieving maximum critical current density and critical temperature of the material, and minimum fault currents. The assembly of such wires into magnets must satisfy the conflicting requirements of magnet size and homogeneity on one hand, and magnetic field strength on the other hand. The general limitation is imposed by the critical current density the superconductive wire can carry, which decreases with rising temperature. State-of-the-art LTS wires in NMR applications carry 200–450 A mm^{-2} and operate at 4 K. The according magnets provide high sensitivity through high magnetic field strengths with flux densities up to 23.5 T along with wide chemical shift dispersion. However, such magnets require constant cooling with continuing supply of liquid cryogenic gases, in

New Developments in NMR No. 5
Mobile NMR and MRI: Developments and Applications
Edited by Michael L. Johns, Einar O. Fridjonsson, Sarah J. Vogt, and Agnes Haber

Published by the Royal Society of Chemistry, www.rsc.org

particular helium and nitrogen, and may deteriorate irreversibly, if cooling is interrupted and a quench occurs.

The increasing shortage of helium pushes the development of so-called 'cryogen-free' magnets. These are superconducting magnets, which still require cryogenic nitrogen and helium, but cooling is achieved by electrically driven heat exchangers, such as Stirling engines and pulse tube refrigerators,[3] or dilution refrigerators combined with gas-recompression.[4] Such systems operate without an active supply of liquid coolants, or even do not require any liquid cryogens at all, if HTS are used. With improving thermo-mechanical stability and spatio-temporal magnetic field stability, this technology is beginning to enter the fields of high-field NMR spectroscopy and medical imaging.[3,5] In NMR spectroscopy and particularly in protein research, the general effort is aimed at achieving increasingly higher magnetic field strengths. In this context, several groups pursue hybrid-systems, which comprise standard superconductive cryomagnets with the addition of 'cryogen-free' inserts or outserts.[6] If successful, magnetic fields up to 47 T may become available in the near future.

In this context, progress in HTS materials research is worth mentioning. In 2008, an organic, fullerene-based HTS superconductor was found with a critical temperature at 38 K,[7] as well as an iron-based superconductor, or oxypnictide superconductor, with a critical temperature at 56 K.[8] Both discoveries bear the prospect of inexpensive superconductors free of rare earth metals and with improved thermo-mechanical properties.

While mature technology exists to build superconducting electromagnets for NMR, the technology to build superconducting permanent magnets for NMR is still in its infancy, because of the difficulty to generate homogeneous magnetic field with these materials. A landmark development is the realization of a 4.7 T MRI magnet with *c*-axis oriented single-domain $EuBa_2Cu_3O_{7-x}$ crystal rings, which have a critical temperature of $T_c = 93$ K (Figure 11.1a).[9] The bulk superconductor is placed in a uniform field of another superconducting magnet and its temperature is lowered past the superconducting transition to trap the field. The magnet operates at around 50 K, which requires only a standard cryogenic refrigerator.[10] The ambient-temperature bore diameter is 23 mm, the other diameter of the vacuum chamber is 88 mm, and the field measured with an 8 mm diameter NMR tube had an RMS homogeneity of 3.1 ppm but a peak-to-peak variation of 37 ppm. This magnet has been shown to be suitable for imaging (Figure 11.1b). The development of such NMR magnets is exciting, because field strengths considerably higher than those of ferromagnetic rare earth magnets can be achieved with such small, permanent HTC magnets. Although the equipment is small enough for desktop operation, it is not yet portable at this stage of development, because the magnet must remain refrigerated at all times.

Portable MR magnets are made from permanently magnetic materials, which are ferromagnetic at room temperature. Commonly, rare-earth metal alloys, such as neodymium–iron–boron (NdFeB) and samarium–cobalt (SmCo), are used to build magnet arrays, which can generate field strengths

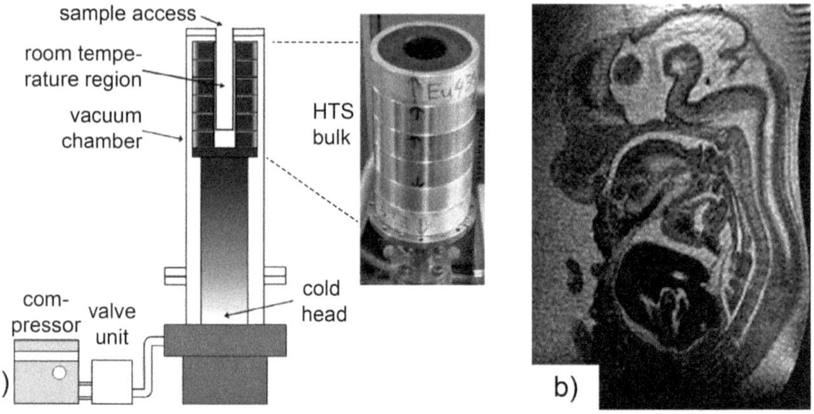

Figure 11.1 High-temperature super-conducting 4.7 T magnet for imaging, adapted from ref. 9 and 10. (a) Schematic of the magnet. (b) Sagittal image of a mouse embryo. Voxel size 50 μm.[3]

of up to 2 T.[11] These alloys are popular because the magnetic energy densities and the Curie temperatures are high, while they are comparatively inexpensive and readily available. Their production usually involves melting the crystalline compounds into ingots, milling of the ingots, sintering the resulting powder, cutting the sinter into shape, galvanic coating of the product, and magnetization *via* a strong electromagnetic pulse. Sintering is a diffusion-controlled process at temperatures higher than 1000 °C, and would require infinite time and repeated milling and sintering to achieve perfect product homogeneity and consistency. Because of this, the magnetic inhomogeneity of the product and material inconsistencies between product batches cannot be avoided.

Research and development of novel bulk permanent magnets has stagnated since the 1980s, when permanent magnet performance met the needs of state-of-the-art electronics, turbines and motors, and the cost of material improvement was seen to outweigh the further benefits gained in such applications. Instead, the focus of magnetic materials research has shifted towards organic magnets and magnetic nano-particles, and their respective applications in bio/medicine and home entertainment electronics. Therefore, the NMR community is unlikely to find stronger inexpensive permanent magnet materials in the near future, and has to keep combating low sensitivity by dedicated hardware designated to meet the needs of specific applications and samples, *e.g.* to maximize filling factors or find new ways of boosting NMR signal amplitudes.

Prominent examples for novel hardware from permanent magnets are small magnets, which produce homogeneous fields for spectroscopy. Different approaches are followed to cope with the field imperfections inherent to sintered rare-earth magnets, to arrive at the field homogeneity sufficient for further improvement by current driven shim coils. Apart from the magnet layout, a first homogenization step is achieved by iron pole shoes (Figure 11.2c and d),

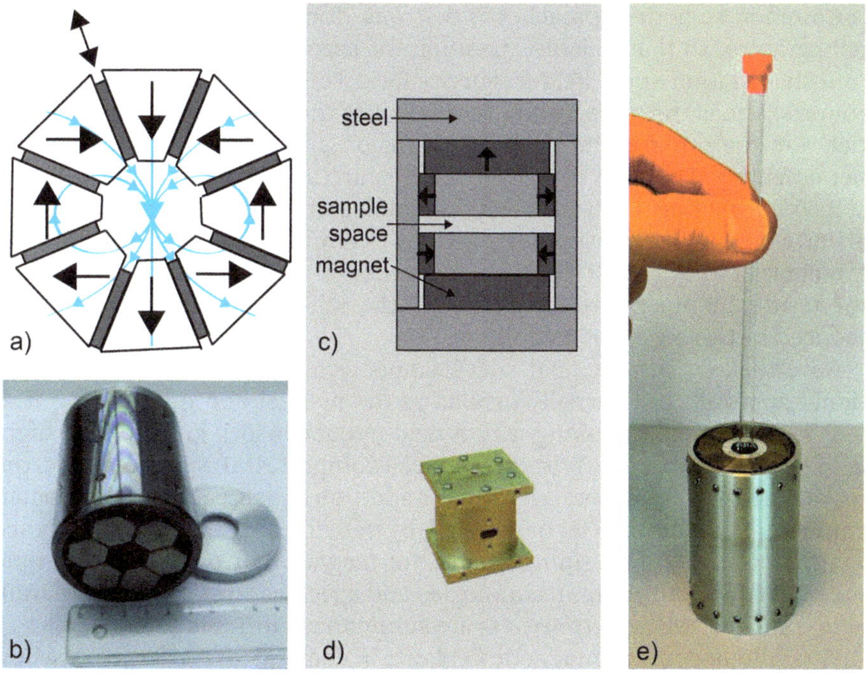

Figure 11.2 Magnets for compact NMR spectroscopy. (a) Principle of a Halbach magnet. Individual magnet blocks are arranged on a ring so that their polarizations trace the field of a dipole. Some magnets can be shimmed with slidable plates. (b) Simple realization of a Halbach magnet from hexagonal cylinder magnets. (c) Encapsulated hybrid magnet from permanent magnet material with pole shoes. (d) Photo of a miniature magnet for NMR spectroscopy with the size of a tennis ball (made by Aster), adapted from ref. 12. (e) Photo of a miniature Halbach magnet for NMR spectroscopy with mechanical shims.[13]

an established technology known from electromagnets, or by positioning the field-defining components of the magnet assembly in a controlled way with high precision (Figure 11.2a and e). The first approach has been taken to build the first small permanent magnet capable of measuring chemical shift-resolved spectra from solution in capillaries which were carefully placed in the region of highest field homogeneity (*cf.* Figure 1.3).[14–16] The second approach has been implemented with a modified small Halbach magnet (Figure 11.2b).[13] Either magnet assembly guides the field lines outside the sample region with magnetic materials so that the stray field outside the magnet is minimized and the field strength in the sample region is maximized. In the dipolar Halbach magnet, the magnet blocks are arranged in such a way that the field lines of a magnetic dipole are traced by the polarizations of the magnet blocks (Figure 11.2a). Such a magnet can be produced in a simple way from identical hexagonal or octagonal cylinder magnets (Figure 11.2b). A better ratio of sample volume to magnet volume, however, is obtained when using individually polarized trapezoidal magnets blocks (Figure 11.2a).

Common to both approaches is a strong drift of the field strength with temperature, so that reliable operation for spectroscopy requires temperature stabilization and a field-frequency lock. This field drift arises from two sources, one is the temperature coefficient of the permanent-magnet material, and the other is the thermal expansion and contraction of the magnet assembly. The field drift from both sources has been suppressed in a self-compensating Halbach array assembled from two magnetic materials arranged with opposite polarizations.[17] While the thermal field drift of the conventional NdFeB Halbach magnet is of the order of 2×10^{-4} T K^{-1}, the residual temperature drift factor of the self-compensating assembly is reduced by two orders of magnitude.

Nevertheless, current field stabilization technology relies on the traditional approach of thermally insulating the magnet and maintaining it at a defined temperature along with a field-frequency lock known from high-field NMR spectroscopy. With commercial compact NMR spectrometers, the lock is either internal, necessitating the addition of deuterated or fluorinated compounds to the analyte, or external, by deriving a signal from a reference sample permanently residing inside the magnet, but outside the sample region. Field stabilization technologies and specifications of currently available compact NMR spectrometers are summarized in Table 11.1.

While homogeneous magnetic fields are in demand for NMR spectroscopy, inhomogeneous fields with uniform gradients are in demand for imaging. In volumetric "inside" magnets, gradient fields are mostly produced by pulsing currents through sets of gradient coils. With single-sided "inside out" magnets, gradient fields are usually time-invariant. The first stray-field imaging (STRAFI) applications made use of a plane outside super-conducting magnets, in which the field gradient is constant and high, of the order of several tesla per meter, which promised challenging applications to solid-state imaging.[18,19] This strong gradient was also employed to study slow diffusion,[20,21] and even a superconducting magnet was constructed for that purpose having a linear field with a high gradient.[22]

Subsequently, permanent magnets came into use for well logging NMR, where relaxation and diffusion are measured from objects outside exposed to the stray-field of the device, and the gradient is used for slice selection.[23–25] These principles were adopted for the NMR-MOUSE,[26] a compact stray-field sensor with a high gradient of the order of 20 T m^{-1}, which has subsequently been refined for imaging,[27] spectroscopy,[28] and high spatial resolution[29] by shimming not the stray field but its gradient to homogeneity or uniformity in a defined volume distant from the sensor surface. Other designs of compact gradient magnets are the GARField magnet,[30] the NMR-MOLE,[31] and the planar Halbach magnet.[30] These designs are summarized in reviews[32–34] and books[11,35,36] together with applications to well logging, materials testing, chemical engineering, biomedicine, and cultural heritage.

Although stray-field NMR methodology for materials research is rather well established, its implementation in compact sensors for materials testing is still relatively new. On the other hand, compact NMR employing

Table 11.1 Overview of commercial bench-top spectrometers and manufacturers' specifications.

Model	^{1}H freq.	Resolution[a] (ppm)	Sample	Nuclei	S/N	Lock	Weight[e] (kg)
Fourier 60	60 MHz	0.03	5 mm NMR tube	^{1}H, ^{19}F, ^{13}C	500[b]	Internal, external	145
NMReady	60 MHz	0.03	5 mm NMR tube	^{1}H, ^{19}F, ^{31}P	Not specified	External	30
picospin 45	45 MHz	0.04	30 μl injection	^{1}H	500[c]	Not specified	5
picospin 80	82 MHz	0.02	40 μl injection	^{1}H	Not specified	Not specified	19
Pulsar	60 MHz	0.02	5 mm NMR tube	^{1}H, ^{19}F	Not specified	Not specified	170
SpinSolve	43 MHz	0.016	5 mm NMR tube	^{1}H, ^{19}F, ^{31}P, ^{13}C	100[d]	Internal	55

[a]Non-spinning resolution.
[b]Single-scan signal-to-noise of 10% solution of ethyl-benzene in deuterated solvent.
[c]Single-scan signal-to-noise of pure water.
[d]Single-scan signal-to-noise of 1% ethyl-benzene in deuterated solvent.
[e]All presented systems combine magnet and spectrometer in one compact device of specified weight.

closed permanent magnets with moderate field homogeneity (Figure 11.1b and d) has a long history, and enjoys great popularity in applied materials research.[34,36,37] For a long time this methodology has been referred to as time-domain NMR or low-field NMR, because NMR instruments with such closed magnet geometries are being used to a great extent without spectroscopic resolution in the food industry to study emulsions, suspensions and gelation. Such magnets are also employed in dedicated MRI scanners for a variety of applications,[38,39] and they were the first to be miniaturized for use in single-chip spectrometers with lab-on-a chip technology for sample preparation.[40–42]

11.2 NMR-on-a-Chip: Miniature Magnets, Coils, Spectrometers and Microfluidics

In NMR, miniaturization does not merely denote the reduction of magnet size. If full mobility and personalized applications are the goal, the optimal NMR instrument, including magnet, detector and spectrometer, should be palm-sized or smaller, should require minimum sample volumes, and minimal power consumption. Miniaturization of NMR hardware has made significant progress in recent years, greatly benefiting from the advent of microfluidics and from the advances in micro-fabrication and lab-on-a-chip (LOC) technology that microfluidics have brought along. Microfluidic platforms provide the means for precise, loss-free handling and processing of micro- to nano-liter sample volumes. LOC operations include mixing, dispersion, separation, fractionation, heterogeneous catalysis, biomedical incubation, and optimal process control *via* tailored flow and stress fields, along with optimization of wettability, residence times, heat exchange and electronic actuation. The precision, efficiency and versatility of such platforms has not only revolutionized biochemical engineering, but has paved the way for micro total analysis systems (μTAS[43]) capable of separating and analyzing fluid or tissue samples into their individual components, or housing cell cultures for biomedical multi-assays.

The original application of microelectronics in LOC technology lies in the engrafting of microfluidic chips with microelectrodes as thermal or electro (osmotic) sensors or actuators. The phenomenal performance of such devices pushed the development of standardized microfluidic microcoil fabrication techniques, and following the pioneering works of Kakuta *et al.*[44] and Wensink *et al.*,[45] who first introduced ultrasensitive microcoil detection on LOC platforms into the NMR community. Implementation of miniature NMR sensors for LOC devices has developed quickly into its own field of research, bringing forth a large variety of new possibilities for spectroscopy and imaging, making use of detection schemes unconventional to standard analytic NMR spectroscopy, such as remote detection,[46] multi-phased arrays[47] and wireless detection,[48] to name a few.[49–51]

While in high-field NMR the use of microcoil technology serves the improvement of detection sensitivity, in low-field applications the low microcoil inductance impairs tuning to low frequencies, unless large capacitance or sacrificial 'tune' inductance are added to the rf circuitry, augmenting resistive losses. As a result, for some time microcoil detection was considered unsuitable for low-field NMR applications, until Sillerud *et al.*[14] were able to demonstrate that theoretically, arbitrarily small microcoils may be inductance-tuned to low detection frequencies without significant losses in signal-to-noise ratio, facilitating low-resolution NMR spectroscopy in low-field NMR microfluidic setups. Simultaneously, they introduced the concept of low-field NMR microfluidic microcoil biosensing of superparamagnetic nanoparticle-tagged biomarkers, in analogy to fluorescence tagging in biomedical assays and applications in AC SQUID biosensing, and its potential for point-of-care applications.

Nanoparticle-tagging relies on antigen–antibody interactions, in the form of either permanent binding or reversible ligand affinity binding, with nanoparticles carrying antibodies that specifically bind to distinct antigens (biomarkers), *i.e.* substances, cells or organisms characteristic for a certain pathogenic, metabolic, genetic or cancerous disease, or even malnutrition. Upon binding, the nanoparticle paramagnetism not only accelerates relaxation of bound biomarker proton signals, but also several nanoparticles linked by biomarkers may agglomerate (clot) and drastically enhance relaxation in all protons in the interstitial space. Both mechanisms facilitate highly sensitive *in vitro* biomarker detection without the need for sophisticated high-resolution spectroscopy in large-scale NMR devices (so long as the magnet used has sufficient homogeneity). Moreover, the specificity of antigen–antibody binding allows for biosensing in largely untreated bodily fluids and fluidized tissue samples, where a large variety of chemically quite similar compounds may be present. If the nanoparticle loading with biomarkers is well defined and constant, flow-through measurements with counting of nanoparticle-tags even allow for quantitative biomarker assessment. The aspects of laymen user-friendliness (no large spectrometers, no spectroscopic analyses) and foolproof specific yes/no detection, or even quantitative measurement, of highly complex compounds in highly complex media, have vast potential for personalized medicine and point-of-care applications in combination with other analytical technology.[36] NMR may become an integral part of devices for medical home diagnostics, complementing the long established fever thermometer and the more recent blood sugar testing devices for diabetics. As a matter of fact, the pioneering work by T2 Biosystems has made desktop NMR biosensors specialized in sensing single distinct biomarkers a valid option in medical diagnostics.

But progress has not stopped at desktop-sized specialized applications. In their remarkable work, Haun *et al.* have extended NMR biosensing to microfluidic multiassays for clinically proven reliable cancer diagnosis, using even smaller magnets.[42] These multiassays are capable of detecting several cancer markers at once, representing a significant step towards NMR-µTAS. This approach does not rely on spectral analysis, but on splitting a sample into sub-volumes which are individually exposed to different biomarker antigens. Moreover, Haun *et al.* have demonstrated that a state-of-the-art mobile phone can easily be modified to operate the NMR biosensor and to analyse the measurement results. In terms of reduction in size and power consumption, this biosensor was only excelled by the consecutive work by Sun *et al.*[52] By use of micro-coil detection on a microfluidic platform in combination with ultrasensitive rf transceivers and CMOS integrated circuits, all mounted on a single chip, they have created the world's smallest NMR biosensor to date, which in its entirety, comprising magnet, rf circuitry, amplifiers and spectrometer, is less than palm-sized and weighs only 100 g, which is less than the weight of an apple.

With the current state of hardware miniaturization NMR has reached a point at which the futuristic StarTrek®Tricorder is merely a stone's throw away. In that context, it is not surprising that companies, such as QualComm have started open competitions for funding the development of Tricorder-like devices for medical healthcare. NMR is envisioned to become an integral part of such a device. By further improvements of spectrometer electronics towards lower noise figures and lower power consumption, an NMR biosensor could be created that is not only operated but also driven by a mobile phone. In addition, single-sided NMR could provide the means for non-destructive *in vivo* testing for biomarkers *via* exogenic non-toxic contrast agents, injected into the patient to target biomarkers released into the blood, if sensitivity can be boosted to allow biomarker detection in subcutaneous blood vessels. Without any need for taking and preparing samples, the gain in time could even shed new light on rapid processes and metabolic changes during cramps, attacks and seizures associated with various conditions and diseases, seeing as the contrast agent could be injected immediately at the onset and would be distributed in the blood stream almost immediately.

Nevertheless, one general limitation of the antigen–antibody recognition/detection principle will always remain: Regardless of the detection method applied, antibodies can only be synthesized to target identified biomarkers, and finding the correct antibody for targeting a certain newly found biomarker will always involve *ex vivo* or *in vitro* trial-and-error combinatorial laboratory assays (see-what-sticks approach). As a result, antigen–antibody recognition will never allow *in vivo* discoveries of novel pathogens, metabolites or toxins, and associated conditions, diseases *etc.*

11.3 Methods

If one sets out to develop an NMR device capable of detecting any change in the genome, transcriptome, proteome or metabolome *in vivo*, while simultaneously being capable of distinguishing significant perturbations indicative of malconditions, from the background noise of regular variations due to daily rhythm, diet, physical activity, stress, sleep *etc.*, then such a device would have to include total analysis *via* NMR spectroscopy, aided by mathematical analysis. The methodology of finding and cataloguing biomarkers (metabonomics) and quantitatively detecting indicative perturbations (metabolomics), mainly in biofluids (blood, lymph, cerebrospinal fluid, urine, sputum and saliva), has been well established in *ex vivo* and *in vitro* high-resolution high-field 1D and 2D NMR spectroscopy (with water suppression), to an extent where data bases of pure metabolite NMR spectra and even malcondition-indicative metabolite profiles are waiting online to be matched to a patient's metabolic NMR spectrum.[53–55]

Equally, the mathematical treatment and analysis of the multivariate biofluid spectra is far advanced. It includes methods for binning of spectral regions and peak-alignment to account for chemical shift displacements between samples due to variation in pH, dilution, viscosity, and salinity. Multivariate analysis, pattern recognition and pattern matching to database profiles mainly rely on principal component analysis (PCA),[56] which, in simple terms, decomposes complex spectra into simpler sub-spectra sorted by magnitude of variance with respect to some variable, such as time, patient's age, or some demographic indicator. PCA has certain limitations as it presumes a linear correlation between variable and observable, and it underestimates low-intensity peak variance, especially in noisy data. Several normalization techniques and non-linear approaches similar to PCA have been successfully demonstrated to account for these issues, and at the current state, mathematical analysis of NMR metabolomics certainly has reached a stage at which compiling the spectral binning or peak-alignment, the multivariate analysis and pattern-matching with online metabolic profiles, would yield a robust, unsupervised and laymen-proof method targeted at detection and monitoring a limited selection of metabolomically well characterized diseases or conditions. Nevertheless, the step towards home applications is speculative at this point, because the sensitivity and resolution of large-scale high-field NMR spectroscopy has not yet found an equal match in inexpensive portable NMR. However, if the improvements in desktop and portable NMR performance continue at their current rate, general home applications in personalized medicine can be expected to become available within this century.

Above all, a further step towards single-sided *in vivo* NMR metabolomics in subcutaneous blood vessels, *i.e.* the NMR Tricorder, is imaginable. Even if the issue of low spectral resolution due to permanent magnet field inhomogeneity finds no solution, the highly homogenous earth's magnetic field may be utilized for this purpose. Here, low sensitivity is a major problem that cannot be addressed by microcoil detection alone. Current developments focus on novel ultrasensitive detection methods, such as optical or atomic magnetometry.[57] These sensors, based on, for instance, vapor cells, NV-defects in diamonds, amorphous wire magnetic impedance sensing, laminated magnetoelectric materials, or superconducting quantum interference devices (SQUIDs),[58] are most capable of sensitive detection in the earth's magnetic field. Their phenomenal sensitivity also facilitates, beyond the detection of nuclear magnetic resonance, magnetoencephalography (MEG), which is sensitive to nervous activity and is utilized in monitoring and imaging of brain, cardiac and gastrointestinal activity. Currently, large effort is put into the miniaturization of magnetometers, which are now (less than) chip-sized,[59–61] and the first desktop-sized magnetometer for small-animal MRI and MEG, based on a miniature AC SQUID, has been introduced recently by Oyama *et al.*[62] When perfected, this technology could revolutionize *ex vivo* desktop or portable NMR spectroscopy and MRI, especially when combined with thermal or electromagnetic prepolarization of samples, the latter of which could easily be miniaturized to nearly arbitrary size.

Subcutaneous *in vivo* spectroscopy in the earth's magnetic field, utilizing either microcoils or magnetometers, could moreover be greatly enhanced by use of the hyperpolarization techniques. Hyperpolarization denotes the defiance of thermal equilibrium polarization, which at ambient conditions and particularly in earth's magnetic field (0.048 mT) is extremely low. Current research in hyperpolarization follows the biosensing approach, and protocols for biomarker hyperpolarization have been established for many important metabolites and contrast agents.[63] In this context, another important aspect is the search for long-lived nuclear spin states to prolong hyperpolarization lifetime in *in vitro* and *in vivo* applications.[64–67]

There are three basic approaches to achieving hyperpolarization. One involves the optical spin excitation of alkaline metal vapour and spin-transfer to large noble gas molecules, commonly ^{129}Xe. ^{129}Xenon, itself having a spin ½ nucleus, can be detected with chemical shift response to its physicochemical environment, but has limited significance/applicability to spectroscopy. Hyperpolarized xenon finds wide application as a gaseous or dissolved contrast agent in medical imaging. By trapping the hyperpolarized gas in cryptophane cages that carry biomarker antigens and injecting those into an organism, the xenon is made to accentuate certain organs or tissues.[68] Dynamic nuclear polarization (DNP) refers to polarization transfer from radical electrons to surrounding protons, ^{13}C or ^{15}N *via* microwave irradiation, assisted by/or spin diffusion processes. DNP is primarily utilized in solid state spectroscopy, and with exogeneous agents it has little to no relevance to biomedical applications, seeing as there are few non-toxic stable radicals available and microwave irradiation is harmful to patients. Current research focuses on chemical or dissolution DNP of endogeneous substances with sufficient relaxation times, *i.e.* ^{13}C or ^{15}N species such as sugars and amino acids, to not only be introduced as contrast agents, but to be metabolized thus highlighting metabolic pathways in living organisms.[69] Finally, there is *para*-hydrogen-induced hyper-polarization (PHIP).[70] This methodology arises from spin-order transfer from *para*-hydrogen to a target molecule. *para*-Hydrogen is a spin-isomer of hydrogen that exists in minuscule amounts at ambient temperature, but may be enriched cryogenically. The spin-order of *para*-hydrogen may be transferred to target molecules by means of hydrogenation reactions or catalyst-mediated spin-order exchange, the first being limited to unsaturated compounds, the latter to nitrogenous targets.

All hyperpolarization methods known to date bear to some extent both the benefits and the disadvantages of target specificity or at least selectivity. The specificity/selectivity has triggered extensive (bio-)sensor research and the development of many novel non-toxic contrast agents. The accumulation of many small successes in these fields has pushed groups and companies world-wide towards the miniaturization of hyperpolarization hardware, though currently we are witnessing merely the beginning of this movement. The disadvantage is the lack of non-selective broad-band signal amplification irrespective of chemical constitution. Of all the methods presented, PHIP can potentially be developed into a broad-band approach, if a non-selective catalyst can be found, which currently is a prime focus of PHIP research. Generally speaking, hyperpolarization technology has not yet reached the stage of becoming an universal and safe, fully laymen-proof technique – partially due to lack of reproducibility of results in some cases, partially due to the sophistication of the method, but mainly due to the costliness and hazardousness of the gases or catalysts involved. Both will be reduced drastically, if miniaturization continues and effectivity keeps on improving.

The last but by far not least point to discuss is the progress and future of NMR in genomics and proteomics. Metabonomics and metabolomics provide a platform mainly for identifying and fingerprinting of diseases, and for statistical analyses on the diagnostic side of medicine and patient care. Proteomics and genomics delve deeper, trying to unravel pathways of metabolism and diseases by elucidating biophysics on the molecular level.[71] Here, the exploration of biomolecular structures, conformations and associated energy-landscapes of physico-chemical interaction is of main concern. Compared to structural analyses of solid crystals by scattering methods such as X-ray diffraction, neutron diffraction or electron beam diffraction, NMR has the enormous advantage of non-invasive, non-destructive exploration of structure, conformation, dynamics (folding) and functionality in an aqueous environment. With this wealth of information NMR has made a significant contribution to the understanding of the function of many membranes, receptors and enzymes, *i.e.* the very building blocks of life, and the origin of life on earth itself.[71,72] Notwithstanding, only a minuscule fraction of the inner workings of the human body has been explored yet, and particularly Alzheimer's disease, AIDS and prions are in the focus of current biophysical research.[73]

NMR structural analyses rely on multidimensional measurements to capture multiple quantum coherences which reveal intramolecular connectivity of atoms or their spatial proximity. Depending on the range of connectivity/proximity and the atomic nuclei to be probed, such multidimensional measurements hitherto have lasted hours to days, or even weeks. This not only limited the timescale on which dynamics could be observed, but equally could lead to degradation and decomposition of the system under investigation. Recently, however, fast multidimensional spectroscopy methods have been introduced that boost sensitivity or speed up acquisition by, for instance, employing parallel receiving technology,[74] sparse sampling of data in combination with appropriate reconstruction of missing data,[75] novel pulse sequences,[76] or dynamic nuclear polarization.[77] Roughly speaking, each of these approaches has reduced measurement time by approximately one order of magnitude and macromolecular dynamics on the timescale of 20 minutes may now be studied in real time. Currently, these methods are applied separately, and we will witness a renaissance in genomics and proteomics if all are combined and standardized for high-resolution multidimensional spectroscopy.

11.4 Applications Beyond Spectroscopy

Shifting the focus from spectroscopy and imaging methodology largely applied in the life sciences, there is a wealth of NMR applications, and particularly low-field NMR applications, in materials research and testing.[32,33,36] They range from on-line quality control in industrial processing, cosmetics and food processing,[78] to research in building materials, wood, paper and cultural heritage,[79,80] well logging,[35,81,82] geohydrology[83] and plant physiology.[78,84–86]

Starting with plant physiology, high-field NMR spectroscopy is a large asset in the research of plant communication and stress response on the one hand, and mass screening for metaboloma and novel drugs, on the other.[85,86] These areas of research have greatly benefited from advances in human NMR metabolomics, and just as much, they would certainly gain from implementation of biosensors and targeted hyperpolarization, in terms of improved sensitivity and real-time monitoring of metabolomic pathways. In the contexts of geohydrology, plant cultivation and industrial farming, MRI has become essential for the monitoring and understanding of plant growth, fruiting and plant hydraulics in optimal and adverse conditions, and for the elucidation of the root structure architecture and its role in soil water uptake and soil hydraulics.[78] Imaging experiments are commonly performed in the laboratory, where plants are removed from their natural environment. Recently, however, several groups have introduced mobile plant tomographs for *in situ* studies of plant water transport and fruiting,[84,87,88] which, if the development continues and spectroscopic performance improves, could be used for *in situ* plant metabolomics.

NMR surveys in industrial processing, well logging, geohydrology, the building and food sector, as well as in cultural heritage, almost exclusively rely on specialized, single-sided low-field NMR sensors, either installed permanently or portable. One-dimensional experiments provide, for instance, a sample's fluid content and moisture distribution, or a sample's relaxation and diffusion signature. In multi-dimensional experiments, correlations between longitudinal, transverse relaxation rate constants and diffusion coefficients are found that can be studied on different timescales. Depending on the application, these signatures in turn are used to identify and quantify individual compounds in liquid or solid mixtures, or they are translated into structural parameters, *e.g.* the porosity and permeability of porous media.[89] Particularly in geohydrology and well logging, noise represents a major obstacle that cannot be solely overcome by hardware design and ultra-sensitive detection alone. Here, data mining techniques, such as principal component analysis, are used to distinguish significant data from noise, and the community continues to surprise with improved methods for adaptive or Bayesian-predictive smoothing of noisy data.

Regarding hardware and methodology, NMR well logging is well established in petrophysics, and advanced, robust logging tools are commercially available. Recently, a prototype of the downhole NMR laboratory has been presented with improved capabilities for underground chemical analyses of fluids. Future developments in the logging sector are likely to be of a pragmatical nature, further optimizing the noise-performance of NMR and the inversion algorithms that are used to derive relaxation and diffusion signatures from NMR data. With improved noise-performance of logging NMR, we may look forward to smaller logging tools with increased vertical resolution, and versatile underground chemical analyses of fluids, as recently introduced by Wu *et al.*[90]

While NMR is an established method in industrial logging, its entry into civil engineering, geohydrology and soil hydraulics is relatively recent. Commercially available logging devices are readily applied to groundwater logging at sufficient depths, if high vertical resolution is not required.[74] Alternatively, large surface coil assemblies can be used to investigate groundwater across large areas.[91] Soil hydraulics in the vadose zone, *i.e.* the part of the earth's crust above the ground water level, are assessed with logging tools that are significantly shorter and employ smaller detection coils.[92] This allows for higher vertical resolution in investigations of soil water transport and soil typing of layered structures. Apart from field applications, NMR has played a significant role in soil organic matter analysis in laboratory studies[93] with primarily employing spectroscopy and solid-state NMR methods.

From the academic viewpoint, the key mathematical and physical achievement, not only in logging, but also in NMR research of porous media in general, would be to gain an analytical understanding of relaxation and diffusion signatures measured in porous media, particularly in chemically and structurally heterogeneous porous media only partially saturated with fluid. This would advance NMR relaxometry from a qualitative science into an *in situ* quantitative porosimetric tool, which, especially when relaxation is correlated with diffusivity or, if possible, chemical shift, would obliterate the common invasive methodologies of measuring sorption isothermals or mercury porosimetry. Equally, pore-space tomography and thin-section imaging may become obsolete if NMR porosimetry proves sufficient to characterize complex porous structures.

The concepts of surface and volume relaxation as implemented in the pioneering 'particle in a box' continuum approach by Brownstein and Tarr[94] have been known for decades and have been applied to manifold systems too numerous to cite. The Brownstein and Tarr model predicts the effective relaxation of fluids trapped in confined spaces, originally biological cells, from the intrinsic relaxation rate and diffusivity of the bulk fluid on the one hand, and on the other hand, from characteristic structural parameters of the confining geometry and the surface relaxivity of the confining material. This model, as commonly applied to porous media, relies on uniform surface relaxivity and free intra-pore diffusion within fluid-saturated spherical or cylindrical pores (capillaries), each being an isolated continuum. Pore throats are considered to be small pores that do not act as conduits. Pore sizes derived using this model are ambiguous, as two or more coupled pores may be mistaken for one large pore. At low fluid saturation, large pores in turn may be mistaken for small ones. Effects of internal magnetic field gradients are neglected altogether. In addition, several groups report anomalous surface diffusion in aqueous, silicate-based porous media that may or may not become dominant at low fluid saturation.[95–97] Nevertheless, valuable information may be derived from NMR relaxometry regardless of absolute pore sizes. For instance, kinetic exchange models have been presented[98,99] to derive exchange rates of fluid between sites with distinct T_2, with the exchange rates relating to pore coupling.[98] Building on that, first steps have been taken towards the discretization of continuous T_2 distributions and parametrization to yield separated micro- and macro-pore diffusion and advection numbers.[100]

The Brownstein–Tarr model is universal in the sense that principally any porous medium may be discretized and parametrized to accommodate multiple, potentially coupled, pore geometries, multiple surface relaxivities, and any diffusion regime. Effects of susceptibility-induced internal magnetic field gradients, although originally not considered within this model, can be accounted for as well. With the increasing complexity of the model, the continuum approach has made way to discrete, pore-space simulations[101,102] that, for instance, use Lattice Boltzmann methods to emulate molecular random walks and can take into account surface wettability. Equally, molecular dynamics simulations have evolved into a reliable tool for the analytical prediction of mass transfer and even NMR signatures of fluids in well-defined nanoporous systems, such as zeolites and metal-organic frameworks, and surface diffusion has been addressed as well.[95,103] Furthermore, first attempts at upscaling pore-scale characteristics, as assessed by NMR, to macroscale material properties[104] have been made using the well-established pore-network models, which abstract a porous medium to a network of nodes and throats (for a comprehensive review of porous media modelling see ref. 105).

Seeing as more and more computational methods are adapted from the porous media modelling community, NMR model equivalents for multiphasic flow or reactive flow will surely follow. If successful, the knowledge gained would be invaluable to understanding and optimizing porous media processes in geohydrology, the building sector and chemical engineering, *i.e.* catalysis, filtration and mixing. The optimization of mass and heat transfer in novel nano-materials, such as metal-organic frameworks (MOFs) or nanostructured electrodes and polyelectrolytes, would greatly benefit from advanced porous models founded on NMR data, seeing as NMR can provide and correlate both chemical and mass transfer information.

11.5 Summary

In general, NMR is already an essential tool in standard laboratory analytics, metabonomics and metabolomics, medical imaging, logging and many areas of materials research and testing. Taken to extremes, one may say that NMR has moved into any field, but not yet the home sector. But in truth, this is only true for first-world countries.

Figure 11.3 Conceptual drawing of a point-of-care NMR analyzer. The palm-sized device contains microfluidic actuators driven by miniature pumps and fluid storage tanks, as well as NMR hardware and a data processor with WiFi uplink to a metabolomics database. The disposable cartridge holds a small sample of blood or saliva. When inserted into the device, microfluidic operations prepare the sample for analysis by separation and possibly hyperpolarization of individual components. Reproduced from ref. 106 with kind permission from the author.

In this context, personalized point-of-care medicine could become the single most significant achievement of this century, and NMR could be the key component. A conceptual drawing of a point-of-care NMR analyzer is shown in Figure 11.3. If the portability, sensitivity and spectral resolution of portable NMR devices continue to improve at their current pace, as well as the price–performance ratio, the demand and popularity will grow, more and more competitors in portable NMR will emerge, and fabrication processes will become increasingly standardized and inexpensive.

The potential of point-of-care NMR and medicine lies not merely in personalized medical applications and finally having nifty futuristic Tricorders at home. The availability of inexpensive mobile NMR analytical platforms would facilitate speedy ambulant analyses, and enable NMR-based research in developing countries. More importantly, the technology would become affordable for destitute communities, and could, much like the cell phone, lead to a drastic increase in living standards and life expectancy in the underprivileged world.

References

1. H. Maeda and Y. Yanagisawa, *IEEE Trans. Appl. Supercond.*, 2014, **24**, 1–12.
2. A. Malozemoff, S. Fleshler, M. Rupich, C. Thieme, X. Li, W. Zhang, A. Otto, J. Maguire, D. Folts and J. Yuan, *Supercond. Sci. Technol.*, 2008, **21**, 034005.
3. A. R. Ackermann, P. Monteur and M. Char, *Very efficient heat-exchanger for cryogen-free MRI magnet, Patent WO2013046129*, Eindhoven, NL, 13 June 2013.
4. G. Batey, A. Casey, M. Cuthbert, A. Matthews, J. Saunders and A. Shibahara, *New J. Phys.*, 2013, **15**, 113034.
5. B. J. Parkinson, R. Slade, M. J. Mallett and V. Chamritski, *IEEE Trans. Appl. Supercond.*, 2013, **23**, 4400405.
6. K. Watanabe, S. Awaji, Y. Hou, H. Oguro, T. Kiyoshi, H. Kumakura, S. Hanai, H. Tsubouchi, M. Sugimoto and I. Inoue, *IEEE Trans. Appl. Supercond.*, 2013, **23**, 4300304.
7. A. Y. Ganin, Y. Takabayashi, Y. Z. Khimyak, S. Margadonna, A. Tamai, M. J. Rosseinsky and K. Prassides, *Nat. Mater.*, 2008, **7**, 367–371.
8. H. Takahashi, K. Igawa, K. Arii, Y. Kamihara, M. Hirano and H. Hosono, *Nature*, 2008, **453**, 376–378.
9. K. Ogawa, T. Nakamura, Y. Terada, K. Kose and T. Haishi, *Appl. Phys. Lett.*, 2011, **98**, 234101.
10. D. Tamada, T. Nakamura, Y. Itoh and K. Kose, *Phys. C: Supercond.*, 2013, **492**, 174–177.
11. F. Casanova, J. Perlo and B. Blümich, *Single Sided NMR*, Springer-Verlag, Berlin, Heidelberg, 2011.
12. V. Demas and P. J. Prado, *Concepts Magn. Reson., Part A*, 2009, **34**, 48–59.
13. E. Danieli, J. Perlo, B. Blümich and F. Casanova, *Angew. Chem., Int. Ed.*, 2010, **49**, 4133–4135.
14. L. O. Sillerud, A. F. McDowell, N. L. Adolphi, R. E. Serda, D. P. Adams, M. J. Vasile and T. M. Alam, *J. Magn. Reson.*, 2006, **181**, 181–190.
15. A. McDowell and E. Fukushima, *Appl. Magn. Reson.*, 2008, **35**, 185–195.
16. A. F. McDowell and N. L. Adolphi, *J. Magn. Reson.*, 2007, **188**, 74–82.
17. E. Danieli, J. Perlo, B. Blümich and F. Casanova, *Phys. Rev. Lett.*, 2013, **110**, 180801.
18. A. Samoilenko, D. Y. Artemov and L. Sibeldina, *JETP Lett.*, 1988, **47**, 417–419.
19. P. McDonald, *Prog. Nucl. Magn. Reson. Spectrosc.*, 1997, **30**, 69–99.
20. R. Kimmich, W. Unrath, G. Schnur and E. Rommel, *J. Magn. Reson.*, 1991, **91**, 136–140.
21. I. Chang, F. Fujara, B. Geil, G. Hinze, H. Sillescu and A. Tölle, *J. Non-Cryst. Solids*, 1994, **172**, 674–681.
22. T. Feiweier, B. Geil, O. Isfort and F. Fujara, *J. Magn. Reson.*, 1998, **131**, 203–208.

23. J. A. Jackson, L. J. Burnett and J. F. Harmon, *J. Magn. Reson.*, 1980, **41**, 411–421.
24. R. L. Kleinberg, in *Encyclopedia of NMR*, ed. D. Grant and R. Harris, Wiley-Liss, New York, 1996.
25. R. L. Kleinberg, A. Sezginer, D. Griffin and M. Fukuhara, *J. Magn. Reson.*, 1992, **97**, 466–485.
26. G. Eidmann, R. Savelsberg, P. Blümler and B. Blümich, *J. Magn. Reson., Ser. A*, 1996, **122**, 104–109.
27. J. Perlo, F. Casanova and B. Blümich, *J. Magn. Reson.*, 2004, **166**, 228–235.
28. J. Perlo, F. Casanova and B. Blümich, *Science*, 2007, **315**, 1110–1112.
29. J. Perlo, F. Casanova and B. Blümich, *J. Magn. Reson.*, 2005, **176**, 64–70.
30. P. McDonald, P. Aptaker, J. Mitchell and M. Mulheron, *J. Magn. Reson.*, 2007, **185**, 1–11.
31. B. Manz, A. Coy, R. Dykstra, C. Eccles, M. Hunter, B. Parkinson and P. Callaghan, *J. Magn. Reson.*, 2006, **183**, 25–31.
32. J. Mitchell, L. Gladden, T. Chandrasekera and E. Fordham, *Prog. Nucl. Magn. Reson. Spectrosc.*, 2014, **76**, 1–60.
33. B. Blümich, J. Perlo and F. Casanova, *Prog. Nucl. Magn. Reson. Spectrosc.*, 2008, **52**, 197–269.
34. J. Mitchell, P. Blümler and P. McDonald, *Prog. Nucl. Magn. Reson. Spectrosc.*, 2006, **48**, 161–181.
35. G. R. Coates, L. Xiao and M. G. Prammer, *NMR Logging: Principles and Applications*, Gulf Professional Publishing, Houston, TX, 1999.
36. B. Blümich, S. Haber-Pohlmeier and W. Zia, *Compact NMR*, Walter de Gruyter, Berlin, 2014.
37. G. Guthausen, *Produkt- und Prozesscharakterisierung mittels NMR-Methoden*, Logos Verlag, Berlin, 2012.
38. K. Kose, T. Haishi and S. Handa, in *Magnetic Resonance Microscopy*, ed. S. Codd and J. Seymour, Wiley VCH, Weinheim, 2009.
39. R. R. Milczarek and M. J. McCarthy, in *Magnetic Resonance Microscopy: Spatially Resolved NMR Techniques and Applications*, ed. S. Codd and J. Seymour, Wiley VCH, Weinheim, 2009.
40. H. Lee, E. Sun, D. Ham and R. Weissleder, *Nat. Med.*, 2008, **14**, 869–874.
41. N. Sun, T.-J. Yoon, H. Lee, W. Andress, R. Weissleder and D. Ham, *IEEE J. Solid-State Circuits*, 2011, **46**, 342–352.
42. J. B. Haun, C. M. Castro, R. Wang, V. M. Peterson, B. S. Marinelli, H. Lee and R. Weissleder, *Sci. Transl. Med.*, 2011, **3**, 71ra16.
43. A. Manz, N. Graber and H. á. Widmer, *Sens. Actuators, B*, 1990, **1**, 244–248.
44. M. Kakuta, D. A. Jayawickrama, A. M. Wolters, A. Manz and J. V. Sweedler, *Anal. Chem.*, 2003, **75**, 956–960.
45. H. Wensink, F. Benito-Lopez, D. C. Hermes, W. Verboom, H. J. Gardeniers, D. N. Reinhoudt and A. van den Berg, *Lab Chip*, 2005, **5**, 280–284.
46. C. Hilty, E. E. McDonnell, J. Granwehr, K. L. Pierce, S.-I. Han and A. Pines, *Proc. Natl. Acad. Sci. U. S. A.*, 2005, **102**, 14960–14963.

47. O. G. Gruschke, N. Baxan, L. Clad, K. Kratt, D. von Elverfeldt, A. Peter, J. Hennig, V. Badilita, U. Wallrabe and J. G. Korvink, *Lab Chip*, 2012, **12**, 495–502.
48. D. Sakellariou, G. Le Goff and J.-F. Jacquinot, *Nature*, 2007, **447**, 694–697.
49. O. Gökay and K. Albert, *Anal. Bioanal. Chem.*, 2012, **402**, 647–669.
50. V. Badilita, R. C. Meier, N. Spengler, U. Wallrabe, M. Utz and J. G. Korvink, *Soft Matter*, 2012, **8**, 10583–10597.
51. A. Webb, *J. Magn. Reson.*, 2013, **229**, 55–66.
52. N. Sun, Y. Liu, L. Qin, H. Lee, R. Weissleder and D. Ham, *Solid-State Electron.*, 2013, **84**, 13–21.
53. A. M. Weljie, J. Newton, P. Mercier, E. Carlson and C. M. Slupsky, *Anal. Chem.*, 2006, **78**, 4430–4442.
54. C. Ludwig and M. R. Viant, *Phytochem. Anal.*, 2010, **21**, 22–32.
55. A. Smolinska, L. Blanchet, L. Buydens and S. S. Wijmenga, *Anal. Chim. Acta*, 2012, **750**, 82–97.
56. K. Pearson, *London, Edinburgh Dublin Philos. Mag. J. Sci.*, 1901, **2**, 559–572.
57. D. Budker and M. Romalis, *Nat. Phys.*, 2007, **3**, 227–234.
58. K. Sternickel and A. I. Braginski, *Supercond. Sci. Technol.*, 2006, **19**, S160.
59. K. Mohri, Y. Honkura, L. Panina and T. Uchiyama, *J. Nanosci. Nanotechnol.*, 2012, **12**, 7491–7495.
60. M. Schmelz, R. Stolz, V. Zakosarenko, S. Anders, L. Fritzsch, H. Roth and H.-G. Meyer, *Phys. C: Supercond.*, 2012, **476**, 77–80.
61. T. Sander, J. Preusser, R. Mhaskar, J. Kitching, L. Trahms and S. Knappe, *Biomed. Opt. Express*, 2012, **3**, 981–990.
62. D. Oyama, J. Hatta, M. Miyamoto, Y. Adachi, J. Kawai, M. Higuchi and G. Uehara, *IEEE Trans. Appl. Supercond.*, 2013, **23**, 1601604.
63. S. Meier, P. R. Jensen, M. Karlsson and M. H. Lerche, *Sensors*, 2014, **14**, 1576–1597.
64. G. Pileio, M. Carravetta and M. H. Levitt, *Proc. Natl. Acad. Sci. U. S. A.*, 2010, **107**, 17135–17139.
65. G. Pileio, S. Bowen, C. Laustsen, M. C. Tayler, J. T. Hill-Cousins, L. J. Brown, R. C. Brown, J. H. Ardenkjaer-Larsen and M. H. Levitt, *J. Am. Chem. Soc.*, 2013, **135**, 5084–5088.
66. Y. Feng, R. M. Davis and W. S. Warren, *Nat. Phys.*, 2012, **8**, 831–837.
67. B. Vuichoud, J. Milani, A. Bornet, R. Melzi, S. Jannin and G. Bodenhausen, *J. Phys. Chem. B*, 2014, **118**, 1411–1415.
68. L. Schröder, T. J. Lowery, C. Hilty, D. E. Wemmer and A. Pines, *Science*, 2006, **314**, 446–449.
69. R. E. Hurd, D. Spielman, S. Josan, Y. F. Yen, A. Pfefferbaum and D. Mayer, *Magn. Reson. Med.*, 2013, **70**, 936–942.
70. S. B. Duckett and R. E. Mewis, *Acc. Chem. Res.*, 2012, **45**, 1247–1257.
71. D. L. Nelson, A. L. Lehninger and M. M. Cox, *Lehninger Principles of Biochemistry*, 5th edn, W. H. Freeman, New York, 2008.
72. *The Handbook of Metabonomics and Metabolomics*, ed. J. Lindon, J. Nicholson and E. Holmes, Elsevier Science, New York, 2006.

73. M. Dhenain, *Magn. Reson. Insights*, 2008, **2**, 75–91.
74. K. J. Donovan, E. Kupče and L. Frydman, *Angew. Chem.*, 2013, **125**, 4246–4249.
75. D. J. Holland, M. J. Bostock, L. F. Gladden and D. Nietlispach, *Angew. Chem.*, 2011, **123**, 6678–6681.
76. E. Rennella, T. Cutuil, P. Schanda, I. Ayala, V. Forge and B. Brutscher, *J. Am. Chem. Soc.*, 2012, **134**, 8066–8069.
77. J. H. Lee, Y. Okuno and S. Cavagnero, *J. Magn. Reson.*, 2014, **241**, 18–31.
78. H. Van As and J. van Duynhoven, *J. Magn. Reson.*, 2013, **229**, 25–34.
79. B. Blümich, F. Casanova, J. Perlo, F. Presciutti, C. Anselmi and B. Doherty, *Acc. Chem. Res.*, 2010, **43**, 761–770.
80. D. Capitani, V. Di Tullio and N. Proietti, *Prog. Nucl. Magn. Reson. Spectrosc.*, 2012, **64**, 29–69.
81. M. D. Hürlimann, *Well logging*, John Wiley & Sons, 2007.
82. D. V. Ellis and J. M. Singer, *Well Logging for Earth Scientists*, Springer, Netherlands, 2007.
83. D. Robinson, C. Campbell, J. Hopmans, B. Hornbuckle, S. B. Jones, R. Knight, F. Ogden, J. Selker and O. Wendroth, *Vadose Zone J.*, 2008, **7**, 358–389.
84. T. Kimura, Y. Geya, Y. Terada, K. Kose, T. Haishi, H. Gemma and Y. Sekozawa, *Rev. Sci. Instrum.*, 2011, **82**, 053704.
85. H. K. Kim, Y. H. Choi and R. Verpoorte, *Trends Biotechnol.*, 2011, **29**, 267–275.
86. J. Schripsema, *Phytochem. Anal.*, 2010, **21**, 14–21.
87. C. W. Windt, H. Soltner, D. v. Dusschoten and P. Blümler, *J. Magn. Reson.*, 2011, **208**, 27–33.
88. M. Jones, P. Aptaker, J. Cox, B. Gardiner and P. McDonald, *J. Magn. Reson.*, 2012, **218**, 133–140.
89. Y.-Q. Song, *J. Magn. Reson.*, 2013, **229**, 12–24.
90. B. Wu, L. Xiao, X. Li, H. Yu and T. An, *Pet. Sci.*, 2012, **9**, 38–45.
91. D. O. Walsh, *J. Appl. Geophys.*, 2008, **66**, 140–150.
92. J. Perlo, E. Danieli, J. Perlo, B. Blümich and F. Casanova, *J. Magn. Reson.*, 2013, **233**, 74–79.
93. C. M. Preston, *Soil Sci.*, 1996, **161**, 144–166.
94. K. R. Brownstein and C. E. Tarr, *J. Magn. Reson.*, 1977, **26**, 17–24.
95. J. Kärger, D. M. Ruthven and D. N. Theodorou, *Diffusion in Nanoporous Materials*, Wiley-VCH, Weinheim, 2012.
96. E. Paciok, A. Habert, M. Van Landeghem and B. Blümich, *Z. Phys. Chem.*, 2012, **226**, 1243–1257.
97. A. Filippov, S. V. Dvinskikh, A. Khakimov, M. Grahn, H. Zhou, I. Furo, O. N. Antzutkin and J. Hedlund, *Magn. Reson. Imaging*, 2012, **30**, 1022–1031.
98. M. Fleury and J. Soualem, *J. Colloid Interface Sci.*, 2009, **336**, 250–259.
99. M. Van Landeghem, A. Haber, J.-B. D'espinose De Lacaillerie and B. Blümich, *Concepts Magn. Reson., Part A*, 2010, **36A**, 153–169.
100. A. M. Olaru, J. Kowalski, V. Sethi and B. Blümich, *J. Magn. Reson.*, 2012, **220**, 32–44.

101. D. Bytchenkoff and S. Rodts, *J. Magn. Reson.*, 2011, **208**, 4–19.
102. N. Klitzsch and O. Mohnke, *Vadose Zone J.*, 2010, **9**, 846–857.
103. *Fluid Transport in Nanoporous Materials*, ed. W. C. Connor and J. Fraissard, Springer, Dordrecht, 2006.
104. V. Guillon, D. Bauer, M. Fleury and M.-C. Neel, *Transp. Porous Media*, 2014, **101**, 251–267.
105. M. Sahimi, *Flow and Transport in Porous Media and Fractured Rock, From Classical Methods to Modern Approaches*, 2nd edn, Wiley-VCH, Weinheim, 2011.
106. C. Pille, Health & Nutrition Advisor, Bachelor thesis, University of Applied Sciences, Münster, 2014.

Subject Index

References to figures are given in *italic* type. References to tables are given in **bold** type.